Current Chinese Economic Report Series

More information about this series at http://www.springer.com/series/11028

Lianshui Li · Zhanyuan Du
Editors

A Research Report on the Development of China's Manufacturing Sector (2016)

Springer

Editors
Lianshui Li
Nanjing University of Information Science and Technology
Nanjing
China

Zhanyuan Du
Ministry of Education of P.R.C
Beijing
China

ISSN 2194-7937 ISSN 2194-7945 (electronic)
Current Chinese Economic Report Series
ISBN 978-981-10-4444-1 ISBN 978-981-10-4445-8 (eBook)
DOI 10.1007/978-981-10-4445-8

Library of Congress Control Number: 2017936646

Printed on acid-free paper

This Springer imprint is published by Springer Nature
The registered company is Springer Nature Singapore Pte Ltd.
The registered company address is: 152 Beach Road, #21-01/04 Gateway East, Singapore 189721, Singapore

Included in the Development Report on Philosophy and Social Sciences by the Ministry of Education in 2013 (13JBG004)

Funded by the China Institute for Manufacture Development (a key research base of the academic institutions of Jiangsu Province for philosophy and social sciences) (2010JDXM028)

Funded by the Research Center for Technological Innovation, Tsinghua University (a key research base for humanities and social sciences of the Ministry of Education)

Funded by the National Natural Science Foundation of China (NSFC) (71173116, 71673145)

Funded by the Preponderant Discipline Development Project of the Colleges of Jiangsu Province.

Preface

Ever since *A Research Report on the Development of China's Manufacturing Sector (2004)* was prepared, this research has focused on new-type manufacturing industries and has explored a road map and way forward for the development of China's manufacturing sector. Thus far, 12 consecutive research reports have been published. This book is the second English edition of the research report. During the research process, we have deeply understood the rapid development of China's manufacturing sector and witnessed the rapid improvement of China's manufacturing sector in terms of economic resourcefulness, technological innovation capability, and resource and environmental protection capability. While China is developing itself into an innovation-oriented country and transforming itself from a large manufacturing country into a strong manufacturing country, this research report is expected to accurately reflect the improvement of China's manufacturing sector in self-dependent innovation capability and to assist in the transformation, upgrade, and innovative development of China's manufacturing sector.

This research report focuses on the topic of technological innovations guiding the development of China's manufacturing sector. Specifically, this report highlights a development road map for China's new-type manufacturing sector, conducts normative research (including comprehensive evaluation, regional research, industry research, and enterprise research), and deepens research on different manufacturing industries, so as to accomplish higher academic value. The unique and innovative points are mainly as follows:

"Comprehensive evaluation" sets forth the connotations of the new-type manufacturing sector, and evaluates the general status of China's new-type manufacturing sector from four aspects, namely, economic resourcefulness, technological innovation capability, energy conservation capability, and environmental protection capability.

"Regional research" sets forth the following points. (1) It evaluates the degree of newness of regional manufacturing industries from four aspects (i.e., economic resourcefulness, technological innovations, resource utilization, and environmental protection). Guangdong and Jiangsu provinces in eastern China take the lead in the comprehensive strength of manufacturing industries and have attained remarkable achievements in self-independent technological innovations. Anhui, Henan, and Hunan in central China have grown rapidly in terms of scale of manufacturing industries, but are confronted with such problems as a sharp increase in pollution discharge and the largest discharge amount of waste gas and waste residue (it is advisable to be vigilant about the well-known old road of "first pollution and then remediation"). Western China has a relatively sluggish manufacturing sector and continues to lag China's developed regions in terms of economic resourcefulness; thus, it is imperative to build an incentive mechanism to promote the clustering and growth of manufacturing industries. (2) The regional research type evaluates the degree of newness of manufacturing industries of China's 30 central cities. Cities in eastern China have outstanding performance in economic resourcefulness (in particular, both Suzhou and Nanjing in Jiangsu rank in the top three), while their environmental protection capability is worrisome (in particular, Suzhou ranks at the bottom). Shenzhen has exceptional performance in both economic resourcefulness and environmental protection capability, providing valuable experience for other manufacturing cities.

"Industry research" sets forth the following points. (1) In light of the connotations and evaluation index system for the new-type manufacturing sector, industry research evaluates and ranks China's manufacturing industries in terms of the degree of newness by using the deviation maximization method. (2) Industry research builds appropriate measuring models and empirically analyzes various issues, for example, the influence of technological innovations on the growth of China's manufacturing sector, the energy efficiency of China's manufacturing sector and its influencing factors, the influence of informatization on the green growth of China's manufacturing sector, and the influence of industry clustering on technological innovations in China's manufacturing sector.

"Enterprise research" analyzes the general development characteristics of China's listed manufacturing enterprises, rebuilds an evaluation and selection system for China's most respectable listed manufacturing enterprises, and selects China's top 50 most respectable enterprises accordingly.

In conjunction with those who are concerned about the development of China's manufacturing sector, we are interested in studying and exploring the development road map for China's manufacturing sector and make our contributions for the attainment of enhanced Chinese development. Given various academic deficiencies, this research study inevitably contains certain errors or imperfections. We sincerely seek your constructive comments and criticisms.

Nanjing, China Lianshui Li
Beijing, China Zhanyuan Du

Contents

Summary

This book is an integral part or outcome of certain major projects, including the following.

- A project in the Development Report on Philosophy and Social Sciences by the Ministry of Education in 2013
- A brainstorm project (03JZD0014) of major research topics on philosophy and social sciences carried out by the Ministry of Education
- Later-stage outcome of Research on the Development Road map of China's Manufacturing Sector Based on Resource Constraints and Self-dependent Innovation (70573045) and Research on the Development of China's Manufacturing Sector under the Pressure of Carbon Emissions Reduction in the Context of Global Warming (70873063) funded by the Natural Science Foundation of China (NSFC)
- A major outcome of the Research on Transformation and Upgrade of China's Manufacturing Sector under Environmental Regulation (71173116) funded by the NSFC
- The second English edition of *A Research Report on the Development of China's Manufacturing Sector (2016)*

First, this book, which is based on related data for 1978–2013, sets forth the characteristics of different stages, typical development patterns, and international status of China's manufacturing sector. Second, it provides an in-depth definition of China's new-type manufacturing sector from the perspective of four aspects, namely economic resourcefulness, technological innovation capability, energy conservation capability, and environmental protection capability. Third, this book accordingly evaluates the status quo of China's new-type manufacturing sector, analyzes the regional development characteristics of China's manufacturing sector, and lists China's top 10 provinces and top 10 cities in terms of their manufacturing sectors. Fourth, this book sets forth the industrial characteristics of China's manufacturing sector and analyzes the influencing factors of development of China's manufacturing sector, as well as the related ranking of regional manufacturing sectors. Finally, this book studies China's listed manufacturing enterprises, further improves the evaluation and selection system for China's most respectable listed manufacturing enterprises, and ranks and briefly evaluates China's top 50 most respectable enterprises.

In summary, this book presents a research study on the development trends of the manufacturing sector, is an annual report on the development progress of China's manufacturing sector, and is also a reference manual that contains rich data about China's manufacturing sector.

This book is intended for all those who show concern for the development of China's manufacturing sector, especially teachers and students in academic institutions, government officials, and managerial personnel in the manufacturing sector and related enterprises.

Chapter 1
China's Manufacturing Sector: Development Characteristics and Typical Patterns Over 35 Years

Caihua Yu, Mingyang Zhang and Feixue Zhou

Over the 35-year period of 1978–2013, China has undergone significant changes, as follows. (1) Institutional innovations have deepened continuously, and social vitality has been stimulated continuously. (2) The social economy has developed rapidly and comprehensive national power has been enhanced significantly. (3) The manufacturing sector has accomplished leap-forward development, and the industrialization process has sped up. In a sense, the reform and opening-up history of the new China is a history during which the development of the manufacturing sector has strengthened rapidly and the industrial system has been optimized and upgraded. The development of China's manufacturing sector not only has taken on distinct stage-specific characteristics, but also has given birth to some typical patterns, which deserve further study.

1.1 Characteristics in Different Stages

Since China carried out its reform and opening-up policy, China's manufacturing sector has undergone three transformations: (1) from labor-intensive industries to capital-intensive and technology-intensive industries; (2) from the original equipment manufacturing and original design manufacturing mode to enhanced self-dependent innovation; and (3) from the development of coastal regions to the development of inland regions. In terms of economic resourcefulness, technological innovation, resource consumption, environmental protection, and social contribution, the development history of China's manufacturing sector can be divided into three stages, as discussed in the following sections.

L. Li and Z. Du (eds.), *A Research Report on the Development of China's Manufacturing Sector (2016)*, Current Chinese Economic Report Series,
DOI 10.1007/978-981-10-4445-8_1

1.1.1 Boom Stage Dominated by Economic Resourcefulness (1978–1994)

From the late 1970s, the floating exchange rate system gradually replaced the fixed exchange rate system under the Bretton Woods system, global economic integration deepened continuously, economic links between different countries became increasingly close, the global economy has begun to emerge from stagnation, and the international economic environment became increasingly flexible. All these provided a favorable external economic environment for the development of China's manufacturing sector.

China's economic reform has continued to deepen. In 1978, the Third Plenary Session of the 11th Communist Party of China (CPC) Central Committee expressly resolved that the CPC's work should refocus on economic development. In 1984, the Central Committee of the CPC promulgated the *Resolutions of the CPC's Central Committee on Reform of the Economic System*, resolving to speed up city-centered reform of the economic system, strengthen enterprises, consciously take advantage of the law of value, and build a reasonable price system and a multiform economic responsibility system. In 1982, the 13th National Congress of the CPC expressly resolved to establish a socialist market economy system. In 1993, the Central Committee of the CPC promulgated the *Resolutions of the CPC's Central Committee regarding Several Issues on Building a Socialist Market Economy System*, resolving to transform the managerial mechanism of state-owned enterprises, and establish a modern enterprise system. All decisions have provided a favorable internal institutional environment for the development of China's manufacturing sector. Since China's reform and opening-up, a sellers' market has existed for several decades due to the great gap between ever-increasing social needs and rapidly developing but backward productivity. This provides a huge sales market and profit space for the development of China's manufacturing sector. Based on the principle that economic development is China's central task, the economic-oriented outlook on political achievement motivates local governments to actively push and even directly become involved in economic affairs, thereby speeding up the development of local manufacturing sectors and creating a new driving force for the development of China's manufacturing sector.

In 1994, the total output value of China's manufacturing sector was 4.48 trillion yuan, which was six times as much as that in 1985. In this stage, China's manufacturing sector was characterized by great economic resourcefulness; certain manufacturing industries (e.g., petroleum processing and coking, manufacture of metal products, transportation equipment manufacturing, manufacture of computers, communication, and other electronic equipment, manufacture of electrical machinery and equipment, and manufacture of chemical fibers) attained above-average growth compared with China's entire manufacturing sector. In 1994, the manufacture of textiles, smelting and pressing of ferrous metals, and manufacture of transportation equipment contributed the largest proportions (11.04, 9.29, and 7.11% respectively) of output value among China's manufacturing industries.

In this stage, China's manufacturing sector attained all-around development, but contributed a very small proportion of output value to the global manufacturing sector, with room for further development. In 1980, the added value of China's manufacturing sector accounted for only 1.5% of that of the global manufacturing sector, or only half that of Brazil's manufacturing sector. However, by 1990, China exceeded Brazil in added value of its manufacturing sector, ranking 8th in the world.

This was the initial stage of China's industrialization, in which the creation of value and profit was the major impetus and the state-owned economy and collective-owned economy played a predominant role. In this stage, China's manufacturing sector was booming, a greater diversity of manufacturing industries was emerging, industrial scale expanded significantly, and manufacturing capability improved rapidly.

1.1.2 Two-Wheel (Economic and Technology) Driving Stage (1995–2010)

From the mid-1990s, the social and economic environment underwent significant changes, both in China and globally. In China, the socialist market economy system took shape in preliminary form, public ownership played a dominant role, with diverse forms of ownership developing side by side in the basic economic system, and essentially, an all-around, multi-field, and multi-level opening-up pattern was formed. Through economic development and market cultivation over a dozen years, a buyers' market became increasingly evident, and competition between enterprises increasingly acute; in addition, China has been confronted with various emerging problems, including irrational economic structure; severe pressure of employment, resources, and the environment; and weak overall economic competitiveness. Across the globe, the manufacturing sector has been confronted with various challenges, the Internet economy is emerging, technological innovations have become a new engine for industrial development, and self-dependent innovations have become conscious behavior for enterprise development and industrial growth. The context of these changes included the Asian financial crisis of 1996, which had a severe impact on the world economy (e.g., international market demand shrunk and the export environment of the manufacturing sector deteriorated); the global economy did not start to recover until the early 21st century.

In the face of these changes in the economic situation and the advent of the Internet economy, more importance was attached to technological innovations in industrial development. Early in 1995, the CPC's Central Committee and the State Council promulgated the *Resolutions on Promoting Scientific and Technological Progress*, resolving to carry out a strategy of "rejuvenating China through science and education", specifically, by enhancing China's scientific and technological strength and capability of transforming scientific and technological achievements into a real productive force; improving the scientific and technological literacy of the whole nation; and developing the national economy by relying on scientific and technological progress and improvements in labor skills. In 2006, the CPC's Central Committee and the State Council promulgated the *Resolutions on Implementing the Scientific and Technological Plan Outline and Strengthening the Self-dependent Innovation Capability*, proposing scientific and technological development guidelines for the new era: self-dependent innovations, a leap forward in key fields, support for economic development, and determining the future. In summary, China has attached increasing importance to the role of scientific and technological progress in economic growth, and China's economic growth has gradually turned to scientific and technological progress and improvement in labor skills.

In 2010, the total output value of China's manufacturing sector was 60.96 trillion yuan, which was 11 times as much as that in 1995. Three manufacturing industries—transportation equipment manufacturing, manufacture of computers, communication, and other electronic equipment, and smelting and pressing of ferrous metals—contributed the largest proportion to output value—9.10, 9.02, and 8.50%, respectively. In this stage, China's manufacturing sector attached great importance to technological innovations. In 2010, China's manufacturing enterprises spent a total of 377.133 billion yuan on research and development (R&D) and 432.4 billion yuan on new product development, and took out 109,721 patents for invention, which were 15.18, 16.72, and 23.45 times as much or many as those in 1995, respectively. In 2010, China's manufacturing sector contributed 7% of the added value of the global manufacturing sector, ranking fourth globally after the US, Japan, and Germany. In 2006, a total of 172 product categories in China's manufacturing sector ranked first globally in terms of annual yield; the added value of China's manufacturing sector amounted to 1.0956 trillion US dollars; and China exceeded Japan in total added value of its manufacturing sector for the first time, becoming the world's second largest manufacturing country. In 2010, China exceeded the US in total added value of its manufacturing sector for the first time, and became the world's largest manufacturing country. This marked a historical breakthrough of China's manufacturing sector, and China's manufacturing sector exceeded the total value added of both Germany and Japan in the early 21st century. Over the last 30 years, the growth rate of China's manufacturing sector has ranked first globally, and the share of China's manufacturing sector in the global manufacturing sector has achieved average annual growth of 8.98%.

In this stage (1995–2010), China's manufacturing sector attached importance to not only the creation of economic value, but also technological innovations; through various platforms (e.g., hi-tech industrial parks and economic development zones), China's small and mid-sized hi-tech enterprises actively introduced, digested, and integrated various innovative achievements, conducted original innovations, improved their production and technological levels, attempted to increase the technological content and added value of their products, and improved their technological innovation capabilities.

1.1.3 New Innovation Driving Stage (2011 to the Present)

In the 21st century, China's manufacturing sector is facing a favorable situation on the whole. Specifically, China's economic strength has been enhanced notably and market reform has advanced to an in-depth level; moreover, different regions are attempting to deepen institutional innovations, intensify the service function of governments, build service platforms actively, create favorable investment environments, introduce foreign capital actively, and take over a new round of international industry transfer. All these factors will promote the development of China's manufacturing sector to a new level. In spite of significant improvement in economic resourcefulness and technological innovation capability, China's manufacturing sector is also confronted with severe challenges. Domestically, China's economic development has stepped into a "new normal" state; specifically, China's economic growth is transforming from high-speed growth to middle and high-speed growth, from extensive scale growth to efficiency-oriented intensive growth, and from a factor-driven and investment-driven mode to an innovation-driven mode. While China's economic growth is characterized by a change in growth rate, structure optimization, and transformation of motivating force, the sustainable development of China's manufacturing sector is confronted with increasing internal pressure. Globally, developed countries, led by the US, are attempting to recapture and retain the commanding height of the manufacturing sector, and to carry out a package of reindustrialization policies. Therefore, China's manufacturing sector is also confronted with more acute global competition. For example, US President Barack Obama signed the US Manufacturing Enhancement Act in 2010; Germany officially promulgated its *Securing the Future of the German Manufacturing Sector: Recommendations for Implementing the Strategic Initiative Sector 4.0* in 2013; France proposed *The New Industrial France* strategy in 2013; and Japan in 2014 promulgated its *3D Printing and Manufacturing Revolution Plan (2014–2019), Supporting Plan for the Upgrade of New Strategic Industrial Base Technologies*, and a *Five-year Plan for Robot Development (2015–2019).*

In this stage (2011 to the present), China has become the world's largest manufacturing country, but there is still room for further progress in smart manufacturing and green manufacturing. It is imperative that China seizes the historical opportunity brought about by the fourth industrial revolution, with a view to driving the development of its manufacturing sector by technological innovations. On May 8, 2015, *Made in China 2025*, signed by Premier Keqiang Li, was officially released by the State Council, and proposed an innovation-oriented development concept. The central idea is to accomplish automation and digitalization of China's manufacturing sector, develop various smart manufacturing modes (e.g., smart devices and smart factories based on cyber-physical systems), and make breakthroughs in such hi-tech fields as three-dimensional printing, mobile Internet, cloud computing, big data, bioengineering, new energy, and new materials.

1.2 Typical Patterns of Development of China's Manufacturing Sector

Over the past 30 years, China's manufacturing sector has given birth to quite a few development patterns with significant regional characteristics. The three regions ofthe Yangtze River Delta, Pearl River Delta, and Circum-Bohai Sea in eastern coastal China as well as the Wuhan and Xi'an regions in middle China have created distinctive development patterns for their manufacturing sectors. The typical development patterns include the Kunshan, Wenzhou, Dongguan, Zhongguancun, Qingdao, Wuhan, and Xi'an patterns.

1.2.1 Kunshan Pattern: Industrialization Driven by Foreign-Capital Economy

Kunshan city is the east portal of Jiangsu province, with its east neighboring Shanghai city and its west neighboring Suzhou city proper. In 1989, Kunshan was administratively upgraded from a county to a city. Kunshan's urban area is 927 km^2. Kunshan has 10 towns under its jurisdiction, and has a registered population of 650,000. Kunshan was traditionally an agricultural county. In the initial period of China's reform and opening-up, Kunshan's manufacturing sector was very underdeveloped, and its comprehensive economic strength ranked at the bottom among the several counties administered by Suzhou city over a long period. Since the 1990s, Kunshan has actively established economic development zones, attempted to improve the investment environment, actively developed a foreign-oriented economy, and coordinated economic and social development with environmental protection. As a result, Kunshan has become an important and highly dense investment destination for foreign capital (especially from China Taiwan's capital). Generally, Kunshan has successfully created the Kunshan pattern, which is characterized by the development of an open economy. Since 1992, Kunshan has attracted investment of 1 billion US dollars annually; in around 2001 especially, quite a few of Taiwan's computer manufacturing enterprises set up shops in Kunshan. In 2013, Kunshan approved a total of 205 foreign-funded projects, attracted total registered capital of 970 million US dollars, received paid-in capital of at least 1.7 billion US dollars, and introduced 10 foreign-funded projects, each of which involved capital of at least 100 million US dollars. By the end of 2013, Kunshan had a total of 58,700 domestic-funded registered enterprises with registered capital of 157.516 billion yuan; in addition, it had 75,100 registered individual businesses with total registered capital of 4.786 billion yuan. Furthermore, the total output value of hi-tech industries amounted to 382.5 billion yuan, a year-on-year increase of 3.3%. In 2014, the city's GDP amounted to 300.102 billion yuan, a year-on-year increase of 7.7%; its total industrial output value amounted to 870.849 billion yuan, a year-on-year decrease of 1.8%; the city's total import and export volume amounted to 84.791 billion US dollars, including the total export volume of 53.577 billion US dollars, (a year-on-year decrease of 4.7 and 2%, respectively). Over the years, Kunshan has cumulatively approved up to 7142 investment projects from 55 countries or regions, and 27 among the world's top 500 enterprises (as measured by Fortune) have invested in Kunshan. Every day, foreign capital of at least 15 million US dollars flows into Kunshan, while it has a daily export volume of at least 10 million US dollars, and daily fiscal revenue of at least 10 million US dollars. With only 0.01% of China's total land area and only 0.05% of China's total population, Kunshan has attracted 2.5% of China's total paid-in foreign capital, 0.32% of China's total import and export volume, and created 0.4% of China's total GDP.

The Kunshan pattern mainly has the following characteristics. (1) It dismantles the shackles of the traditional economic system, and the economic system and mechanism have been vigorously innovated. For example, Kunshan has set up economic development zones at its own expense, intensified the government's service function, scientifically planned and implemented urban construction, sped up the urbanization process, improved transportation infrastructure, advanced the informatization process, constantly optimized the investment environment, developed a modern service sector system, improved the supporting capability for the manufacturing sector, and actively developed industrial clusters. (2) The Kunshan pattern carries out an export-driven strategy and develops an externally oriented economy. Kunshan has taken full advantage of its geological location, set up various economic development zones and industrial parks, actively attracted foreign entrepreneurs and capital through diverse channels (e.g., intermediary agents, agencies and the Internet), and determinedly introduced the world's top 500 enterprises, well-known Chinese enterprises, hi-tech projects, and investment projects that can play a significant driving role in economic development. (3) Kunshan has increased the supply of non-agricultural land by means of land reclamation and established farmer households or farmer cooperatives by signing contracts between collectives and farmer households. Local village collectives have obtained non-agricultural construction land by means of land reclamation first, and then, through inviting bids from cooperative economic organizations (e.g., investment cooperatives and residents' wealth-creation cooperatives) formed by local farmer households or farmers. Bid-winning cooperatives jointly constructed standard factory buildings, shops, or worker dormitories, for renting out to external investors. (4) The Kunshan pattern develops the local private economy by endogenously driving foreign-capital. By taking full advantage of the developed foreign-capital economy, Kunshan has actively developed its local private economy, for example, as follows: ① through the restructuring of enterprises and policy support, it has developed a number of local private enterprises; ② it has incubated a number of private enterprises in different ways (e.g., shareholding, business cooperation, and supporting services), and has promoted harmonious coexistence between foreign-capital enterprises and private enterprises; and ③ it has encouraged foreign-capital enterprises to develop the service sector and thereby extend the industrial chain, so that domestic-capital enterprises can share the fruits created by the foreign-capital economy. (5) Taiwan-funded enterprises account for a large proportion of Kunshan's total foreign-capital enterprises, and mainly invest in the processing and assembling sector. Taiwan's capital plays an important role in Kunshan's foreign-capital economy. Among at least 1300 foreign capital enterprises in operation, enterprises funded by China Taiwan's capital account for at least 60%. China Taiwan-funded enterprises mainly invest in the manufacturing sector, specifically, IT, precision machinery manufacturing, and fine chemicals. Of products manufactured by China Taiwan-funded enterprises, 90% are exported, and thus, China Taiwan-funded enterprises are mainly foreign-oriented, but have a limited link to Kunshan's local economy. (6) Kunshan has improved the living standard of its residents, and promoted coordinated development of both rural and urban areas. In order to improve the living standard of urban and rural residents, Kunshan began to carry out a residents' wealth-creation project on a large scale in 2000. Kunshan has actively developed its private economy, so as to increase employment opportunities, and has carried out the "Three-haves Project" (in which every resident should have skills, have a job, and own real estate), so as to increase the income of residents. In addition, Kunshan has set aside fiscal revenue of nearly 100 million yuan to local residents, built a quintuple social security system for local residents, comprising rural endowment insurance, rural medical insurance, home relocation compensation, basic livelihood security for rural residents, and minimum livelihood security for urban and rural residents.

What can be learnt from the Kunshan pattern? The following measures are advised: take full advantage of geological location, and actively accept the industrial radiation and industrial transfer from the modern manufacturing sector's center; define the role of governments and markets clearly, divide the work in definitive ways, and ensure governments are vigorous in providing public goods; develop scarce resources (e.g., scare land), increase the total supply of economic factors, innovate in operation modes, and cater for the development needs of the manufacturing sector; drive the development of the local private economy through foreign investment, and encourage the foreign-capital economy to root itself deeply in the investment destination; build commanding heights for the IT industry, intensify the supporting capability for the IT industry, and expand the scale of industrial cluster; improve the livelihoods of people, accelerate the urban-rural integration process, and attain perfect integration of industrialization, informatization, marketization, urbanization, and environmental protection.

For the Kunshan pattern, it is advisable to pay attention to the following points. (1) Attracting industrial capital through cheap land runs counter to the sustainable development of the manufacturing sector and regional economy. (2) The acceleration of urbanization and narrowing of urban-rural gaps should not be accomplished at the expense of agricultural land; land is an important production factor for both the agricultural and manufacturing sectors, and thus, land planning should be undertaken from a long-term perspective. (3) Both industrial and enterprise brands should be emphasized and original design manufacture and original brand manufacturer modes should be developed actively.

1.2.2 *Wenzhou Pattern: Industrialization Incubated by Private Economy*

Located in mountainous areas in the southeast of Zhejiang province, Wenzhou is far from large and medium-sized industrial cities and national-level markets, and thus, transportation expenses and information acquisition costs are very high. The per-capita cultivated land area of Wenzhou is less than 0.5 mu (1 mu equals 0.0667 ha), and the soil quality and irrigation conditions of Wenzhou are far worse than those of rural areas of south Jiangsu province. Moreover, Wenzhou's agriculture is underdeveloped, and its rural collective economy is very weak. In the early period of China's reform and opening-up, Wenzhou lacked special support via national policy, and Wenzhou's economy was hardly shackled by China's traditional economic system. However, Wenzhou natives have kept up the tradition of engaging in the household handicraft industry. The Wenzhou pattern originated in the early 1980s. A number of Wenzhou natives left their hometowns to engage in small-scale business and the supply and sales of commodities; then, the household handicraft industry and joint household enterprises were gradually established and developed. On this basis, some large-scale specialized markets, which were characterized "One Product in One Village and One Trade in One Township," were established successively. *Jiefang Daily* published a report in May 1985, *Viewing Township Industry from South Jiangsu Province and Viewing Household Handicraft Industry from South Zhejiang Province—330,000 Wenzhou People Engaged in the Household Handicraft Industry*. This report introduced Wenzhou's household handicraft industry, and first used the term "Wenzhou pattern." Subsequently, "Wenzhou pattern" was used in official documents and was in wide use in the mid-1990s.

In short, the Wenzhou pattern is defined as follows. Wenzhou actively develops non-agricultural industries in the form of the household handicraft industry and specialized markets, thereby forming a development pattern characterized by petty commodities and large markets. Here, petty commodities refers to those commodities characterized by small production scale, low technical content, and low transportation costs, while large markets refer to the countrywide market networks established by the people of Wenzhou.

The Wenzhou pattern mainly has the following characteristics. (1) In the ownership structure, the private economy plays a dominant role. In the early stage, under constraints of both capital and technologies, Wenzhou's manufacturing sector mainly relied on private capital; specifically, the manufacturing sector started from the individual handicraft sector, and made full use of various civil resources, including private capital and civil organizations. (2) Small and mid-sized household handicraft workshops played a dominant role in the initial stage of the pattern in the form of enterprise organizations. Because collective resources were very limited, Wenzhou's enterprises existed mainly in the form of small and mid-sized household handicraft workshops. In addition, a large part of the labor force left their hometowns to work in other places, thereby facilitating not only the acquisition of market information and technologies, but also primitive accumulation of private capital. (3) In the industrial structure, the labor-intensive petty commodities industry played a dominant role in the initial stage of the pattern. Household handicraft workshops were usually very small, and local labor resources were very abundant and cheap. Therefore, Wenzhou mainly manufactured labor-intensive petty commodities (e.g., buttons and zippers) in the early stage. In recent years, Wenzhou's equipment manufacturing industries (e.g., transportation equipment manufacturing, manufacture of general purpose machinery, and manufacture of measuring instruments) as well as its port-related industries (e.g., shipbuilding and petroleum processing) have been developing rapidly, and traditional manufacturing industries (e.g., manufacture of leather shoes, manufacture of plastic products, and manufacture of furniture) have been developing at a steady pace. (4) With regard to the conditions of economic development, Wenzhou's agriculture is very underdeveloped. Wenzhou's per-capita cultivated land area is very small, its soil quality and irrigation conditions are far inferior to those of the countryside of south Jiangsu province, and its rural collective economy is very weak. Therefore, Wenzhou's people usually leave their farmlands and hometowns to participate in various non-agricultural industries. (5) In the driving force of development, the Wenzhou pattern mainly relies on the market mechanism and trade associations. The Wenzhou pattern is summarized as a "market-oriented pattern," "spontaneous development pattern," and "self-organizing pattern," and its core is to develop the economy spontaneously and endogenously through market forces. The function of governments is mainly to provide an institutional environment and public goods, and accordingly, the administrative function of governments is weak. In addition, various trade associations are an important driving force of the Wenzhou pattern because they have played a noteworthy role to engender self-reliance, self-support, and self-governance.

Based on the Wenzhou pattern, the following conclusion can be made (1) a region with underdeveloped agriculture and a weak collective economy can accomplish industrialization and urbanization by developing its manufacturing sector; (2) the private economy has a natural link with the market economy, and can enhance the vitality of economic development; (3) according to Hymer's monopolistic advantage theory, the advantages of knowledge-based assets (including market information and technologies) are a basic precondition for the Wenzhou pattern; (4) the labor-intensive manufacturing sector can take full advantage of cheap labor resources; (5) market demand is a decisive factor of industrial development; (6) trade associations provide important support for the Wenzhou pattern; (7) extension of the industrial chain and enhancement of supporting capability work in favor of the development of industrial clusters. Meanwhile, the following points should be noted: (1) industrial development calls for strong support by the capital market; (2) long-term industrial development will be constrained by small industrial size and small enterprise size; (3) in response to market changes and technological progress, the industrial system should be adjusted, optimized, and upgraded constantly; (4) high technological innovation capability provides powerful support for industrial development, and improvement in self-dependent innovation capability is an important means to enhance core industrial competitiveness.

1.2.3 Dongguan Pattern: Industrialization Driven by Foreign-Capital Processing Sector

Located in the economic corridor along HongKong, Shenzhen, and Guangzhou, Dongguan began to develop a foreign-oriented economy very early. Since China's reform and opening-up, Dongguan has attracted more than 10,000 foreign-capital enterprises, actively taken over industries transferred from HongKong and Taiwan, and subsequently, from Japan, South Korea, Europe, and the US, and built a foreign-oriented economy dominated by processing trade, as well as a corresponding economic framework. Over a long period, Dongguan has ranked in the top three among China's cities by total export volume.

The Dongguan pattern is defined as follows: Foreign-capital enterprises provide capital while Dongguan provides land and factory buildings; by taking full advantage of the Chinese mainland's rich labor resources, Dongguan has actively developed its manufacturing sector (e.g., manufacture of electronic products, clothing, toys, and furniture), and has exported these goods to global markets (e.g., the EU, US, and Japan) through the export channels of Hong Kong and Shenzhen. Owing to booming foreign-capital enterprises and private Chinese enterprises, Dongguan has established an economic pattern characterized by one industry driving the development of one town, and one industry creating prosperity for one village. Starting with processing, some towns and villages successfully transformed themselves into highly competitive industrial clusters, such as, Humen's clothing industry, Dalang's wool industry, Houjie's furniture industry, and Shijie's electronic industry. Over the past 30 years, Dongguan's manufacturing sector has gradually transformed from labor-intensive industries to capital-intensive and technology-intensive industries. Private enterprises have gradually extended their business from commerce and catering to manufacturing industries, real estate, and agriculture, and have even entered emerging industries, such as, hi-tech industries and IT industries. In 2013, Dongguan had 174,300 registered enterprises, among which state-owned and collective-owned enterprises accounted for 1%, and private enterprises accounted for the vast majority. Dongguan's total industrial output value was 243.614 billion yuan, and industrial enterprises above a designated size (revenue of more than 20 million yuan) attained an added value of 242.562 billion yuan. The added value of Dongguan's hi-tech manufacturing industries had a year-on-year increase of 19.6% (specifically, a 20.8% rise in the electronic and telecommunications equipment manufacturing industry and an 8% rise in the manufacture of computers and office equipment industry). In 2014, Dongguan's annual fixed investment amounted to 142.711 billion yuan, among which foreign-capital enterprises contributed 15.66%, public enterprises 21.7%, and private enterprises 66.55%.

The Dongguan pattern mainly has the following characteristics. (1) The foreign-oriented economy plays a major role. Dongguan is adjacent to Hong Kong and Macau. By taking full advantage of its geological location and wide network of Chinese living overseas, Dongguan has actively taken over international industrial transfer and developeda foreign-oriented manufacturing economy characterized by the "three-plus-one" tradingmix (customized manufacturing with materials, designs, or samples supplied and compensation trade). As a result, the three-plus-one tradingmix plays a decisive role in Dongguan's economic development. (2) The local private economy has developed with the help of the foreign-capital economy. A number of local private enterprises started and grew rapidly by providing various services for foreign-capital enterprises, and thus, have close links to foreign-capital enterprises. A number of astute and market-conscious laborers have gradually grown into excellent entrepreneurs, by participating in various affairs of foreign-capital enterprises (e.g., operation management, technology development, production, and on-the-job training). In addition, they have fully digested and absorbed knowledge and skills in production, marketing, and business management acquired from foreign-capital enterprises, and they have founded a number of private enterprises in reality. (3) The manufacturing sector has an excellent digestion and absorption ability. By learn from and participating in the business operations of foreign-capital enterprises, local private enterprises have grown rapidly, and continued to improve their digestion and absorption ability. After many years of development, local private enterprises have gradually replaced foreign-capital enterprises in such industries as clothing, shoemaking, and wool, keeping pace with foreign-capital enterprises in such industries as plastics, printing, and furniture, and are catching up with foreign-capital enterprises in electronic and information product manufacturing. (4) The overall development level of Dongguan's enterprises is not high. The Dongguan pattern is faced with various problems, for example, enterprise scale is universally small, technical content and added value of products are not high, there are very few brands with proprietary intellectual property or large-scale, well-known enterprises, and enterprises are still deficient in the ability to withstand market risks and core competitiveness. All these factors have severely restricted the further development of Dongguan's private economy. (5) Products are increasingly hi-tech while labor is simplified. Dongguan is a processing and manufacturing base of foreign-capital enterprises. Assembling industries (e.g., assembly of telecommunications equipment, computers, and other electronic equipment) play a dominant role in Dongguan's economic development, and such products contain rich technical content. However, there is a low technical content for the productive labor required for these industries because the production pattern is dominated by assembly-line work and processing of semi-finished products.

The valuable experience of the Dongguan pattern can be summarized as follows. By taking full advantage of geological location, Dongguan has actively developed the "three-plus-one" tradingmix and has taken over international industrial transfer.

Meanwhile, it is recommended that Dongguan should acquire self-independent innovation capability, develop proprietary brands, encourage preponderant manufacturing enterprises to be deeply rooted in Dongguan, and adhere to resource conservation, environmental protection, and sustainable development.

1.2.4 Shunde Pattern: Development Driven by Large-Scale Manufacturing

Located in the middle of the Pearl River Delta, Shunde is under the jurisdiction of Foshan city and is adjacent to Hong Kong and Macau. Since 1978, Shunde has actively developed township enterprises, foreign-funded enterprises (including Chinese-foreign equity joint ventures, Chinese-foreign contractual joint ventures, and wholly foreign-owned enterprises), and individual enterprises. As a result, Shunde has gradually established an industrial system dominated by the manufacture of textiles and other light industries (e.g., household appliance manufacturing). Within only a dozen years, Shunde has successfully turned itself into a newlyemerging industrial city from a traditional agricultural county known for its mulberry fishponds. Today, Shunde is considered one of the "Four Tigers" of Guangdong Province, including Nanhai, Shunde, Dongguan, and Zhongshan

In the Shunde pattern, collective economy, the manufacturing sector, and supporting enterprises play a dominant role. Shunde's manufacturing sector mainly includes such industries as household appliance manufacturing, transportation equipment manufacturing, electronic product manufacturing, telecommunications cable manufacturing, and ironware manufacturing. In 2014, the total industrial added value of Shunde amounted to 227.991 billion yuan, and the added value created by industrial enterprises above a designated size increased year-on-year by 8.3%. With respect to enterprise ownership, the added value increased as follows: there was a 4.9% rise in state-owned and state-holding enterprises, a 1.7% rise in collective enterprises, a 9.7% rise in joint-stock enterprises, a 6.3% rise in foreign-funded, Hong Kong-funded, Macau-funded, and Taiwan-funded enterprises, and a 10.2% rise in private enterprises. By industries, the added value increased as follows: there was a 4.5% rise in mining industries, a 9.4% rise in manufacturing industries, and a 3.2% rise in public-sector industries (including power supply, heat supply, gas supply, and water production and supply). Among the industrial enterprises above a designated size, the added value increased as follows: there was a 7.7% rise in the processing of food from agricultural products, a 6.7% rise in the manufacture of textiles, a 9.1% rise in the manufacture of general purpose machinery, a 6.9% rise in the manufacture of special purpose machinery, an 11.8% rise in the manufacture of automobiles, a 12.2% rise in the manufacture of computers, communication, and other electronic equipment, and a 9.4% rise in the manufacture of general purpose machinery. The total added value created by six highly energy-consuming industries increased year-on-year by 7.5%; specifically, there was a 9.3% rise in the manufacture of non-metallic mineral products, a 10.3% rise in the manufacture of raw chemical materials and chemical products, a 12.4% rise in the smelting and pressing of non-ferrous metals, a 6.2% rise in the smelting and pressing of ferrous metals, a 2.2% rise in the generation and supply of electric power and heat, and a 5.4% rise in the processing of petroleum, coking, processing of nuclear fuel. The added value created by hi-tech manufacturing industries increased year-on-year by 12.3%, accounting for 10.6% among all industrial enterprises above a designated size. The added value created by the manufacture of equipment increased year on year by 10.5%, accounting for 30.4% among all industrial enterprises above a designated size.

The Shunde pattern mainly has the following characteristics. (1) It emphasizes large-scale manufacturing and the pursuit of the scale effect. Large-scale manufacturing is an inevitable requirement for modern enterprise development. In 1991, 5 out of China's top 10 township enterprises were from Shunde. (2) The pattern intensifies brand building, attaches importance to brand investment, and earns brand benefits. Household appliance manufacturing, furniture manufacturing, and floriculture are Shunde's three dominant industries, which are well known in both China and abroad. For both foreign and domestic capital, Shunde is the preferred destination of investment in these three industries. Shunde's manufacture of household electrical appliances boasts the largest number of national well-known trademarks in China, accounting for 31%. (3) With the Shunde patter, a strong supporting capability for household appliance manufacturing was developed. Shunde is a manufacturing base for household white appliances, and has a large number of manufacturing enterprises of fittings for household appliances (from raw materials through to core parts). Today, Shunde is one of the regions in China in which are based a large number of manufacturers of fittings for household appliances and that has a strong supporting capability for the household appliance manufacturing industry. The output value created by the fitting manufacturing industry accounted for 5% of Shunde's total industrial output value, and 10% of the output value of Shunde's household appliance manufacturing industry. (4) The Shunde pattern promotes industry upgrade in response to consumption upgrade. Shunde's household appliance manufacturing industry is highly developed. Today, the low-end household appliance market is almost saturated and the living standards of people have improved continuously. Accordingly, Shunde's household appliance manufacturers actively carry out industrial and product upgrades, and launch various hi-tech household appliance products, so as to increase industrial profits.

The valuable experience of the Shunde pattern can be summarized as follows. Industrial clusters and scale effect are vitally important for the development of the manufacturing sector, and proprietary brands should be developed actively, so as to earn brand benefits. Meanwhile, it is advisable to exploit and utilize land resources and protect the ecological environment during the development of the manufacturing sector, to improve the self-dependent innovation capability, and to develop middle and high-end products, so as to increase the added value.

1.2.5 Zhongguancun Pattern: Development Driven by Technological Innovations

Zhongguancun Science Park is China's first national-level hi-tech development zone. Zhongguancun Science Park possesses the most intensive resources in technologies, brainpower, and information. Over a dozen years, Zhongguancun Science Park has maintained an annual economic growth rate of at least 30%, and has built an industrial system dominated by the software industry, information service industry, IT manufacturing industry, optical-electronic machinery integration industry, new material industry, bioengineering and medicine industry, and environmental protection industry. Today, Zhongguancun Science Park has become an important economic growth source for Beijing. The electronic information industry of Zhongguancun Science Park mainly focuses on the computer industry (including computer hardware, computer software, and information services), and plays a leading role in China's hi-tech industries. The hi-tech products launched by Zhongguancun Science Park have a large market share in domestic Chinese markets, and have anauthoritative market share in international markets.

The Zhongguancun pattern mainly has the following characteristics. (1) Market forces played a dominant role in the early stage of the pattern, and significant government support was provided in later stages. Zhongguancun Science Park is blessed with a number of colleges and universities, scientific research institutes, and high-caliber professional and technical personnel, as well as innovative entrepreneurs and educators. In the context of the reform and opening-up, Zhongguancun Science Park has gradually turned itself into a unique hi-tech industrial park. In some respects, the Zhongguancun pattern is similar to Japan's Tsukuba pattern; specifically, the preferential policies provided by governments play a driving role in economic development. (2) Zhongguancun Science Park is backed by colleges, universities, and a scientific research institute, and technology-based enterprises spring up successively. Zhongguancun is blessed with a large number of scientific research institutes, thereby providing powerful intellectual resource support for the development of technology-based enterprises. The initiators of the newly emerging hi-tech enterprises are mostly from the Chinese Academy of Sciences, colleges, and universities. By relying on the powerful technical support of scientific research institutes, colleges, and universities, the enterprises in the Zhongguancun Science Park actively develop new products. In the early 1980s, China generally lagged behind overseas advanced countries in computer technologies. In a few specific areas, however, China has established its own technological advantages and accumulated technological reserves. Due to the shackles of the traditional economic system, these technological achievements were set aside, and thus, could not be commercialized. For example, Xuan Wang, a professor of Beijing University, began to explore laser phototypesetting technology early in the 1970s, which provided technological reserves for the advent of the Founder Group. Moreover, the rich technological reserves acquired by the Institute of Computing Technology in the Chinese Academy of Sciences, especially the significant merits of Guangnan Ni, were vital for the advent of Lenovo. Commercializing technological reserves quickly was an important motivating force behind the rapid development of newly emerging enterprises in the 1980s. (3) Zhongguancun Science Park pursued a technology-industry trade integration pattern (e.g., supporting technological development via trade, and supporting the manufacturing sector via trade). Developed in a closed environment, China's electronic and computer industries lagged far behind those of foreign advanced countries in terms of technological level, manufacturing scale, and degree of commercialization. In that situation, it was imperative that advanced technologies and products were introduced from developed countries. Due to the shackles of the traditional economic system and lack of related professional personnel and experience, the old-brand electronic enterprises and traditional trade companies could do nothing to introduce foreign advanced technologies and products. This presented a valuable opportunity for the newly emerging enterprises of Zhongguancun. Zhongguancun was blessed with a large number of computer professionals, and the Chinese Academy of Science was actively exploring a new development pattern. As a result, many newly emerging enterprises established themselves in trade and agency services, and embraced rapid development due to flexibility of their business operations.

After many years of development, the hi-tech enterprises have acquired a certain financial strength and marketing competence, and have built their sales networks. Furthermore, Legend, Founder, and Stone Group have successfully established themselves in the computer sector. Zhongguancun Science Park carries out a market-oriented development strategy, and has established a technology-industry trade development pattern that integrates R&D, production, sales, and services.

The Zhongguancun pattern can be summarized as follows. It takes full advantage of intensive intellectual resources, and allocates resources and promotes economic development through the market mechanism. However, there are very few brand-name enterprises and products that originated in Zhongguancun Science Park, and it has not established an appropriate mechanism and environment in which technological innovation capability can be transformed continuously into economic resourcefulness.

1.2.6 Qingdao Pattern: Brand-Name Enterprises Play a Dominant Role

Qingdao is an important industrial base in China's modern history. Historically, Qingdao was one of China's three textile industry bases of Shanghai, Qingdao, and Tianjin. Since China's reform and opening-up, Qingdao's manufacturing sector has embraced rapid development, and is the pillar of Qingdao's economy. Since the 12th Five-year Plan period, Qingdao's secondary sector has maintained an annual growth rate of at least 39%. In 2014, the total output value created by industrial enterprises above a designated size was 1.6761 trillion yuan, a year-on-year increase of 9.6%; the output value of Qingdao's hi-tech industries increased year-on-year by 16.7%, accounting for 40.7% of the total output value created by the industrial enterprises above a designated size. The industrial upgrade involved speeding up growth in the following industries: household appliances and electronic products, machine manufacturing, petrochemicals, and food and beverages. Today, Qingdao has become an important manufacturing sector base in China and even in the world. The manufacturing sector is the primary driving force of Qingdao's economic development. Among China's 42 major economic center cities, Qingdao ranks 9th by total industrial output value, 8th by industrial added value and 10th by industrial profit. Today, Qingdao is among China's main cities with rapid development of the manufacturing sector and significant economic vitality.

The Qingdao pattern mainly has the following characteristics. (1) The transportation industry is highly developed. Qingdao Port has a throughput capacity of at least 100 million tons, and ranks in the top 20 globally by international container throughput capacity. Qingdao Airport operates 40 domestic air routes and 9 international air routes, ranking in the top 12 in China by throughput capacity. Various newly built expressways (e.g., Jinan-Qingdao Expressway, Weifang-Laiyang Expressway, and Qingdao-Yinchuan Expressway) and the Jiaozhou-Xinyi Railway have improved the transportation conditions of Qingdao further. With rapid development of modern information technologies, China is linked with the world more closely. Certain water conservancy projects (e.g., water diversion from the Yellow River into Qingdao and the South-to-North Water Diversion) and expansion of power plants has provided strong guarantees for water and power supplies. (2) Qingdao boasts a number of brand-name enterprises and products, thereby enhancing the overall competitiveness of its manufacturing sector. Qingdao boasts a number of domestic well-known trademarks (including Haier, Qingdao Beer, Hisense, Aucma, and Double Star). Among the first of 57 well-known Chinese brands, Qingdao has 8 well-known brands; up to 73 product categories have been awarded the title of "Shandong's Well-known Brands." By carrying out a well-known brand strategy, a number of excellent enterprises have stood out. For example, Qingdao's top 10 enterprise groups play a leading role, and its top 10 enterprises and top 10 hi-tech enterprises play an important role. As a result, a ladder development pattern has taken shape. Headed by large enterprises, Qingdao has actively carried out intensive construction, logistics development, and informatization of industrial parks, and has built a new framework for a modern manufacturing sector, thereby enhancing the competitiveness of its manufacturing sector significantly. (3) Economic openness has increased further, and a number of foreign-funded manufacturing enterprises (Chinese-foreign equity joint ventures, Chinese-foreign contractual joint ventures, and wholly foreign-owned enterprises) have invested in Qingdao. More than 40 of the world's top 500 enterprises have invested in Qingdao, including Lucent, Rhone-Poulenc, Mitsubishi Heavy Industries, Panasonic, and HP. The fields of their investment have extended from labor/resource-intensive industries (e.g., manufacture of food and beverages, manufacture of textile wearing and apparel) to technology-intensive industries (e.g., electronic product manufacturing, telecommunications equipment manufacturing, machine manufacturing, and chemicals). Such foreign-funded enterprises have become an important part of Qingdao's manufacturing sector.

The Qingdao pattern can be summarized as follows. It is centered on supporting enterprises, and establishes preponderant manufacturing enterprise groups with a certain industrial concentration ratio; it aims to achieve international segmented markets; it builds world-class brands and enterprises and continuously enhances international fame and competitiveness. Meanwhile, the Qingdao pattern is mainly faced with the following problems. The manufacturing sector has very few monopolistic technologies with proprietary intellectual property and private manufacturing enterprises develop somewhat slowly.

1.2.7 Tianjin Binhai Pattern: Development Driven by Port Transportation Trade

Centrally located in the Circum-Bohai Sea region, Tianjin Binhai New Area has a remarkable location advantage, solid industrial foundation, and huge economic growth potential. It is backed by the Beijing-Tianjin-Hebei region, serves the Circum-Bohai Sea region, has a radiating influence on north China, northwest China, and northeast China, and faces the northeast Asia region. In particular, it has the potential to play a demonstrative and leading role in the old industrial base of northeast China. The hardware and software facilities of its port, wharfs, and industrial park are all constructed using high standards, and meet advanced international levels. This provides a basic guarantee for the rapid development of its manufacturing sector.

The Tianjin Binhai pattern mainly has the following characteristics. (1) Tianjin Binhai New Area has a locational advantage. It is located at the crossing point between the Beijing-Tianjin urban belt and the Circum-Bohai Sea urban belt; it is a transportation hub that connects China with the world, and has a radiating influence on north China and northwest China; it has the largest trade port in the northern part of China; it has a wide hinterland, operates a large number of shipping lines, and has trade contacts with more than 300 ports of at least 170 countries or regions; it is the starting point of the Eurasian Continental Bridge in the northern part of China. (2) Tianjin Binhai New Area boasts high economic openness, and is the only region in China blessed with a port, an economic development zone, a bonded zone, a hi-tech development zone, and a large industrial base. After 10 years of development and construction, Tianjin Binhai New Area has built an industrial system that covers such industries as marine chemicals, petrochemicals, metallurgy, machinery, electronics, biochemistry, and food and beverages. Therefore, a good manufacturing foundation has been formed. (3) Tianjin Binhai New Area has a solid industrial foundation, good infrastructure, and a strong technical force. Some industrial clusters have a considerable influence in both China and abroad. Such industrial clusters include the metallurgical industry (represented by steel tube enterprises), the marine petrochemical industry (represented by Dagang Oilfield, Bohai Oilfield, Tianjin Petrochemical, and Bohai Chemical), the electronic and computer industry (represented by Motorola and Samsung), the machinery industry (represented by Germany SEW and VICTOR Taichung Machinery Works), and the food industry (represented by Novo Nordisk and Smith Kline). By way of industrial clustering, Tianjin Binhai New Area has established pillar industries, including the electronic information industry, machine manufacturing industry, bio-pharmaceutical industry, chemical industry, and food and beverages industry. The electronic information industry (headed by Motorola and Samsung) and the machine manufacturing industry (headed by Toyota and related enterprises), which have become industrial clusters with considerable scale.

The Tianjin Binhai pattern can be summarized as follows. It seized the opportunity fora Chinese manufacturing sector to become integrated into the global economy, making the best of its location advantage; reduced overall business costs, and gave full play to its comparative advantages. Meanwhile, the Tianjin Binhai pattern is faced with the following problems: (1) the industrial level is somewhat low, and the synergy and complementarity between different leading industries are weak; and (2) related service industries have developed very slowly and cannot provide effective support for the rapid development of leading industries.

1.2.8 Wuhan Pattern: Clustering Development of the Optoelectronic Information Industry

Wuhan's economic development zones started very early and developed very rapidly. The Donghu Hi-tech Development Zone, Zhuankou Development Zone, and Dongxihu Development Zone complement each other industrially, and constitute a locational corner. Today, they play amore notable driving role in the economic development of Wuhan. Specifically, Donghu Hi-tech Development Zone has become a hi-tech industrial park admitted by Asia-Pacific Economic Cooperation. Together with the Wujiashan Cross-Strait Hi-tech Industrial Park and College Software Park, they jointly constitute a multi-park and multi-zone pattern. Generally, Wuhan has met the industrial conditions for developing the optoelectronic industry and building the Wuhan Optics Valley. In recent years, some rapidly developing and influential optoelectronic enterprises have stood out, constituting an optoelectronic industry chain. In 2014, the selling profits earned by the optoelectronic information enterprises in Wuhan Optics Valley increased year-on-year by at least 10%, and were 10 times as much as those of Wuhan's manufacturing sector; the optoelectronic information enterprises in Wuhan Optics Valley contributed 37.37% of the total profits of Wuhan's manufacturing sector. In Donghu Hi-tech Development Zone alone, there are nearly 100 enterprises engaged in R&D and production in optoelectronic information. Wuhan ranks first in China by the production capacity of fibers and optical cables, and has adefinitive production capacity in optoelectronic devices and optoelectronic communication equipment.

The Wuhan pattern is mainly characterized by industrial clustering. In addition to optoelectronic-related resources, Donghu Hi-tech Development Zone has an abundant variety of clustered resources. In addition to the optoelectronic sector, Wuhan Optics Valley has developed a variety of other hi-tech industries. Therefore, the industrialization pattern of Wuhan Optics Valley should be industrial clustering, with a view to integrating the resources of the Donghu Hi-tech Development Zone and Wuhan city fully, and allowing full play to the overall growth effect and role of Wuhan Optics Valley. Specifically, the information industry should play a dominant role, and Wuhan Optics Valley should host a variety of hi-tech industrial clusters that are capable of sustainable development and industrial upgrade, such as, IT, life science, new energy, new materials, space technologies, marine technologies, and environmental protection. Among the diverse industrial clusters, the optoelectronic industry should play a core and dominant role and should be a driving force for the development of other industrial clusters.

The Wuhan pattern can be summarized as follows. It takes full advantage of Wuhan's technological and human resources, develops the optoelectronics sector cluster, and allows full play to scale and cluster effects. Meanwhile, it is recommended that Wuhan should give importance to both domestic and international markets; develop a series of products characterized by deep processing, high added value, and knowledge intensiveness; cooperate with colleges, research institutes, and enterprises in both China and abroad; expand industrial scale based on core advantages; and attempt to enhance international competitiveness.

1.2.9 Xi'an Pattern: Distinctive Development of the Equipment Manufacturing Sector

Xi'an is a manufacturing base of China with strategic significance, and has significant strength in the manufacturing of electric power equipment, military equipment, airplanes, spaceflight equipment, and machine tools. In addition, Xi'an is an important scientific research base in the fields of spaceflight, aviation, weapons, machinery, electronics, measuring instruments, and agriculture. In 2014, the total industrial added value of Xi'an amounted to 152.312 billion yuan, a year on year increase of 10.7%; and the added value created by industrial enterprises above a designated size amounted to 119.528 billion yuan, a year on year increase of 11.1%. Among the industrial enterprises above a designated size, the added value created by light-industry enterprises amounted to 29.358 billion yuan, a year on year increase of 8.9%; and the added value created by heavy-industry enterprises amounted to 90.17 billion yuan, a year on year increase of 11.8%. Among the industrial enterprises above a designated size, the added value increased as follows: there was a 140% rise in the manufacture of computers, communication, and other electronic equipment, an 11.9% rise in the manufacture of metal products, a 13.7% rise in the manufacture of automobiles, and a 21.9% rise in the manufacture of railway, ship, aerospace, and other transportation equipment. The total added value created by six highly energy-consuming sectors increased year on year by 9.4%, or specifically, there was a 10.8% rise in the manufacture of non-metallic mineral products, a 35.1% rise in chemical materials and chemical products manufacturing, a 39.2% rise in the smelting and pressing of non-ferrous metals, a 20.8% decline in the smelting and pressing of ferrous metals, a 5.5% rise in electric power and heat production and supply, and an 11.1% decline in processing of petroleum, coking, processing of nuclear fuel.

The Xi'an pattern is mainly characterized by the development of the distinctive equipment manufacturing industry. In recent years, large and medium-sized state-owned equipment manufacturing enterprises of Xi'an exhibited significant strength in China's new round of economic growth, embraced rapid development, and became leading enterprises in China's equipment manufacturing industry. Generally, Xi'an has a significant advantage in the development of a distinctive equipment manufacturing industry. In this regard, typical enterprises include Xi'an Aircraft Industry (Group) Co., Ltd., Xi'an Electric Power Machinery Manufacturing Co., Ltd., and Shaanxi Blower (Group) Co., Ltd.

The Xi'an pattern can be summarized as follows. Backed by the rich scientific research resources and favorable software/hardware environments, Xi'an provides key support to the equipment manufacturing industry, key component manufacturing industry, and core spare parts industry, thereby promoting the leap-forward development of the equipment manufacturing industry of Xi'an and driving the development of the new material industry and mechanical-electrical industry. Meanwhile, it is recommended that Xi'an optimize the organizational structures of enterprises, foster large-scale supporting enterprises, intensify specialized division of labor, develop industrial clusters, improve self-dependent innovation capability, and drive the scientific development of other manufacturing industries via the distinctive equipment manufacturing industry.

1.3 Position of China's Manufacturing Sector in the World

Ever since the 1990s, China's manufacturing sector has maintained continuous rapid development and has been the main driving force for the continuous rapid development of China's economy. Today, China is transforming itself into a new-generation manufacturing center in the world, and is playing an increasingly important role in the division of the global manufacturing system.

1.3.1 Largest Global Scale

According to statistical data from the World Bank, in 2009, the added value of China's manufacturing sector surpassed that of the US's manufacturing sector, ranking China first in the world. In 2013, the added value of China's manufacturing sector accounted for 23.2% of the total added value of the global manufacturing sector, and the yields of at least 220 product categories each ranked first in the world; meanwhile, in 2013, the added value of the US's manufacturing sector (ranked second in the world) declined to 17.2% of the world's manufacturing sector total (data source: United Nations Statistics Division, May 2015) (Table 1.1).

According to the *International Trade Statistics 2013* published by the World Trade Organization, (1) in 2012, the export volume of the world's manufacturing sector amounted to 11.49 trillion US dollars; (2) the export volume of the US's manufacturing sector amounted to 1.102 trillion US dollars, accounting for 9.6% of the global manufacturing sector total (compared with 13.8% in 2000), indicating that the export capacity of the US's manufacturing sector was on a downward trend; (3) the export volume of Japan's manufacturing sector amounted to 710 billion US dollars, accounting for 6.2% of the global manufacturing sector total (compared with 9.6% in 2000; 4) the export volume of China's manufacturing sector amounted to 1.925 trillion US dollars, accounting for 16.8% of the global manufacturing sector total. Specifically, steel accounted for 11.1% of the Chinese total, textiles 33.4%, clothing 37.8%, office and telecom equipment 32.3%. In addition, the following industries had the largest proportion of the world's total export volume: electronic data processing and office equipment industry (41.2%), telecommunications equipment (36.1%), and integrated circuits and electronic components (17%). The export volume of China's manufacturing sector accounted for 94% of China's total export volume. The abovementioned statistical data show that China's manufacturing sector has powerful export capacity, and assumes a very important position in the global manufacturing sector.

Table 1.1 Export volumes of the manufacturing sectors of them in countries/regions of the world

Country/region	Export volume (Unit: 1 billion US dollars)					Proportion of total export volume (%)	
	1990	2000	2010	2011	2012	2005	2012
US	290	646	944	1047	1102	81	71.3
Japan	275	450	680	725	710	91.9	88.9
South Korea	61	155	412	473	463	90.8	84.4
China	44	220	1477	1772	1925	91.9	94
Hong Kong	76	192	373	408	423	95.8	85.9
Taipei	62	141	245	271	262	86.5	87
The world	2391	4691	9986	11,523	11,490	71.2	64.1

Data source: *International Trade Statistics 2013* published by the World Trade Organization

1.3.2 Rapid Improvement in Self-dependent Innovation Capability

In order to become a strong manufacturing country, the manufacturing sector of a nation must acquire world-class technologies. The UK, US, Japan, and Germany were all among the world's scientific and technological centers in their historical stages. In the decade before World War II, the US's manufacturing sector not only enjoyed a leading edge in scale, but also acquired an unrivalled technological level. For example, 60% of global technological innovations were made in the US. A country's technological innovation capability can be measured by the increase in scientific and technology personnel, R&D expenditures, and patent grants.

1. China's manufacturing sector has the largest scale of R&D personnel globally

In 2013, Chinahad 3.533 million R&D personnel, ranking first in the world and far above many countries or regions in the world. By contrast, the US had 1.68 million R&D personnel (2011 estimation), Japan had 866,000 R&D personnel, and Germany had 591,000 R&D personnel. Relative to China's huge employed population, however, the proportion of China's R&D personnel per 10,000 employees (2013) is not large at 45.9, compared with 160.4 in South Korea (2012) and 133.5 in Japan (2013). Figure 1.1 shows the related statistics of certain typical countries.

2. Patent applications of China's manufacturing sector rank first globally by growth rate, and third globally by total quantity

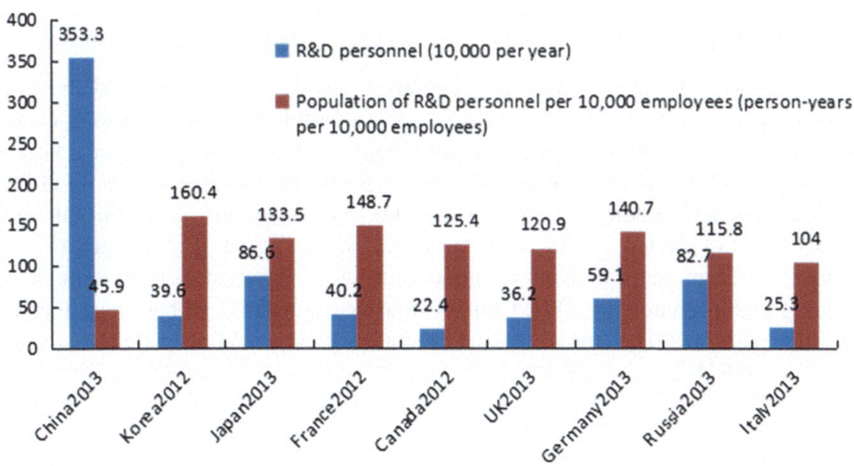

Fig. 1.1 Number of R&D personnel in some countries. Data source: *Main Science & Technology Indicators 2015/1* (OECD), provided by the Ministry of Science and Technology, China

While technology transfer and R&D become increasingly active in the world, transnational companies have attached more importance to intellectual property protection. According to statistical data published by the World Intellectual Property Organization, a total of 205,000 international patent applications were submitted in 2013 via the Patent Cooperation Treaty (PCT), a year on year increase of 5.1% from 195,331 in 2012. Specifically, the US (57,239 PCT patent applications, accounting for 27.9% of the global total) continuously kept far ahead of other countries or regions; Japan (43,918 PCT patent applications, accounting for 21.4% of the global total) continuously ranked second; while China submitted 21,516 PCT patent applications, an increase of 15.5% from the previous year and accounting for 10.5% of the global total. Although China lagged far behind the US and Japan, China surpassed Germany (17,927 PCT patent applications) for the first time to rank third globally. In addition, China was granted a total of 1851 triad patents, an increase of 11.6% over the previous year, to rank sixth globally. These data show that China's self-dependent innovation capability is being improved significantly.

3. R&D expenditures of China's manufacturing sector rank second globally

In 2013, China's R&D expenditures amounted to 191.2 billion US dollars (equivalent to 1.18466 trillion yuan), surpassing Japan's R&D expenditures (170.9 billion US dollars), and ranking second globally. However, China's R&D expenditures were still less than half that of the US (about 42%). China's R&D expenditure input intensity was 2.01%, still lagging behind Japan (3.49%), Germany (2.94%), and the US (2.81%), but the figure increased by 0.32% points over 2010, was above the average level of the EU's 28 member countries (1.92%), and was equivalent to the R&D expenditure input intensity of moderately developed countries (Fig. 1.2).

4. Breakthrough by China's manufacturing sector in building global brands

The world's strong manufacturing countries usually occupy high-end links of value chains (e.g., marketing and brands), keep control of global marketing channels, and own well-known global brands. As the world's No. 1 manufacturing country, the US has the largest number of proprietary well-known brands. In 2015, the US owned 52 among the world's top 100 brands, slightly lower than the 53 brands it had in the top 100 in 2005, but remaining far ahead of other countries or regions. In recent years, China's manufacturing sector has made a breakthrough in building global brands. In 2004, China's Huawei was rated a global brand for the first time; and in 2005, Lenovo was selected among the world's top 100 brands (as listed in Table 1.2).

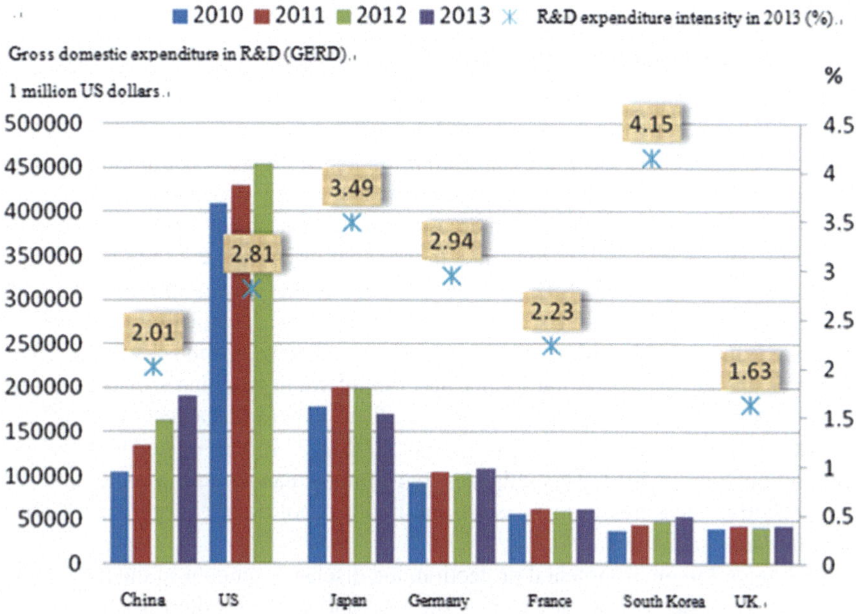

Fig. 1.2 R&D expenditures of certain countries. Data source: *Statistics Report on Science and Technology* (March 2015), published by the Division of Innovative Development in the Ministry of Science and Technology

Table 1.2 Quantity of brands among the world's top 100 brands by different countries

Country item	Quantity of top brands (2005)	Quantity of top brands (2015)
US	53	52
Japan	7	6
Germany	9	10
UK	4.5	6
France	8	7
Italy	4	2
Switzerland	5	2
Holland	3.5	3
South Korea	3	3
Spain	1	2
China		2
Finland	1	
Sweden	1	2
Mexico		1
Canada		1
Denmark		1

Data source: *The 100 Best Global Brands* released by Inter brand on October 6, 2015 (http://interbrand.com/best-brand)

1.3.3 Problems with Resource Utilization and Environmental Protection

The rapid growth of China's manufacturing sector mainly relies on large quantities of resource inputs, and the energy consumption and resource consumption per unit product are far above the world average level. The energy consumption of China's manufacturing sector continues to increase year by year; specifically, there was an increase from 1.43051 billion tons of standard coal in 2006 to 2.22528 billion tons of standard coal in 2013. Likewise, the electric power consumption of China's manufacturing sector continues to increase year by year; specifically, there was an increase from 1.5371 trillion kWh in 2006 to 2.8638 trillion kWh in 2013. However, the energy consumption per unit output value in China's manufacturing sector continued to decline year by year; specifically, there was a decline from 5210 tons of standard coal per 100 million yuan in 2006 to 2340 tons of standard coal per 100 million yuan in 2013. This indicates that China's manufacturing sector actively takes effective measures in response to the national policy of energy conservation and emissions reduction.

With regard to environmental protection, the discharge amount of the three types of industrial waste (i.e., wastewater, waste gas, and waste residue) per unit output value generally tended to decline slowly during 2006–2013. Specifically, the discharge amount of wastewater per 100 million yuan output value declined from 60,700 tons in 2006 to 21,500 tons in 2013; the discharge amount of waste gas per 100 million yuan output value declined from 79 million standard cubic meters in 2006 to 49 million standard cubic meters in 2013; and the discharge amount of waste residue per 100 million yuan output value declined from 13.95 tons in 2006 to 0.73 tons in 2013. Meanwhile, the total output value by comprehensive utilization of the three types of industrial waste generally tended to increase; specifically, there was an increase from 91.631622 billion yuan in 2006 to 223.7174245 billion yuan in 2013.

In summary, China has made a breakthrough in both economic development and technological innovation, and has gained a distinctive comparative advantage. Meanwhile, China has made definitive progress in resource utilization and environmental protection, but is still expected to make further efforts.

Chapter 2
Connotations and Prospects of the New-Type Manufacturing Sector

Lijie Zhang, Wei Sun and Changkai Wang

In the 21st century, China is being confronted with not only various urgent challenges (including industrialization, urbanization, modernization, and informatization), but also various pressures (including energy/resource shortages and environmental pollution). In this context, it is imperative for China to carry out new-type industrialization characterized by high technological content, high economic benefits, low resource consumption, minimal environmental pollution, and full utilization of human resources. This study sets forth connotations for new-type manufacturing sector from four aspects: economic resourcefulness, technological innovation capability, energy conservation capability, environmental protection capability. Then, this study forecasts the impacts of the new-type manufacturing sector on social and economic development. Finally, this study proposes an evaluation index system for the new-type manufacturing sector, which comprises four indexes and 31 subindexes.

2.1 Connotations of the New-Type Manufacturing Sector

Research on the connotations and evaluation of the new-type manufacturing sector mainly has been conducted in China. Li and Du (2004) were the first to define the connotations of the "new-type manufacturing sector," pointing out that the new-type manufacturing sector was capable of improving technological innovations, reducing energy consumption and environmental pollution, increasing employment, improving economic benefits, strengthening competitiveness, and achieving sustainable development. The evaluation index system for the new-type manufacturing sector covered three major indexes, namely, economic resource-

L. Li and Z. Du (eds.), *A Research Report on the Development of China's Manufacturing Sector (2016)*, Current Chinese Economic Report Series,
DOI 10.1007/978-981-10-4445-8_2

fulness, technological competitiveness, and resource and environment protection capability, as well as 20 subindexes. Using the abovementioned evaluation index system, Li and Zhou (2005) used the principal component analysis method to conduct comparative analysis and clustering analysis for the degree of newness of manufacturing industries in China's 30 regions, and summarized the development characteristics of manufacturing industries in these regions. Based on the above-mentioned philosophy of a new-type manufacturing sector, Li and Zang (2008) used a three-dimensional time-series method to conduct comparative research on the manufacturing industries of China and main developed countries from three aspects, namely, economic resourcefulness, technological innovations, and resource and environment protection. The research results were as follows: (1) China was far below the US and Japan in the output value of its manufacturing sector, and lagged far below developed countries in technological innovation inputs and outputs; (2) China's manufacturing sector was notably characterized by high energy consumption and high pollution, but insufficient inputs in pollution abatement; (3) China's manufacturing sector had a certain comparative advantage in growth rate, product yield, employment, and labor costs. From the perspective of environmental protection, Xu (2010) established an index evaluation system for the new-type manufacturing sector based on three yardsticks, that is, low energy consumption, low pollution, and low emissions, and accordingly analyzed the degree of newness of Jiangsu province's manufacturing sector. Using specific industries as research samples, some scholars built an evaluation index system for the new-type manufacturing sector, and analyzed the newness of specific industries accordingly. Liu (2006) put forward the concept of the new-type equipment manufacturing industry, and built a competitiveness evaluation system for this new-type equipment manufacturing industry; using this competitiveness evaluation system, Liu (2006) used a mathematical statistics method to quantitatively analyze the competitiveness of Liaoning province's new-type equipment manufacturing industry based on four yardsticks, namely, industrial technological competitiveness, economic benefit competitiveness, human resource competitiveness, and regional technological competitiveness. Lv (2014) built an evaluation system for Liaoning province's new-type equipment manufacturing sector, which comprised seven major indexes, namely, informatization, resource utilization, technological content, economic benefit, environmental protection, human resource utilization, and openness and 20 subindexes; using the analytic hierarchy process, Lv (2014) built a comprehensive evaluation model for the new-type equipment manufacturing industry. Wang (2007) conducted systematic evaluation for the evolutionary level of the equipment manufacturing industry by using three yardsticks, including economic benefits, technological potential, and environmental harmony; specifically, economic indexes were used to measure the current contributions made by the equipment manufacturing industry to the national economy, technological indexes were used to measure the future competitiveness of the equipment manufacturing industry, and environmental indexes were used to measure the development sustainability and long-term benefits of the equipment manufacturing industry. Zhang (2009) studied how to evaluate the technological innovation capability of China's manufacture of computers, communication, and other electronic equipment; specifically, the technological innovation capability for this industry was broken down into four primary indexes, that is, innovation environment, potential innovation resources, innovation inputs, and innovation outputs, and 15 secondary indexes. From the perspective of industrial security, Zheng (2010) built an evaluation index system for China's manufacture of medicines, which comprised five indexes, namely, self-dependent innovation capability, development environment, international competitiveness, industrial controlling power, and external dependence.

Regarding the connotations and evaluation of the new-type manufacturing sector, abundant research findings have been attained. Such findings are of high theoretical and practical significance. The evaluation index system mainly focuses on economic resourcefulness, technological innovation capability, and resource and environment protection capability, but gives very little consideration to social contribution capability, including employment and taxes. While the mass of people pay more attention to enterprises' social responsibilities (e.g., employment and taxes), it is necessary to redefine the concept of the new-type manufacturing sector.

This research study considers that the connotations of the new-type manufacturing sector should include not only economic resourcefulness, technological innovation capability, energy conservation capability, and environmental protection capability, but also social contribution capability. Therefore, the new-type manufacturing sector can be redefined as the manufacturing sector that is capable of improving technological innovation capability, reducing energy consumption and environmental pollution, improving economic benefits, increasing employment opportunities and tax contributions, and attaining sustainable development.

1. Economic resourcefulness

Economic resourcefulness is an important part of the new-type manufacturing sector, and is an important yardstick for the degree of newness of the manufacturing sector. Economic resourcefulness is particularly important for a country that is in the process of industrialization. Only by creating great economic benefits, can China's manufacturing sector obtain the impetus for sustainable development and provide material support for developing science and technologies, improving economic efficiency, increasing employment opportunities, enhancing the taxpaying capability, and protecting the environment.

2. Technological innovation capability

Technological innovation capability is an important part of the new-type manufacturing sector. In the process of transforming from extensive traditional manufacturing sector to the intensive new-type manufacturing sector, science and technologies are of vital importance. Only by making the best of modern science and technologies, conducting technological innovations, and taking full advantage of China's human resources can China's manufacturing sector improve efficiency and benefits, reduce environmental pollution, develop hi-tech industries, and be transformed from the "Made in China" mode to the "Created in China" mode.

3. Energy conservation capability

Nowadays, China's traditional manufacturing sector adopts an extensive production mode characterized by low benefits, high consumption, and high pollution. This has aggravated China's resource shortages and ecological deterioration. The resources herein mainly refer to natural resources that are closely linked to the development of China's manufacturing sector and other biological resources used as industrial raw materials. These resources provide the material basis for the production activities of China's manufacturing sector, and many of them are non-renewable. Unreasonable utilization of resources will lead to resources waste and environmental deterioration. Reasonable and efficient utilization of resources is increasingly important for the sustainable development of the social economy.

4. Environmental protection capability

Environmental protection capability refers to the comprehensive capability to solve real or potential pollution problems encountered in the production activities; coordinate the relationship between economic activities and environmental protection; and ensure sustainable development of the social economy. Environmental harmony and ecological protection are prerequisites for sustainable social and economic development. Environmental protection attained by technological progress is increasingly important for the sustainable development of China's manufacturing sector.

2.2 Motivating Force and Prospects of the New-Type Manufacturing Sector

The 16th National Congress of the CPC put forth the concept of "new-type industrialization," and creatively resolved that China should follow a new-type industrialization path characterized by high technological content, high economic benefits, low resource consumption, minimal environmental pollution, and full utilization of human resources. In 2015, China adopted the *Made in China 2025* development program. New-type industrialization is mainly embodied in innovations in China's manufacturing sector, and the enhancement of overall national strength largely relies on the development of the manufacturing sector. In addition, the manufacturing sector is the foundation and pillar of a country's production capacity and national economy. This has been proved by the practices of developed countries. It should be noted that China's manufacturing sector is faced with various bottlenecks (e.g., environmental, resource, and population bottlenecks). Thus, China's manufacturing sector should take a new-type industrialization path, and China should actively develop the new-type manufacturing sector by seizing the development opportunity created by the knowledge economy era and global economic integration. As a representative force of industrial progress, the new-type manufacturing sector is gradually taking shape and the theoretical system for it has been improved continuously. The new-type manufacturing sector adheres to the philosophy of "human-oriented, self-independent innovation that is environmentally friendly and future oriented." In light of this philosophy and the related findings in *A Research Report on the Development of China's Manufacturing Sector* (2004–2014), this study further examines the motivating forces and prospects for the new-type manufacturing sector.

Economic motivation: The core of traditional industrialization is to develop a large-scale and high-efficiency manufacturing sector, and thereby drive the development of agriculture and service sectors, with a view to providing abundant material products to the society. The production process largely relies on the massive consumption of capital, labor, and energy, thereby causing a variety of severe problems, including environmental pollution, resource shortages, and a significant wealth gap. As a type of industrialization driven by informatization, new-type industrialization emphasizes the role of knowledge and technologies in production activities, and has changed the constitution of conventional production factors and the relative importance of different production factors. The new-type manufacturing sector is a prerequisite and foundation for new-type industrialization. The new-type manufacturing sector will drive China's new-type industrialization, or more specifically, the new-type manufacturing sector is an inevitable requirement for accelerating the adjustment of economic structure and promoting the transformation and upgrade of industrial structure. The development of the new-type manufacturing sector provides a guarantee for the enhancement of international competitiveness of the traditional manufacturing sector. The new-type manufacturing sector can transform the traditional manufacturing sector through various modern technologies (especially information technologies). The new-type manufacturing sector serves to change the extensive growth mode and attain sustainable economic development. Characterized by high added value, high growth, high efficiency, low energy consumption, and low pollution, the new-type manufacturing sector is an inevitable choice for new-type industrialization and sustainable development. In addition, enterprise development is mainly measured by economic performance, enterprise survival entails low transaction costs, and profit seeking is a natural instinct of entrepreneurs. Therefore, economic performance directly determines the result of enterprise development. As an important part of China's market economy system, manufacturing enterprises play an important role in attaining harmonious and coordinated development of the society, promoting economic development, creating employment opportunities, facilitating technological progress and innovations, and invigorating the market economy. The rapid growth of manufacturing enterprises will directly promote industrial upgrade and industrial structure optimization, and will provide a basic impetus for the development of the new-type manufacturing sector.

Technology motivation: With the global integration of the manufacturing sector and advent of the knowledge economy, the level of technological development has gradually become the decisive factor in increasingly fierce global economic competition. Since the 1990s, several developed countries have successively transferred their low-end technologies and low-tech industries to other countries, encouraged their transnational companies to invest across the world, and attempted to reach the commanding heights of global economic competition through various means (e.g., capital, financial, public opinion, and military resources). As a result, developed countries with strong technological innovation capabilities have gradually dominated the global production system and profit distribution system. Seemingly, the financial crisis in Southeast Asia and the economic crisis in Latin America were caused by these regions' economic systems and management systems. However, at a more in-depth level, such crises can be ascribed to excessive reliance on foreign technologies, markets, and capital; put another way, their innovation capability was not sufficient. The *11th Five-year Plan on National Economic and Social Development* formulated by the Central Committee of the CPC pointed out that in order to carry out a complete scientific outlook on development, China must improve its self-dependent innovation capability. China's long-term and sustained development should rely on technological progress and improvement of labor competences. More specifically, China should ensure deep implementation of strategies; take enhanced self-dependent innovation capability as the cornerstone for scientific and technological development and as a central task for adjusting industrial structure and transforming the economic growth pattern; and actively improve its original innovation capability, integrated innovation capability, and its introduction, digestion, absorption, and re-innovation capabilities. The technological concept of the manufacturing sector lies in a variety of technological innovations, which include not only original innovations (represented by self-dependent innovations), integrated innovations, and re-innovations based on introduction, digestion, and absorption, but also application innovations (represented by knowledge innovations) and sustained innovations. Therefore, technology motivation is manifested as a balance of innovations and utilizations.

Resource motivation: China has an abundant variety of natural resources, and is blessed with both conventional and scarce resources. In addition, China ranks at the top globally in reserves of main industrial raw materials (e.g., petroleum, iron ores, and copper ores). This provides the natural resource conditions required for development of the new-type manufacturing sector. Meanwhile, such ideas as ecological civilization, resource conservation, and environmental protection have been deeply rooted in the hearts of people. Overall, China has made great progress in environmental protection inputs, efficient resource utilization, energy conservation and emissions reduction, and resource exploitation. Moreover, information technologies are profoundly affecting the IT application competence of China's labor forces, and for new employees, it has become a necessary skill to handle their daily work using modern information technologies. This provides powerful human resource support for enhancing the level of informatization in the new-type manufacturing sector and for increasing the added value in the IT sector.

2.3 Four-Yardstick Evaluation Index System for China's Manufacturing Sector

Based on the abovementioned analyses of the new-type manufacturing sector, this annual research report proposes a four-yardstick evaluation index system for the new-type manufacturing sector.

2.3.1 Design Principles for the Evaluation Index System

The connotations of the new-type manufacturing sector provide the guidelines for its evaluation index system. Such connotations cover four main aspects, that is, economic resourcefulness, technological innovation capability, energy conservation capability, and environmental protection capability, each of which covers diverse points. In order to measure the degree of newness of the manufacturing sector systematically and accurately, a reasonable evaluation index system should be built based on appropriate principles. In this research study, we consider that the evaluation index system for the new-type manufacturing sector should follow the principles of scientificity, systematicness, comparability, and operability.

1. Scientificity

Can the new-type manufacturing sector be evaluated accurately and reasonably? This largely depends on whether the evaluation indexes, evaluation criteria, and evaluation methods are selected scientifically. In evaluating the new-type manufacturing sector, the scientificity principle for the evaluation index system mainly covers accuracy and completeness. Accuracy has the following requirements: (1) the evaluation indexes should be defined accurately and clearly, so as to avoid subjective assumptions or misconceptions; (2) the evaluation index system should be structured and hierarchized reasonably, and the different indexes should be well coordinated to serve the entire evaluation index system. In addition, the completeness of the evaluation indexes places the following requirements: (1) the evaluation index system should reflect the evaluation objects completely around the evaluation purpose; (2) the evaluation index system should highlight the key points and take account of all aspects without omitting any essential aspect.

2. Systematicness

The new-type manufacturing sector involves diverse aspects, each of which is denoted by appropriate indexes. The selected indexes should cover as many aspects as possible, have a certain representativeness, indicate the main connotations, characteristics, current status, and evolutionary process of the new-type manufacturing sector, reflect the inherent connections between different aspects of the connotations, and be hierarchized clearly. Systematicness requires that the evaluation index system is not a simple pile-up of indexes, but an organic whole.

3. Comparability

Statistical indexes may vary among different manufacturing industries. The selection of evaluation indexes should take full account of the differences in statistical indexes between different manufacturing industries, and ensure consistency in meaning, statistical caliber, and statistical range, so as to ensure the comparability of the evaluation indexes. The selection of evaluation indexes should take account of the differences in statistical standards and statistical calibers between different provinces. In selecting evaluation indexes, the statistical indexes and statistical calibers should be consistent between different manufacturing sectors and between different manufacturing regions, so that the evaluation index system and evaluation criteria have definite comparability among different sectors and regions.

4. Operability

In addition to scientificity, systematicness and comparability, the evaluation index system for the new-type manufacturing sector should follow the operability principle; specifically, the required data for the evaluation index system should be easy to acquire, process, and operate, and the evaluation results should be easy to utilize, for example, guiding the development practice of the manufacturing sector.

2.3.2 Evaluation Index System for the New-Type Manufacturing Sector

Based on the abovementioned connotations, the evaluation index system for the new-type manufacturing sector mainly comprises four major indexes, namely, economic resourcefulness, technological innovation capability, energy conservation capability, and environmental protection capability. Economic indexes are used to measure the contributions made by the manufacturing sector to the national economy, technological indexes are used to measure the future competitiveness of the manufacturing sector, energy indexes are used to measure the degree to which the development of the manufacturing sector relies on energy consumption, and environmental indexes are employed to measure the degree to which the manufacturing sector affects and damages the environment. Accordingly, this research study proposes an evaluation index system for the new-type manufacturing sector, which comprises four major indexes and 31 subindexes (as described in Table 2.1).

1. Economic indexes

Economic resourcefulness is mainly measured in terms of output value, profits, efficiency, market performance, employment, and tax. In Table 2.1, A1 and A2 are output value subindexes, which are used to reflect the output capacity of the manufacturing sector and its contributions to the national economy. A3 and A4 are profit subindexes, which are used to reflect the total profit and per-capita profit in the manufacturing sector. A5 is an efficiency subindex, which is used to reflect the

Table 2.1 Evaluation index system for the new-type manufacturing sector

General index	No.	Major index	No.	Subindex	
Evaluation index system for the new-type manufacturing sector	A	Economic index	A1	Output value	Total output value of the manufacturing sector (unit: 100 million yuan)
			A2		Ratio of total output value of the manufacturing sector to total industrial output value (%)
			A3	Profit	Total profit of manufacturing enterprises (unit: 100 million yuan)
			A4		Per-capita profit in the manufacturing sector (unit: yuan per capita)
			A5	Efficiency	Labor productivity of the manufacturing sector (unit: 10,000 yuan per capita)
			A6	Market	Product sales rate of the manufacturing sector (%)
			A7	Employment	Employed population in the manufacturing sector (unit: 10,000)
			A8		Ratio of employed population in the manufacturing sector to total employed population (%)
			A9	Tax	Total taxes of manufacturing enterprises (unit: 100 million yuan)
			A10		Per-capita tax in the manufacturing sector (unit: 10,000 yuan per capita)
	B	Technological index	B1	R&D	R&D expenditure in the manufacturing sector (unit: 10,000 yuan)
			B2		Full-time equivalent of R&D personnel in the manufacturing sector (unit: person-year)
			B3		R&D expenditure intensity in the manufacturing sector (%)
			B4		Ratio of R&D personnel to total employed population in the manufacturing sector (%)
			B5	Product development	Quantity of new product development projects in the manufacturing sector
			B6		New product development expenditures in the manufacturing sector (unit: 10,000 yuan)
			B7	Patent	Quantity of patent applications in the manufacturing sector
			B8		Quantity of patent grants in the manufacturing sector

(continued)

Table 2.1 (continued)

General index	No.	Major index	No.	Subindex	
			B9	Technological achievement transformation	Output value of new products in the manufacturing sector (unit: 10,000 yuan)
			B10		Rate of output value of new products in the manufacturing sector (%)
			B11		Input–output coefficient of technological innovations in the manufacturing sector
	C	Energy index	C1	Gross consumption	Energy consumption in the manufacturing sector (unit: 10,000 tons of standard coal)
			C2		Energy consumption per unit output value in the manufacturing sector (unit: 10,000 tons of standard coal per 100 million yuan)
			C3	Electric power consumption	Electric power consumption in the manufacturing sector (unit: 100 million kWh)
	D	Environmental index	D1	Wastewater	Discharge amount of wastewater in the manufacturing sector (unit: 10,000 tons)
			D2		Discharge amount of wastewater per unit output value in the manufacturing sector (unit: 10,000 tons per 100 million yuan)
			D3	Waste gas	Discharge amount of waste gas in the manufacturing sector (unit: 100 million standard cubic meters)
			D4		Discharge amount of waste gas per unit output value in the manufacturing sector (unit: 100 million standard cubic meters per 100 million yuan)
			D5	Waste residue	Discharge amount of waste residue in the manufacturing sector (unit: ton)
			D6		Discharge amount of waste residue per unit output value in the manufacturing sector (unit: tons per 100 million yuan)
			D7	Comprehensive utilization	Output value by comprehensive utilization of three types of industrial waste (unit: 10,000 yuan)

labor productivity of manufacturing enterprises. A6 is a market subindex, which is used to reflect the product sales performance of the manufacturing sector, as well as the degree to which the manufactured products cater to social needs. A7 is an aggregate employment subindex, which is used to reflect the capacity of the manufacturing sector to provide employment opportunities. A8 is a relative subindex about employment, which is used to reflect the proportion of employed population in the manufacturing sector to the total employed population. A9 and A10 are tax subindexes, which are used to reflect the tax contributions made by manufacturing enterprises. The following section describes the calculation methods for the abovementioned subindexes.

(1) Total output value of the manufacturing sector is equal to $\sum_{j=1}^{30} \mathrm{TVP_j}$, where $\mathrm{TVP_j}$ indicates the total industrial output value of the j-th manufacturing industry, and $j = 1, 2, \ldots, 30$. For details on the manufacturing industries, refer to *Classification and Code Standard of National Economy Industry* (GB/T4754-2002).
(2) Ratio of total output value of the manufacturing sector to total industrial output value is equal to $\frac{\sum_{j=1}^{30} \mathrm{TVP_j}}{\mathrm{TP}} \times 100\%$, where, $\sum_{j=1}^{30} \mathrm{TVP_j}$ indicates the total output value of the manufacturing sector, and TP indicates the total industrial output value.
(3) Total profit of manufacturing enterprises (denoted by S) equals the sum of the profits earned by all manufacturing enterprises, and is determined by adding the profits earned by 30 manufacturing sectors at the two-digit SIC level.
(4) Per-capita profit in the manufacturing sector equals $\frac{\mathrm{S} \times 10{,}000}{\mathrm{L}}$, where S indicates the total profit of manufacturing enterprises (unit: 100 million yuan), and L indicates the total employed population of manufacturing enterprises (unit: 100,000).
(5) Labor productivity of the manufacturing sector equals $\frac{\sum_{j=1}^{30} \mathrm{TVP_j}}{\mathrm{L}} \times 100\%$.
(6) Product sales rate of the manufacturing sector equals $\frac{\sum_{k=1}^{30} \mathrm{SR_k}}{\sum_{j=1}^{30} \mathrm{TVP_j}} \times 100\%$, where $\mathrm{SP_k}$ indicates the product sales income of the k-th manufacturing industry, $\mathrm{k} = 1, 2, \ldots, 30$, $\mathrm{TVP_j}$ indicates the total industrial output value of the j-th manufacturing industry, and $\mathrm{j} = 1, 2, \ldots, 30$.
(7) Employed population in the manufacturing sector is denoted by L.
(8) The ratio of employed population in the manufacturing sector to total employed population equals $\frac{\mathrm{L}}{\mathrm{L_q}} \times 100\%$, where, L indicates the employed population in manufacturing enterprises, and $\mathrm{L_q}$ indicates the regional employed population.
(9) Total tax of manufacturing enterprises is denoted by T.

(10) Per-capita profit in the manufacturing sector equals $\frac{\mathrm{T}\times 10{,}000}{\mathrm{L}}$, where, T indicates the total profit of manufacturing enterprises (unit: 100 million yuan), and *L* indicates the total employed population of manufacturing enterprises (unit: 100,000).

2. Technological indexes

Technological innovation capability is mainly measured in terms of R&D expenditures, product development, patents, and technological achievement transformation. In Table 2.1, B1, B2, B3, and B4 are R&D subindexes of the manufacturing sector, which reflect the total R&D expenditure and R&D expenditure intensity of manufacturing enterprises. B5 and B6 are product development subindexes, which can objectively reflect the expenditures and expenditure intensity of manufacturing enterprises on new product development. B7 and B8 are patent subindexes, which reflect the input–output performance of technological innovations made by manufacturing enterprises. B9, B10, and B11 are subindexes on technological achievement transformation, which reflect the capability of manufacturing enterprises to transform and apply technological achievements. These 11 subindexes reflect the technological innovation capability of the manufacturing sector from the perspective of R&D inputs, new product development, technological outputs, and technological achievement transformation and application. The following section describes the calculation methods for these subindexes.

(1) R&D expenditure in the manufacturing sector refers to the sum of R&D expenditures inputted by all manufacturing enterprises.
(2) Full-time equivalent of R&D personnel in the manufacturing sector refers to the sum of full-time R&D personnel in all manufacturing enterprises.
(3) R&D expenditure intensity in the manufacturing sector equals $\frac{R\&D}{GDP} \times 100\%$, where R&D indicates the R&D expenditures of the manufacturing sector, and GDP indicates the gross domestic product.
(4) Ratio of R&D personnel to total employed population in the manufacturing sector equals $\frac{L}{L'} \times 100\%$, where L′ indicates the R&D personnel in the manufacturing sector, and *L* indicates the total employed population in the manufacturing sector.
(5) Quantity of new product development projects in the manufacturing sector refers to the total quantity of new product development projects carried out by all manufacturing enterprises.
(6) New product development expenditures in the manufacturing sector refer to the total expenditures of new product development inputted by all manufacturing enterprises.
(7) Quantity of patent applications in the manufacturing sector refers to the total quantity of patent applications submitted by all manufacturing enterprises.
(8) Quantity of patent grants in the manufacturing sector refers to the total quantity of patent grants owned by all manufacturing enterprises.

(9) Output value of new products in the manufacturing sector refers to the total output value of new products created by all manufacturing enterprises.
(10) Rate of output value of new products in the manufacturing sector equals $\frac{NPV}{\sum_{j=1}^{30} TVP_j} \times 100\%$, where NPV indicates the output value of new products in the manufacturing sector (unit: 100 million yuan), TVP_j indicates the total industrial output value of the j-th manufacturing industry (unit: 100 million yuan), and $j = 1, 2, \ldots, 30$.
(11) Input–output coefficient of technological innovations in the manufacturing sector equals $= \frac{\mathrm{NPV}}{NPR} \times 10,000$, where NPV indicates the output value of new products in the manufacturing sector (unit: 100 million yuan), and NPR indicates the expenditures on new product development (unit: 10,000 yuan).

3. Energy indexes

Energy conservation capability is mainly measured in terms of total energy consumption and electricity consumption. In Table 2.1, C1 and C2 are subindexes on total energy consumption, which reflect the total energy consumption and energy intensity, respectively, in the manufacturing sector. C3 is an electricity consumption subindex, which reflects the degree to which the manufacturing sector relies on electricity. The three subindexes jointly reflect the dependency relationship between the development of the manufacturing sector and energy consumption, and comprehensively embody the energy conservation capability of the manufacturing sector. The following describes the calculation methods for these subindexes.

(1) Energy consumption in the manufacturing sector (denoted by CC) refers to the total amount of energy consumed by all manufacturing enterprises.
(2) Energy consumption per unit of output value in the manufacturing sector equals $\mathrm{CC}/\sum_{\mathrm{j=1}}^{30} \mathrm{TVP_j}$, where CC again indicates the energy consumption of the manufacturing sector.
(3) Electricity consumption in the manufacturing sector (unit: 100,000,000 kWh).

4. Environmental indexes

Environmental protection capability is mainly measured in terms of discharge of wastewater, discharge of waste gas, discharge of waste residue, and comprehensive utilization of the three types of industrial waste. D1 and D2 reflect the total discharge amount and discharge intensity of wastewater, respectively, D3 and D4 reflect the total discharge amount and discharge intensity of waste gas, respectively, D5 and D6 reflect the total discharge amount and discharge intensity of waste residue, respectively, and D7 refers to the output value by comprehensive utilization of the three types of industrial waste, and is a specific embodiment of the circular economy in the manufacturing sector. These subindexes reflect the environmental protection capability of the manufacturing sector in different ways. The following describes the calculation methods for these subindexes.

(1) Discharge amount of wastewater in the manufacturing sector refers to the amount of wastewater discharged by all manufacturing enterprises.
(2) Discharge amount of wastewater per unit output value in the manufacturing sector is equal to $\frac{\text{WWD}}{\sum_{j=1}^{30} \text{TVP}_j}$, where WWD indicates the discharge amount of wastewater in the manufacturing sector during the report period.
(3) Discharge amount of waste gas in the manufacturing sector refers to the amount of waste gas discharged by all manufacturing enterprises.
(4) Discharge amount of waste gas per unit of output value in the manufacturing sector is equal to $\frac{\text{WGD}}{\sum_{j=1}^{30} \text{TVP}_j}$, where WGD indicates the discharge amount of waste gas in the manufacturing sector during the report period.
(5) Discharge amount of waste residue in the manufacturing sector refers to the amount of waste residue discharged by all manufacturing enterprises.
(6) Discharge amount of waste residue per unit of output value in the manufacturing sector is equal to $\frac{\text{WSD}}{\sum_{j=1}^{30} \text{TVP}_j}$, where WSD indicates the discharge amount of waste residue in the manufacturing sector during the report period.
(7) Output value by comprehensive utilization of the three types of industrial waste refers to the total output value created by the comprehensive utilization of the three types of waste in all manufacturing industries.

References

Li, L.S., and Z.Y. Du. 2004. *A research report on the development of China's Manufacturing Sector 2004*. Beijing: Science Press.

Li, L.H., and Y. Zhou. 2005. Empirical analysis of the status of the "new industrial model" in China's manufacturing sector–an assessment based on the manufacturing sectors of 30 Chinese regions. *Management World* 6:76–81.

Li, L.S., and Z.P Zang. 2008. The international Comparison of new style manufacturing industry three-dimensional of time series. In *Collection of papers on the Academic Annual Conference of the Fourth Term of the China Science and Technology Policy Studies: II.*

Liu, Z.B. 2006. Developing modern producer service and adjusting and optimizing Structure of manufacturing, *Journal of Nanjing University (Philosophy, Humanities and Social Sciences)* 5:36–44.

Lv, F.B. 2014. Research of industrial development environment and technological innovation problem of the equipment manufacturing industry, illustrated by the case of the technological innovation of technology, bases small and micro enterprises in Liaoning. *Scientific management research*, 32(5):71–73.

Wang, Q. 2007. Public policy choices to TNCs' M&A practices in China's equipment manufacture enterprises. *Journal of Beijing Normal University (Social Sciences)* 1:115121

Xu, X.C. 2010. The new-style of analyses of Jiangsu manufacturing sector based on the visual angle of environment protection, industrial and Science Tribune 9:7.

Zhang, X.G. 2009. On the driving functions of designing and creating industry in China's economic transformation. *Journal of Henan University of Technology (Social Science Edition)* 5(4):43–46,50.

Zheng, R.G. 2010. Evaluation of the Capacity of Industrial Self-Innovation in China's Manufacturing Industry. *Contemporary Finance & Economics*,1:89–95.

Chapter 3
The Comprehensive Evaluation of China's Manufacturing Sector

Wei Sun, Zaiwu Gong and Changkai Wang

Based on data about the economic, technological, energy, and environmental indexes of China's manufacturing sector, this section evaluates and ranks the economic resourcefulness, technological innovation capability, energy conservation capability, and environmental protection capability of China's manufacturing sector by using the method of multi-attribute deviation maximization decision making. Then, it conducts a comprehensive evaluation for China's new-type manufacturing sector accordingly.

Multiple indexes are involved in the evaluation and ranking of the comprehensive development of China's manufacturing sector. Therefore, this is a multi-attribute decision-making problem. Multi-attribute is also referred to as multi-criteria decision making, whose core and key issue is to determine the weight of each index. To this end, this study resorts to the deviation maximization method, which is a completely objective evaluation method that not only eliminates human-made influences in the subjective evaluation method, but also is clearly defined and easy to use. In practice, this evaluation method is widely applied.

$A = |A_1, A_2, \ldots, A_n|$ indicates the solution set for a multi-criteria evaluation problem, $G = |G_1, G_2, \ldots, G_m|$ indicates an index set, Y_{ij} ($i = 1, 2\ldots, n$; $j = 1, 2\ldots, m$) indicates the index value of the A_i solution with respect to the G_j index, and $Y = (y_{ij})_{n \times m}$ indicates the attribute matrix of the A solution set with respect to the G index set, that is, the evaluation matrix.

Indexes can usually be classified into four types based on their nature: benefit indexes, cost indexes, fixed indexes, and interval indexes. Dimensions and dimensional units vary with evaluation indexes. Therefore, evaluation indexes are made to be dimensionless (normalized processing), thereby preventing the incommensurability arising from the disparity in dimensions and dimensional units.

L. Li and Z. Du (eds.), *A Research Report on the Development of China's Manufacturing Sector (2016)*, Current Chinese Economic Report Series,
DOI 10.1007/978-981-10-4445-8_3

This section involves only benefit indexes and cost indexes. The larger a benefit index is, the better it is; the smaller a cost index is, the better it is. The normalized processing methods are as follows.

For the cost indexes,

$$Z_{ij} = \frac{y_j^{max} - y_{ij}}{y_j^{max} - y_j^{min}} \quad i = 1, 2, \ldots, n; \quad j = 1, 2, \ldots, m. \tag{3.1}$$

For the benefit indexes,

$$Z_{ij} = \frac{y_{ij} - y_j^{min}}{y_j^{max} - y_j^{min}} \quad i = 1, 2, \ldots, n; \quad j = 1, 2, \ldots, m. \tag{3.2}$$

where, y_j^{min} and y_j^{max} indicate the minimum value and maximum value, respectively, of the index G_j.

$Z = (Z_{ij})_{n\times m}$ indicates the evaluation matrix obtained after the evaluation indexes are made to be dimensionless. Obviously, the larger the Z_{ij} value is, the better it is. $w = (w_1, w_2 \ldots, w_m)^T \succ 0$ indicates the weight vector of an evaluation index. Meanwhile, the following unitization constraint condition should be met:

$$\sum_{j=1}^{m} w_j^2 = 1 \tag{3.3}$$

After obtaining the weight vector w, we construct the following evaluation matrix:

$$c = \begin{array}{c} \\ A_1 \\ A_2 \\ \cdots \\ A_n \end{array} \begin{array}{c} \begin{array}{cccc} G_1 & G_2 & \cdots & G_m \end{array} \\ \begin{bmatrix} w_1 z_{11} & w_2 z_{12} & \cdots & w_m z_{1m} \\ w_1 z_{21} & w_2 z_{22} & \cdots & w_m z_{2m} \\ \cdots & \cdots & \cdots & \cdots \\ w_1 z_{n1} & w_2 z_{n2} & \cdots & w_m z_{nm} \end{bmatrix} \end{array} \tag{3.4}$$

Then, we obtain the multi-index comprehensive evaluation score of the A_i solution by the simple arithmetic mean weighting method, as described in Eq. (3.5):

$$D_i(w) = \sum_{j=1}^{m} z_{ij} w_j, \quad i = 1, 2, \ldots, n. \tag{3.5}$$

Likewise, the larger the $D_i(w)$ value is, the better the A_i solution is. When the weight vector w is known, the A_i solution can be evaluated and ranked according to Eqs. (3.1)–(3.5).

Subsequently, we further analyze how to determine the weight vector w. if an index G_j has no effect on the final evaluation score and rank of the solution A_i, we assign 0 to the weight of G_j. If an index G_j can affect the final evaluation score and rank of solution A_i significantly, we assign a high weight value to G_j. For index G_j, $v_{ij}(w)$ indicates the deviation between solution A_i and other solutions; then,

$$v_{ij}(w) = \sum_{k=1}^{n} \left| w_j z_{ij} - w_j z_{kj} \right|, \quad i = 1, 2, \ldots, n, \quad j = 1, 2, \ldots, m. \tag{3.6}$$

$$v_j(w) = \sum_{i=1}^{n} v_{ij}(w) = \sum_{i=1}^{n} \sum_{k=1}^{n} \left| z_{ij} - z_{kj} \right| w_j, \quad j = 1, 2, \ldots, m. \tag{3.7}$$

$v_j(w)$ indicates the sum of the deviations between solution A_i and other solutions under index G_j. The selected weight vector w needs to maximize the sum of deviations between all indexes and all solutions; therefore, we construct the following objective function:

$$\max F(w) = \sum_{j=1}^{m} v_j(w) = \sum_{j=1}^{m} \sum_{i=1}^{n} \sum_{k=1}^{n} \left| z_{ij} - z_{kj} \right| w_j \tag{3.8}$$

Then, the problem of how to determine the weight vector w is equal to a non-linear programming problem.

$$\begin{cases} \max F(w) = \sum\limits_{j=1}^{m} v_j(w) = \sum\limits_{j=1}^{m} \sum\limits_{i=1}^{n} \sum\limits_{k=1}^{n} \left| z_{ij} - z_{kj} \right| w_j \\ s.t. \quad \sum\limits_{j=1}^{m} w_j^2 = 1 \end{cases} \tag{3.9}$$

We solve this non-linear programming problem, and normalize the weight vector w. Then, we obtain the following equation:

$$w_j^* = \frac{\sum\limits_{i=1}^{n} \sum\limits_{k=1}^{n} \left| z_{ij} - z_{kj} \right|}{\sum\limits_{j=1}^{m} \sum\limits_{i=1}^{n} \sum\limits_{k=1}^{n} \left| z_{ij} - z_{kj} \right|} \quad j = 1, 2, \cdots, m. \tag{3.10}$$

In summary, we perform the following three steps to solve the multi-index evaluation and ranking problem by using the deviation maximization method.

(1) We process the benefit indexes and cost indexes and obtain the normalized evaluation matrix $Z = (Z_{ij})_{n \times m}$.

(2) We use the deviation maximization method to determine the optimal weight vector $w^* = (w_1^*, w_2^*, \ldots, w_m^*)^T$; accordingly, we determine the comprehensive evaluation score $D_i(w), \quad i = 1, 2, \ldots, n$ for solution A_i.
(3) According to the comprehensive evaluation scores of different solutions, we conduct reasonable evaluation and ranking analysis for a multi-index problem.

3.1 Comprehensive Evaluation of Economic Resourcefulness

We obtain data about China's manufacturing sector for 2006–2013 from the *China Statistical Yearbook*, and select the indexes that can objectively and comprehensively reflect the economic resourcefulness of China's manufacturing sector, such as, the total enterprise profit and total output value. We calculate the weight of each index by using the deviation maximization method, and determine the comprehensive evaluation score for the economic resourcefulness of China's manufacturing sector during 2006–2013 according to the normalized values of the related indexes.

1. The comprehensive evaluation score for the economic resourcefulness of China's manufacturing sector continued to increase during the years 2006–2013

As shown in Fig. 3.1, the comprehensive evaluation score of China's manufacturing sector increased from 0.0333 (2006) to 0.8736 (2013), indicating a continuous improvement in the economic resourcefulness of China's manufacturing sector. During 2006–2013, the economic resourcefulness of China's manufacturing sector improved year by year; in particular, this improvement accelerated very significantly from 2009 to 2010 (the comprehensive evaluation score rapidly increased from 0.4117 to 0.7944).

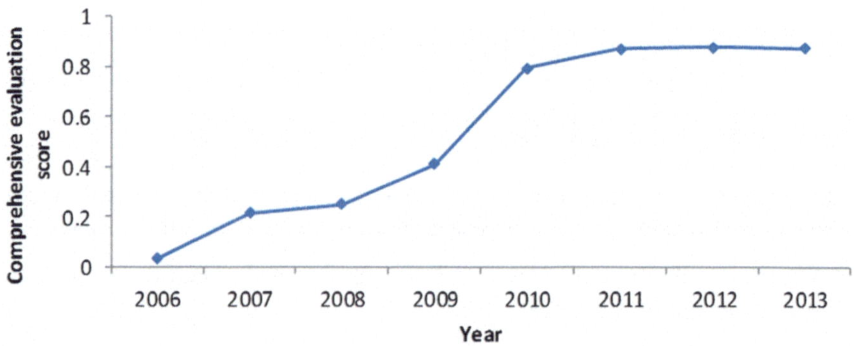

Fig. 3.1 Comprehensive evaluation of economic resourcefulness of China's manufacturing sector from 2006 to 2013

2. Analyses of single indexes on economic resourcefulness of China's manufacturing sector

(1) The total output value of China's manufacturing sector increased very rapidly, especially during 2009–2011. The total output value assumed an upward trend over the 8 years analyzed, reflecting the development trend of China's manufacturing sector to a certain extent. Among the indexes for China's manufacturing sector, the weight of total output value of the manufacturing sector was 0.1489. The total output value of the manufacturing sector reflects the development status of a country's manufacturing sector. As shown in Fig. 3.2, the total output value of China's manufacturing sector assumed a steady upward trend. In particular, this index value increased very significantly, from 47.9 to 73.4 trillion yuan.

(2) The total enterprise profit of the manufacturing sector enjoys a high weight; overall, the total enterprise profit of China's manufacturing sector faced a slowdown in growth. The total enterprise profit of the manufacturing sector refers to the surplus earned by manufacturing enterprises after various costs are deducted from various incomes during production operations. This economic index reflects the total profit or loss in the period of the study. To a certain extent, this index reflects the profitability of China's manufacturing enterprises. Among the indexes for China's manufacturing sector, the weight of total enterprise profit is as high as 0.1854. As shown in Fig. 3.3, the total enterprise profit of China's manufacturing sector assumed an upward trend on the whole. In particular, total enterprise profit increased very dramatically during 2009 and 2010. Evidently, China's manufacturing sector was not affected in these 2 years by the financial crisis of 2008. As the European debt crisis continued for 3 years, the EU economy was still embroiled in the crisis, the economic recovery of the world's main economic entities (such as the US and Japan) was very weak, and the economic development of emerging economies slowed in 2013. In summary, the world's real economy was severely affected. In the context of sluggish external demand and domestic economic downturn, the total enterprise profit of China's manufacturing sector slowed in 2013.

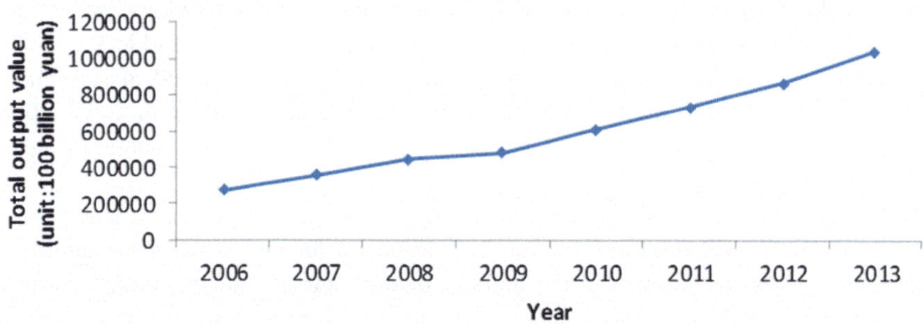

Fig. 3.2 Total output value of China's manufacturing sector from 2006 to 2013

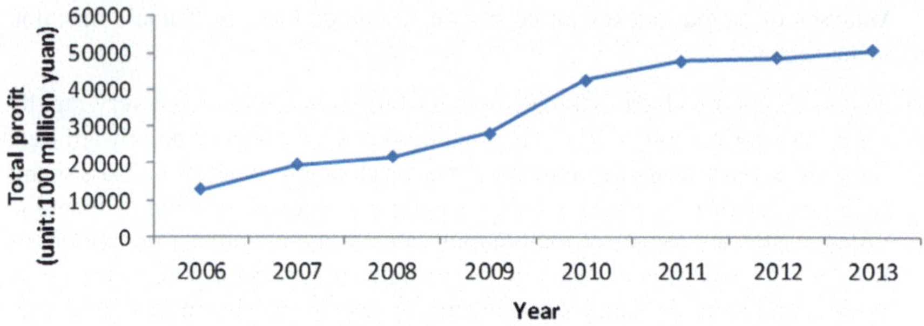

Fig. 3.3 Total enterprise profit of China's manufacturing sector from 2006 to 2013

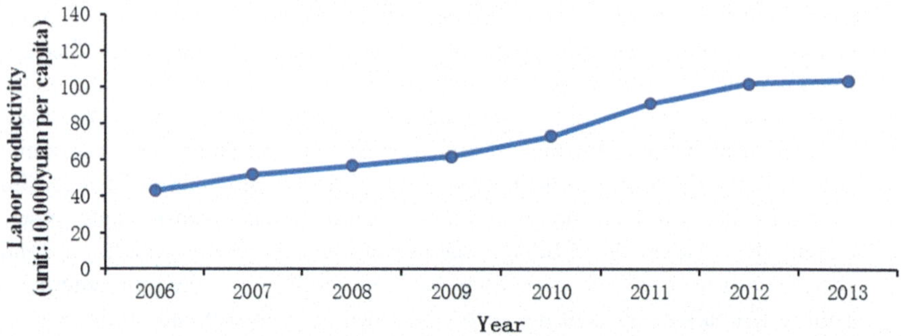

Fig. 3.4 Labor productivity of employees in China's manufacturing sector from 2006 to 2013

(3) The labor productivity of employees in China's manufacturing sector assumed an upward trend, and in particular, this index value increased very dramatically in 2011. As shown in Fig. 3.4, the index value assumed an upward trend during 2006–2013, reflecting favorable development status of China's manufacturing sector to a certain extent. During 2006–2010, labor productivity essentially increased at the same rate; during 2009–2013, the growth rate of labor productivity increased somewhat (demonstrated by the slope rate of the broken line). It is not difficult to establish that China's hyper-conventional economic stimulus package (e.g., the 4 trillion yuan economic stimulus plan) carried out at the end of 2008 had a certain effect in helping China to tackle the 2008 financial crisis. It was forecast by Shusong Ba that a new round of economic growth would start in 2011 after China's economy bottomed. In fact, the labor productivity of employees in China's manufacturing sector increased very significantly in 2011.
(4) Product sales in China's manufacturing sector are still in early stage, and overall sales still need to ramp up. The product sales rate of the manufacturing sector refers to the ratio of the amount of product sales in the manufacturing industry to the total output value of the manufacturing sector in the period of the study. This index can be used to reflect the proportion of products in the

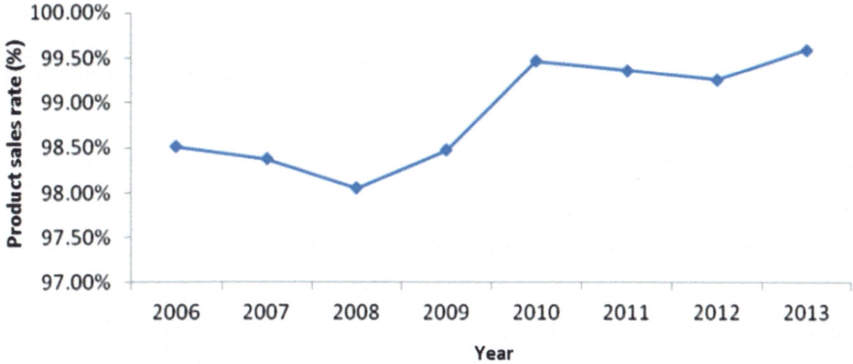

Fig. 3.5 Product sales rate of China's manufacturing sector from 2006 to 2013

manufacturing sector that have been sold, to analyze the link between sales and production in the manufacturing sector, and to study the extent to which manufacturing sector products can satisfy social needs. The index visually reflects the sales status of products, thereby helping the distributors increase their profits. Figure 3.5 shows the change trend of the product sales rate of the manufacturing sector. During 2006–2013, the product sales rate changed in a wave-like manner (i.e., alternately increased and decreased). During 2006–2008, the product sales rate assumed a downward trend, and rebounded in 2008. Among the six indexes for China's manufacturing sector, the product sales rate enjoyed the highest weight of 0.1905. According to previous statistical data, China's exports to the US continued to grow rapidly by at least 24.9% during 2002–2006, according to statistical data of China Customs. After August 2007, the growth of China's exports to the US fell sharply, and declined to 6.8% in December 2007, according to Chinese statistical data. Therefore, the weakness in the US economy arising from the subprime lending crisis played a significant role in the declining product sales rate of China's manufacturing sector. The product sales rate enjoyed certain growth during 2009–2010, but growth dropped slightly in 2012 and rose significantly in 2013, and the product sales rate fluctuated dramatically. This shows that product sales of China's manufacturing sector were still in the early stage during 2006–2013. The repeated wave-like changes of this index reflect the existing problems in the overall sales pattern and social needs of the products of China's manufacturing sector.

(5) The employed population of the manufacturing sector reflects the total employment scale of the manufacturing sector in a region during a specific period. Figure 3.6 shows the change trend of this index. During 2006–2013, the employment scale of China's manufacturing sector continued to expand. Specifically, the employed population was 33.516 million in 2006, and peaked at 52.579 million in 2013. Over the 8 years of the sample period, the employed population of China's manufacturing sector increased by 19 million. During

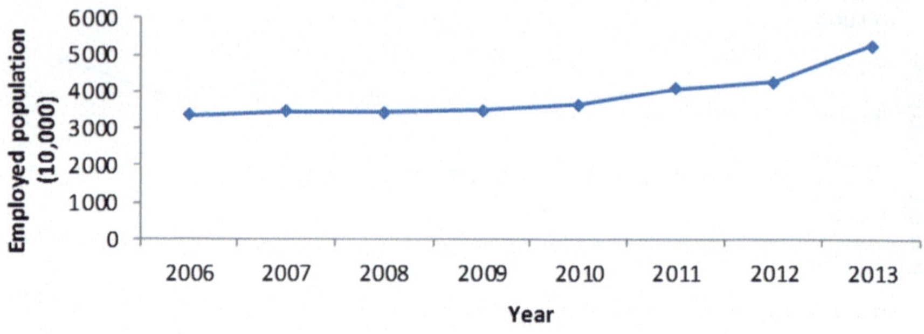

Fig. 3.6 Employed population of China's manufacturing sector

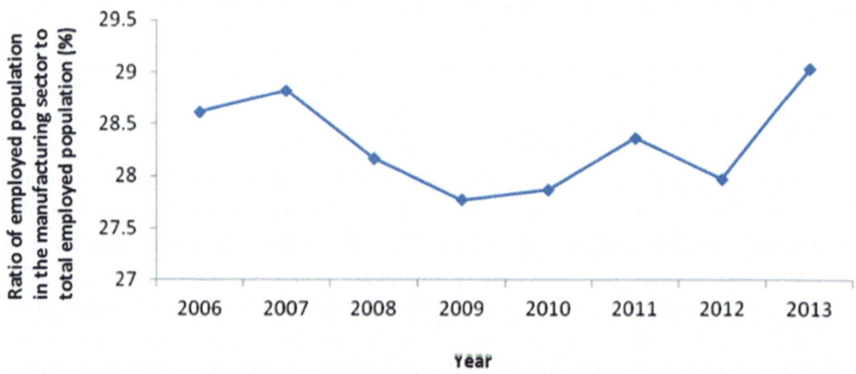

Fig. 3.7 Ratio of employed population in the manufacturing sector to total employed population

2011–2013, the employed population of China's manufacturing sector began to rise rapidly, to 40.883 million in 2011, 42.622 million in 2012, and 52.579 million in 2013. The rise in the employed population might have been due to China's favorable employment policy.

(6) The proportion of the employed population in the manufacturing sector to the total employed population reflects the capacity of the manufacturing sector to create employment opportunities, and is an important index that measures the social service capability of the manufacturing sector. Figure 3.7 shows the change trend of this index.
During 2006–2013, there were obvious fluctuations in this index. Specifically, the index was 28.61% in 2006, rose slightly to 28.82% in 2007, declined to 27.77% in 2009 (the lowest level in the sampling period), and then rose with fluctuations to reach 29.04% in 2013 (the peak in recent years). According to various statistical data, China's manufacturing sector remains the main force for the creation of employment opportunities.

(7) Total taxes of manufacturing enterprises reflect the capacity of manufacturing enterprises to pay taxes. Figure 3.8 shows the change trend of this index. Over 2006–2013, this index rose year by year. Specifically, it rose sharply by 270%

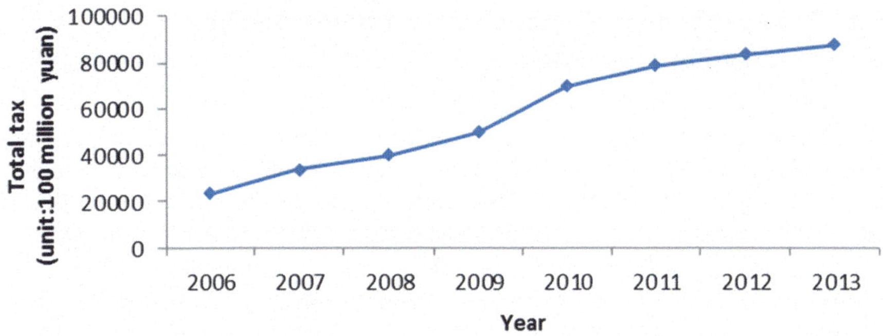

Fig. 3.8 Total taxes of manufacturing enterprises

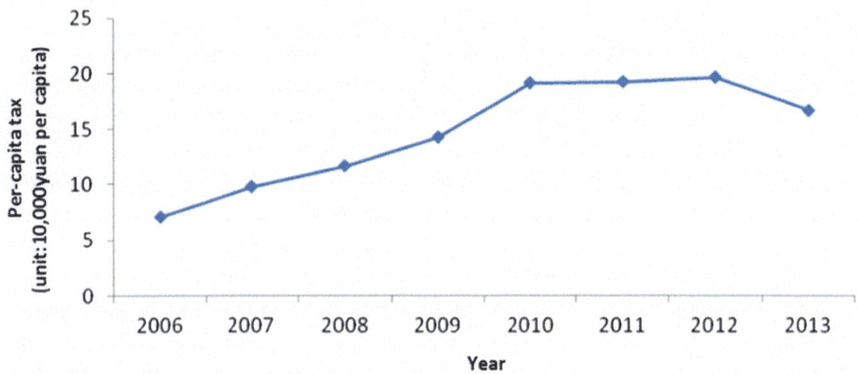

Fig. 3.9 Per-capita tax in the manufacturing sector

from 2.37 trillion yuan in 2006 to 8.76 trillion yuan in 2013; the average year-on-year rate of increase was as high as 21.37%. This index rose most rapidly in 2007 (42.87%), and followed by 2010 (39.91%), and it rose most slowly in 2012 (only 4.74%).

(8) Per-capita tax in the manufacturing sector reflects the capacity of manufacturing enterprises to pay taxes on a per-capita basis. Figure 3.9 shows the change trend of this index. During 2006–2010, this index rose continuously from 70,609 yuan (2006) to 191,321 yuan (2010), an increase of 171%. During 2010–2012, the index rose slowly, while it declined from 196,327 yuan in 2012 to 166,687 yuan in 2013. Evidently, China's manufacturing sector improved production technologies year by year, while labor productivity improved slowly year by year.

3.2 Comprehensive Evaluation of Technological Innovation Capability

Manufacturing sector firms are usually involved with a variety of complex and advanced technologies. Currently, people attach more importance to the technological innovation capability of the manufacturing sector than ever. About 70% of R&D personnel and expenditures in manufacturing enterprises are from western developed countries. By output value, China's manufacturing sector ranked first globally in 2010, but is weak overall in its technological innovation capability. This has become a major obstacle to the transformation of China's manufacturing sector. In this situation, it has become an urgent problem to evaluate the technological innovation capability of China's manufacturing sector, identify the key barriers to China's technological innovation capability, forge a new path in the technological innovation capability of China's manufacturing sector, and transform China's manufacturing industries into a technology-intensive sector.

Based on data on technological innovations of China's manufacturing sector from the *China Statistical Yearbook* and *China Statistical Yearbook on Science and Technology* (2007, 2008, 2009, 2010, 2011, 2012, 2013, 2014), this study selects 11 indexes that can reflect the technological innovation capability of China's manufacturing sector objectively and comprehensively, uses the deviation maximization method to calculate the weight of each index, and determines a comprehensive evaluation score of the technological innovation capability of China's manufacturing sector during 2006–2013 according to the normalized values of the indexes. Based on the original data, normalized data and weight of each index, comprehensive evaluation score of innovation capability, and their ranking results, this study evaluates the technological innovation capability of China's manufacturing sector. Table 3.1 lists the original data for the different indexes on the technological innovation capability of China's manufacturing sector.

Based on Eq. (3.2), we construct the normalized data for the different technological indexes of China's manufacturing sector during 2006–2013, as described in Table 3.2.

We calculate the weight of each technological index of China's manufacturing sector during 2006–2013, and comprehensively evaluate the technological innovation capability of China's manufacturing sector during these years, as described in Table 3.3.

1. Overall evaluation of technological innovation capability of China's manufacturing sector

According to Table 3.3, the comprehensive evaluation score of technological innovation capability continued to increase during 2006–2013; specifically, this value gradually increased from 0.0932 in 2006 to 0.8925 in 2015. Evidently, the technological innovation capability of China's manufacturing sector improved year by year, and the technological innovation capability in 2013 improved significantly from 2006.

Table 3.1 Statistical data on technological indexes of China's manufacturing sector from 2006 to 2013

	R&D expenditure (unit: 10,000 yuan)	Full-time equivalent of R&D personnel (unit: person-year)	R&D expenditure intensity (%)	Ratio of R&D personnel to total employed population (%)	Quantity of new product development projects
2006	15513883.9000	621991.3300	0.7172	1.8558	98,040
2007	20095640.5000	777570.0000	0.7560	2.2438	109,305
2008	25463701.8000	922832.8900	0.8108	2.6871	116,679
2009	30142350.8000	1207549.7000	0.8842	3.4581	147,778
2010	37713266.9000	1275445.3180	0.9393	3.5067	155,072
2011	56923791.5000	1823783.3000	1.2032	4.4610	261,564
2012	68408380.5000	2126589.1000	1.3182	4.9894	316,883
2013	79502287.0000	2368205.0000	1.3976	4.5041	351,682

	New product development expenditure (unit: 10,000 yuan)	Quantity of patent applications	Quantity of invention patent grants	Output value of new products (unit: 10,000 yuan)	Rate of output value of new products (%)	Input–output coefficient of technological innovations
2006	18335256.0000	67,227	28,168	318862362.0000	11.6131	17.3907
2007	24025786.0000	93,576	4455	422591301.0000	11.9501	17.5891
2008	30284884.2000	118,048	54,223	510857359.4000	11.5747	16.8684
2009	35503978.8000	162,694	78,905	578707859.3000	12.0765	16.2998
2010	43240022.2000	192,661	109,721	729658232.1000	11.9703	16.8746
2011	67234362.1000	374,112	196,521	988687652.3000	13.4702	14.7051
2012	78372872.8000	468,831	270,841	1085763794.0000	13.7052	13.8538
2013	90561522.0000	534,927	327,989	1265454673.0000	14.1047	13.9734

Data source: *China Statistical Yearbook on Science and Technology* (2007, 2008, 2009, 2010, 2011, 2012, 2013)

Table 3.2 Normalized data on technological indexes of China's manufacturing sector from 2006 to 2013

	R&D expenditure (unit: 10,000 yuan)	Full-time equivalent of R&D personnel (unit: person-year)	R&D expenditure intensity (%)	Ratio of R&D personnel to total employed population (%)	Quantity of new product development projects
2006	0	0	0	0	0
2007	0.0716	0.0891	0.0570	0.1238	0.0444
2008	0.1555	0.1723	0.1376	0.2653	0.0735
2009	0.2286	0.3353	0.2454	0.5113	0.1961
2010	0.3469	0.3742	0.3264	0.5268	0.2249
2011	0.6471	0.6882	0.7143	0.8314	0.6447
2012	0.8266	0.8616	0.8833	1	0.8628
2013	1	1	1	0.8451	1

	New product development expenditure (unit: 10,000 yuan)	Quantity of patent applications	Quantity of invention patent grants	Output value of new products (unit: 10,000 yuan)	Rate of output value of new products (%)	Input–output coefficient of technological innovations
2006	0	0	0	0	0.0152	0.9469
2007	0.0788	0.0563	0.0477	0.1096	0.1484	1
2008	0.1654	0.1087	0.0869	0.2028	0	0.8071
2009	0.2377	0.2041	0.1692	0.2745	0.1983	0.6548
2010	0.3448	0.2682	0.2720	0.4340	0.1564	0.8087
2011	0.6770	0.6562	0.5615	0.7076	0.7492	0.2279
2012	0.8312	0.8587	0.8094	0.8102	0.8421	0
2013	1	1	1	1	1	0.0320

Data source Author's calculation

As shown in Fig. 3.10, the growth rate of 10 indexes (excluding the input–output coefficient of technological innovation) in 2011 was significantly different from that in other years. The output value of China's manufacturing sector reached 1.955 trillion yuan (*China Statistical Yearbook* 2014), ranking first globally. IHS Global Insight released a report in 2011 and pointed out that the per-capita output value of China's manufacturing sector was only one-eighth of that of the US's manufacturing sector, indicating that "Made in China" surpassed "Made in the USA" only in quantitative terms. In order to improve its technological innovation capability, China's manufacturing sector gave more importance to and inputs into R&D, new product development, and patent grants in 2011. Therefore, it is easy to observe that the technological innovation capability of China's manufacturing sector improved significantly in 2011. In addition, China's hi-tech manufacturing enterprises that are above a designated size created a total output value of

Table 3.3 Rank of technological innovation capability of China's manufacturing sector from 2006 to 2013

Weight coefficient	0.0890	0.0896	0.0931	0.0886	0.0923
	R&D expenditure (unit: 10,000 yuan)	Full-time equivalent of R&D personnel (unit: person-year)	R&D expenditure intensity (%)	Ratio of R&D personnel to total employed population (%)	Quantity of new product development projects
2006	0	0	0	0	0
2007	0.0716	0.0891	0.0570	0.1238	0.0444
2008	0.1555	0.1723	0.1376	0.2653	0.0735
2009	0.2286	0.3353	0.2454	0.5113	0.1961
2010	0.3469	0.3742	0.3264	0.5268	0.2249
2011	0.6471	0.6882	0.7143	0.8314	0.6447
2012	0.8266	0.8616	0.8833	1	0.8628
2013	1	1	1	0.8451	1

Weight coefficient	0.0892	0.0915	0.0888	0.0876	0.0934	0.0969		
	New product development expenditure (unit: 10,000 yuan)	Quantity of patent applications	Quantity of invention patent grants	Output value of new products (unit: 10,000 yuan)	Rate of output value of new products (%)	Input–output coefficient of technological innovation	Evaluation score Di (w)	Rank No.
2006	0	0	0	0	0.0152	0.9469	0.0932	8
2007	0.0788	0.0563	0.0477	0.1096	0.1484	1	0.1715	7
2008	0.1654	0.1087	0.0869	0.2028	0	0.8071	0.2008	6
2009	0.2377	0.2041	0.1692	0.2745	0.1983	0.6548	0.2976	5
2010	0.3448	0.2682	0.2720	0.4340	0.1564	0.8087	0.3727	4
2011	0.6770	0.6562	0.5615	0.7076	0.7492	0.2279	0.6433	3
2012	0.8312	0.8587	0.8094	0.8102	0.8421	0	0.7755	2
2013	1	1	1	1	1	0.0320	0.8925	1

Data source Author's calculation

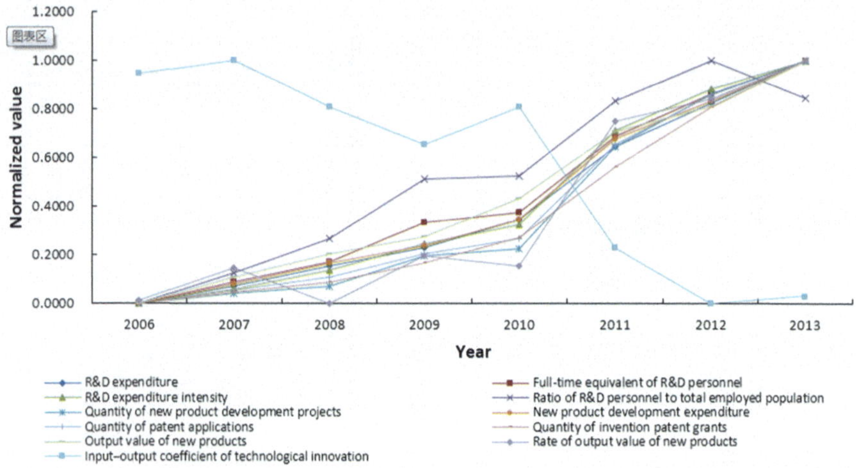

Fig. 3.10 Change trend of economic indexes of China's manufacturing sector from 2006 to 2013

9.2 trillion yuan in 2011, ranking second globally. This shows that China's manufacturing sector attached more importance to the improvement of technological innovation capability after it began to rank first globally by output value in 2010.

The following section evaluates the technological innovation capability of China's manufacturing sector from four aspects: R&D, new product development, patent grants, and new product benefits:

(1) Analyses of R&D input

During the 8 years of the sample period, the values of the four indexes related to R&D all continued to increase at a similar rate. The scale and intensity of R&D activities reflects the technological innovation capability and core competitiveness of an enterprise, an industry, and even a country. As shown in Fig. 3.11, the ratio of

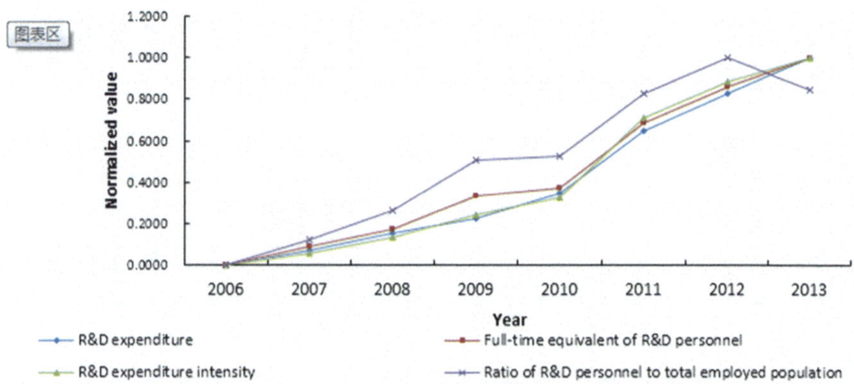

Fig. 3.11 Change trend of R&D input of China's manufacturing sector from 2006 to 2013

R&D personnel to total employed population in the manufacturing sector declined slightly in 2013 from 2012, while the values of four indexes related to R&D all continued to increase at a similar rate during 2006–2013. This shows that R&D expenditure input was in direct proportion to R&D labor input, and China's manufacturing sector attached more importance to R&D, thereby investing more labor and expenditure in R&D.

R&D expenditure intensity is an index of wide concern. Notwithstanding the continuous upward trend in R&D input, the R&D expenditure intensity of China's manufacturing sector increased from 0.7172% in 2006 to 1.3976% in 2013, increasing by only 0.68% during the 8 years. Although this index is favorably affected by the rapid growth of China's GDP, R&D expenditure input could hardly keep pace with the growth of GDP in the short term. This shows that China's R&D expenditure input grew rapidly in quantitative terms, but there was room for further improvement in qualitative terms. In recent years, the R&D expenditure intensity of China's manufacturing sector has maintained a growth rate of 5–10% (an average annual growth rate of 10.23%); in 2011, the growth rate of this index increased significantly (as high as 28.10%). This might be ascribed to a national target in the 11th Five-year Plan (2006–2010), that the R&D–GDP ratio should reach 2%.

Jiabao Wen, former premier of China, pointed out in 2011 that the R&D–GDP ratio was more important than GDP; in addition, the *12th Five-Year Plan for National Economic and Social Development* stipulates a target of 2.2% for China's R&D–GDP ratio by 2015. This shows that China's government has attached significant importance to R&D expenditure intensity. To a certain extent, this led to the rapid growth in R&D expenditure intensity of China's manufacturing sector in 2011.

The ratio of R&D personnel to total employed population in the manufacturing sector increased very slowly. As shown in Fig. 3.11, this index enjoyed a very low weight among the 11 indexes. During 2006–2013, the index was on an upward trend; however, while it increased obviously in 2009 and 2011, it did so only very slowly in other years (the average annual increase was 14.3%). In summary, the R&D labor input somewhat increased during the 8 years of the sample period, but the rate of increase was very low.

(2) Analyses of new product development

There has been a significant increase in both the quantity of new product development projects and new product development expenditures. New product development is intended to respond to diverse and ever-changing social needs. In order to survive acute market competition, enterprises have to embark on timely launches of new products that are suited to mass needs. During the 8 years of the sample period, the quantity of initiated new product development projects increased from less than 100,000 in 2006 to 350,000 in 2013. Such a dramatic increase clearly shows that China's manufacturing enterprises have attached significant importance to new product development.

New product development cannot do without financial support. With the increase in the quantity of new product development projects, new product development expenditures increased from 183,352 billion yuan in 2006 to 905.615 billion yuan in 2013. A short product lifecycle brings about increasingly high operation costs and risks to enterprises. In order to win greater market share and open up new business fields, enterprises have to spend increasingly more funds on new product development.

(3) Analyses of patent grants

In China's manufacturing sector, the quantity of patent grants has risen annually by at least 30%. During the 8 years of the sample period, the quantity of patent applications in China's manufacturing sector increased dramatically from 67,227 (2006) to 534,927 (2013). This rapid increase could not have proceeded without the effective administration and protection of intellectual property, including patents, by China's government. Since the 11th Five-year Plan period, China has revised various laws and regulations on patents repeatedly. In 2007, the 17th National Congress of the CPC expressly resolved that China should carry out an intellectual property strategy. In 2008, China began its *Outline of National Intellectual Property Strategy*. Such laws and regulations provide a powerful assurance for the accomplishment of patented inventions and creations, and stimulate the enthusiasm of China's manufacturing sector for R&D in patents.

Invention patents are the type of patent that best reflects the degree of technological innovations and original innovation capability. The quantity of invention patent grants in China's manufacturing sector increased rapidly from 28,168 (2006) to 327,989 (2013), indicating that China's manufacturing enterprises are taking part in technological innovations more and more actively, and this has borne great fruit. In recent years (excluding 2011), the rate of increase of invention patent grants has ranged from 21 to 51%. In 2011, the rate of increase of this index was as high as 79%, a rise of 40% points from 2010. By the end of 2011, the quantity of China's effective invention patent grants was as high as 351,288, for the first time surpassing the quantity of foreign invention patents granted in China; in addition, the quantity of invention patent grants per 10,000 Chinese mainlanders reached 2.37. Why was China able to achieve this? In 2011, a target of "3.3 invention patent grants per 10,000 people" was incorporated into the outline of the 12th Five-year Plan for the first time, and different regions of China successively promulgated favorable policies to support R&D in patents.

(4) Analyses of new product benefits

In China's manufacturing sector, the output value of new products increases year by year, the rate of output value of new products varies unsteadily, and the change trend of the input–output coefficient of technological innovations is abnormal. According to the normalized data shown in Fig. 3.12, there is a significant difference in the change trend between the three indexes on new product benefits during 2006–2013. The output value of new products increased year by year, indicating

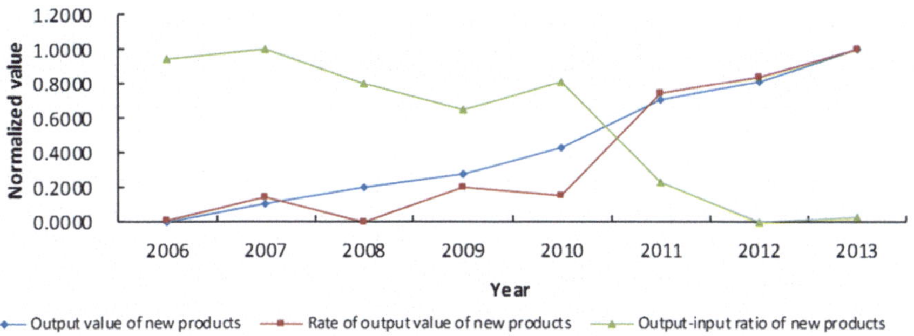

Fig. 3.12 Change trend of new product benefits of China's manufacturing sector from 2006 to 2013

that considerable benefits were created by new product development. The rate of output value of new products decreased in 2008 and 2010, but for the 8 years sample period, this index peaked in 2013. As shown in Fig. 3.12, the rate of output value of new products had a high weight among the 11 indexes, indicating that this index varied greatly and was the most unstable of the 11 indexes, and that the contribution made by the output value of new products to the total industrial output value was very instable.

It is noteworthy that the change trend of the input–output coefficient of technological innovations is very abnormal. The index was very high in 2006 and 2007, fell short of the average level of 2006 and 2007 during subsequent years, and reached a low for the 8 years sample period in 2012. The comprehensive evaluation score of technological innovation capability peaked in 2013 for the 8 years sample period, but the input–output coefficient of technological innovations in 2013 declined 3.4173 from the 2006 level of 17.3907. The input–output coefficient of technological innovations is the ratio of the benefits of technological innovations to the costs of technological innovations in the manufacturing sector. It should be highlighted that this index declined almost continuously during the 8 years. With the constant development of modern technologies, it has become increasingly difficult for the manufacturing sector to obtain benefits through technological innovations; the same expenditure input cannot lead to the same benefits obtained previously. This also implies that technological innovations need more financial support. Put differently, it implies that the expenditure input of China's manufacturing sector is very inefficient economically.

3. Summary and suggestions

For China's manufacturing sector, both the input intensity and output of technological innovations increased significantly during 2006–2013. Except for one or two indexes not being ideal (e.g., the input–output coefficient of technological innovations), the technological innovation capability of China's manufacturing sector reached a new level, and there is room for further improvement. Based on the abovementioned analyses of the technological innovation capability of China's manufacturing sector, this study makes the following suggestions.

(1) China should increase expenditure input in R&D and new product development continuously. China's manufacturing sector has increased expenditure input in technological innovations year by year, and consistently made certain achievements in patent grants and output value of new products. However, China's expenditure input in R&D is not high; in particular, China's R&D–GDP ratio is far below the average level (2–3%) of developed countries. In order to improve the technological innovation capability of China's manufacturing sector further, it is inevitable that China should increase expenditure input.
(2) China should strengthen the development of innovative human resources. R&D personnel are the main driving force of technological innovations, and are the key driving force for improving the technological level of China's manufacturing sector. In recent years, both the population of R&D personnel and R&D personnel input intensity have increased in China's manufacturing sector; however, the R&D expenditure intensity of manufacturing enterprises remains relatively low. Effective labor input and HR development are the key factors to prevent any waste of R&D expenditures. Therefore, it is very urgent to increase the R&D labor input and build highly qualified R&D teams, so as to transform China's manufacturing sector into a technology-intensive sector.
(3) China should improve the transformation of technological achievements, and strengthen the connections between technological R&D and market applications. China's manufacturing sector has made great progress in patent applications, patent grants, and output value of new products, but the input–output coefficient of technological innovations has continued to decline in recent years. This has a great deal to do with the way in which China evaluates technological innovation capability, that is, China overemphasizes R&D input, quantity of patent grants, and quantity of published articles. Innovations will not be actualized unless they can compete in the market. Therefore, it is imperative that China's manufacturing sector conduct market-oriented technological innovations and transform technological achievements into real productivity, thereby changing the "Made in China" pattern into "Intelligent Manufacturing in China."

3.3 Comprehensive Evaluation of Energy Conservation Capability

The scale and level of a country's manufacturing sector reflect its comprehensive national strength. Since China's reform and opening-up, China's manufacturing sector has made great progress in both scale and level, and the added value of

China's manufacturing sector always accounts for at least one-third of China's GDP. Judging by the growth rate, it is strongly anticipated that China will become a large manufacturing country and even the world's manufacturing center. The manufacturing sector is one of the main fields of international economic competition, and many industrial countries rely on the output of manufactured products to promote their economic development. Since China's reform and opening-up, the structure of China's export products has undergone great changes: textile products have replaced petroleum as the main export product. This implies that China's main export products have become labor-intensive instead of being natural resources, while capital-intensive and technology-intensive products are accounting for an increasing proportion of China's total export products year by year. Therefore, the manufacturing sector has become a major driving force for China's economic development.

In the process of China's industrialization, the contradiction between energy consumption and energy supply has become increasingly severe since 2002. The supply of main energy (e.g., coal, electricity, and petroleum) has come under great pressure. Energy shortages have become a major constraint to the sustainable development of China's national economy. Considering the severe situation of energy supply, both governments and enterprises have to attach great importance to energy conservation and emissions reduction.

The manufacturing sector is not only the mainstay sector of China's national economy, but also the largest energy-consuming sector. In order to reduce China's total energy consumption, the key is to reduce the energy consumption of China's manufacturing sector. Thanks to the adjustment of industrial structure and improvement in energy efficiency, the energy consumption intensity of China's manufacturing sector continues to decline. Therefore, it is very important to evaluate the energy efficiency of China's energy consumption.

This section mainly studies the energy consumption of China's manufacturing sector, and calculates and analyzes related data from the *China Statistical Yearbook* (2006–2013). Regarding the main indexes on energy consumption from 2006 to 2013, this section uses the deviation maximization method to calculate the weight of each index and thereby comprehensively evaluates the trend of energy consumption of China's manufacturing sector.

Table 3.4 lists the normalized data on energy consumption of China's manufacturing sector during 2006–2013, and Table 3.5 lists the comprehensive evaluation scores of energy consumption of China's manufacturing sector during 2006–2013.

According to the normalized data on energy consumption during 2006–2013, we draw a line chart on its linear distribution trend (Fig. 3.13).

1. Comprehensive evaluation of resource/energy consumption of China's manufacturing sector

The comprehensive evaluation score of energy consumption of China's manufacturing sector continued to decline during 2006–2013. As described in Table 3.5, the comprehensive evaluation score declined continuously from 0.6677 in 2006 to 0.3323 in 2013. While substantial energy consumption exerts increasing pressure on China's sustainable development, the energy consumption intensity of China's manufacturing sector has continued to decline in recent years by adjusting the industrial structure and improving energy efficiency.

2. Single-index evaluation of energy consumption of China's manufacturing sector

The following section evaluates the resource/energy consumption of China's manufacturing sector from three aspects: energy consumption, electricity consumption, and energy consumption per unit of output value.

(1) The total energy consumption of China's manufacturing sector continued to increase during the 8 years of the sample period. As shown in Fig. 3.14, the total energy consumption of China's manufacturing sector increased continuously from 1.43 billion tons of standard coal in 2006 to 2.225 billion tons of standard coal in 2013. Evidently, most of China's manufacturing industries are highly energy-consuming and energy-dependent, and will become increasingly energy-dependent with further growth. In order to achieve sustainable development over the long term and to strengthen its competitiveness, it is imperative that China's manufacturing sector takes various measures to reduce energy consumption.
(2) The electricity consumption of China's manufacturing sector assumed an upward trend in the 8 years sample period. Specifically, as shown in Fig. 3.14, it increased from 1.5371 trillion kWh in 2006 to 2.8638 trillion kWh in 2013. Evidently, China's manufacturing industries are highly reliant on electricity, and will be increasingly reliant on electricity with further growth. Therefore, it is imperative that China's manufacturing sector reduces the intensity of electricity consumption.
(3) The energy consumption per unit of output value in China's manufacturing sector continued to decline in the 8 years sample period. As shown in Fig. 3.15,

Table 3.4 Normalized data on energy consumption of China's manufacturing sector from 2006 to 2013

Year	Energy consumption in the manufacturing sector	Energy consumption per unit of output value in the manufacturing sector	Electricity consumption in the manufacturing sector
2006	1	0	1
2007	0.8343	0.2760	0.7939
2008	0.6344	0.4568	0.7575
2009	0.5276	0.5021	0.6748
2010	0.4282	0.7380	0.4348
2011	0.2784	0.8641	0.2346
2012	0.2122	0.9108	0.1369
2013	0	1	0

Data source Author's calculation

Table 3.5 Comprehensive evaluation of energy consumption of China's manufacturing sector from 2006 to 2013

Weight year	0.3222	0.3323	0.3455		
	Energy consumption in the manufacturing sector (unit: 10,000 tons of standard coal)	Energy consumption per unit of output value in the manufacturing sector (unit: 10,000 tons of standard coal per 100 million yuan)	Electricity consumption in the manufacturing sector (unit: 100 million kWh)	Evaluation score [$D_i(w)$]	Rank No.
2006	1	0	1	0.6677	1
2007	0.8343	0.2760	0.7939	0.6348	2
2008	0.6344	0.4568	0.7575	0.6179	3
2009	0.5276	0.5021	0.6748	0.5700	4
2010	0.4282	0.7380	0.4348	0.5334	5
2011	0.2784	0.8641	0.2346	0.4579	6
2012	0.2122	0.9108	0.1369	0.4183	7
2013	0	1	0	0.3323	8

Data source Author's calculation

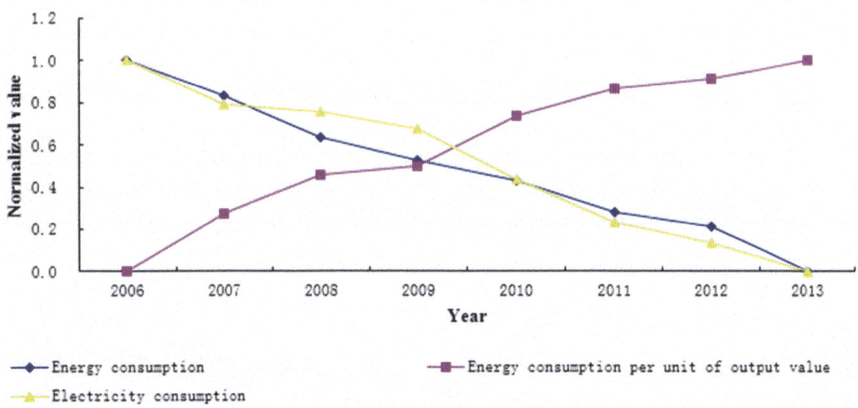

Fig. 3.13 Data normalization trend of energy consumption

the energy consumption per unit of output value declined from 5210 tons of standard coal per 100 million yuan output value in 2006 to 2340 tons of standard coal per 100 million yuan output value in 2013.

Obviously, China's manufacturing sector actively responds to the national policy on energy conservation and emissions reduction, and takes effective measures to improve energy efficiency.

3. Summary and suggestions

China's manufacturing sector is increasingly energy-dependent; both total energy consumption and electricity consumption increase year by year. Therefore, China should attach greater importance to energy conservation and emissions reduction and respond appropriately. The year-by-year decline in energy consumption per unit of output value demonstrates that China's manufacturing sector has attached increasing importance to energy conservation and energy efficiency.

China's economic development cannot proceed without strong support from the manufacturing sector. In order to strengthen the international competitiveness of China's manufacturing sector and the entire Chinese economy, it is imperative to attain harmony between the manufacturing sector and the development of resources and energy, to reduce the intensity of resource/energy consumption, and to improve the efficiency of resource/energy utilization. In view of this, China's national energy development strategy should give sufficient importance to the manufacturing sector and resource/energy development. Based on these analyses, this study makes the following suggestions.

(1) China's manufacturing sector needs to adjust its industrial structure. While maintaining an appropriate growth rate, China's manufacturing sector should pay more attention to the quality of development, reasonably adjust its industrial structure, transform its growth pattern to reduce resource/energy consumption, restrict and even prohibit the development of highly energy-consuming manufacturing industries, encourage the development of highly energy-efficient manufacturing industries, and provide significant support for the development of green and environmentally friendly manufacturing industries.
(2) China's manufacturing sector needs to adjust its industrial structure. China's manufacturing sector should enhance technological innovations, eliminate outdated technologies, develop hi-tech products, reasonably allocate enterprise

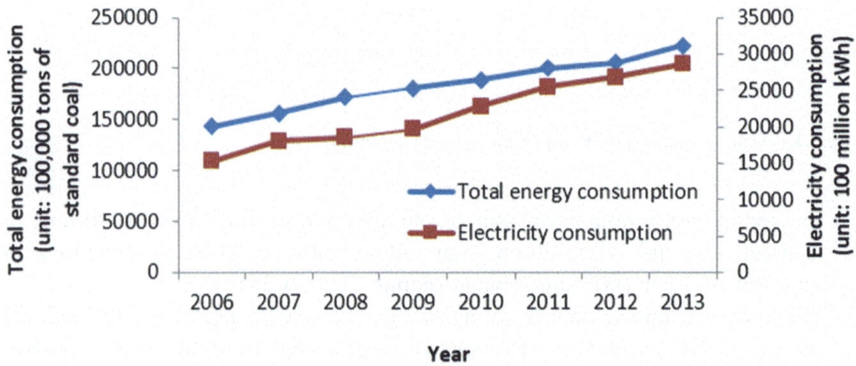

Fig. 3.14 Energy and electric power consumption of China's manufacturing sector from 2006 to 2013

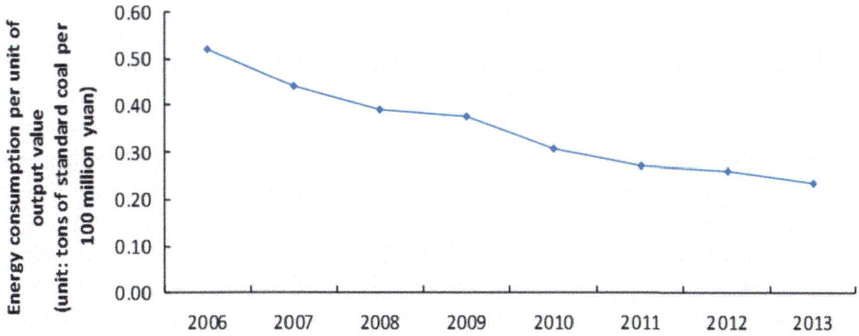

Fig. 3.15 Energy consumption per unit output value in China's manufacturing sector from 2006 to 2013

resources to hi-tech products, transform traditional high energy-consuming industries into highly energy-efficient and hi-tech industries, and improve the efficiency of energy/resource utilization.

(3) China's manufacturing sector needs to increase production efficiency. It should improve production technologies, increase production efficiency and output value, and reduce energy consumption per unit of output value. On the one hand, inefficient enterprises should either improve their productivity or be eliminated, so as to improve the productivity of the whole manufacturing sector; on the other hand, highly efficient enterprises should be encouraged to maintain high production efficiency and help other enterprises improve production efficiency, so as to increase the output value of the whole manufacturing sector and raise resource efficiency per unit of output value.

3.4 Comprehensive Evaluation of Environmental Protection Capability

Based on statistical data for 2006–2013, this section analyzes the change trend of four aspects, namely, discharge amount of wastewater, waste gas, and waste residues, and comprehensive utilization of the three types of waste, and the reasons therefore; in addition, it evaluates the achievements and deficiencies of these four aspects, with a view to providing academic support for pollution abatement of China's manufacturing sector. By using the deviation maximization method, this study normalizes the related indexes, conducts a comprehensive comparison between the normalized data, and analyzes the change trend of environmental efficiency and the contribution of discharge and comprehensive utilization of the three types of industrial waste to environmental efficiency, so as to propose effective measures for improving environmental efficiency.

1. Data processing and calculation

The statistical data on environmental indexes for 2006–2013 used in this section are mainly from the *China Statistical Yearbook* for these years. Specifically, the statistical data on the discharge amount of waste gas of China's manufacturing sector for 2006–2013 and the statistical data on the output value by comprehensive utilization of the three types of industrial waste are available from the *China Statistical Yearbook on Environment*; the output values of China's manufacturing sector for 2012 and 2013 are derived from the sales values of China's manufacturing sector contained in the *China Industry Statistical Yearbook* (the two values approximate each other, and thus, this substitution has no significant effect on the analysis and evaluation results); the discharge amount of waste residue and the output value by comprehensive utilization of the three types of industrial waste for 2011–2013 are not available in the main statistical yearbooks, and thus, associated data for the two indexes in the *China Statistical Yearbook* (2005 to 2010) are used as the original data. Then, a grey GM (1,1) prediction model is used to derive their values, as described in Table 3.6.

Using Eqs. (3.1) and (3.2), this study obtains dimensionless data for the environmental indexes of China's manufacturing sector, as listed in Table 3.7.

Table 3.8 lists and ranks the comprehensive evaluation scores of the environmental indexes of China's manufacturing sector

2. Analyses of environmental influence by China's manufacturing sector

(1) Discharge of wastewater of China's manufacturing sector

During 2007–2012, the discharge amount of wastewater by China's manufacturing sector assumed a downward trend; the discharge amount of wastewater in 2012 declined to the level of 2006, but in 2013, it rose to almost double the amount in 2012. Figure 3.16 shows the discharge of wastewater of China's 30 manufacturing industries: the total discharge amount of wastewater of China's manufacturing sector was lowest in 2006, but increased sharply from 2012 to 2013 due to the remarkable increase in the discharge amount of wastewater in the smelting and pressing of ferrous metals and the manufacture of raw chemical materials and chemical products. In 2013, the annual operating costs of industrial wastewater treatment facilities in smelting and pressing of ferrous metals were equal to only 0.8% of those incurred in 2012, the amount of treated industrial wastewater in this industry was equal to only 48% of that in 2012, and the amount of discharged industrial wastewater was 21.86335 billion tons (this was 20.6 times as much as that in 2012, and accounted for 55% of the total discharge amount of wastewater in China's manufacturing sector). Smelting and pressing of ferrous metals replaced the manufacture of paper and paper products as the industry that discharged the most wastewater. In 2013, the annual operating costs of industrial wastewater treatment facilities in the manufacture of raw chemical materials and chemical products were equal to only 0.2% of those incurred in 2012, and the amount of discharged industrial wastewater was 4.09677 billion tons, accounting for 10% of the total discharge amount of wastewater in China's manufacturing sector. The manufacture

Table 3.6 Statistical data on environmental indexes of China's manufacturing sector

Year	Discharge amount of wastewater (unit: 10,000 tons)	Discharge amount of wastewater per unit of output value (unit: 10,000 tons per 100 million yuan)	Discharge amount of waste gas (unit: 100 million standard cubic meters)	Discharge amount of waste gas per unit of output value (unit: 100 million standard cubic meters per 100 million yuan)	Discharge amount of waste residue (unit: tons)
2006	1666639.00	6.07	217628.00	0.79	3881210.00
2007	1810570.29	5.12	254326.90	0.72	2821850.00
2008	1791223.00	4.06	273702.00	0.62	2140000.00
2009	1747242.57	3.87	281675.00	0.62	1708070.00
2010	1766872.10	2.66	328155.00	0.54	997860.00
2011	1730429.00	2.36	463004.00	0.63	853141.16
2012	1702421.00	2.15	420576.00	0.53	630946.31
2013	4006729.00	4.47	435474.00	0.49	547257.86

Year	Discharge amount of waste residue per unit of output value (unit: tons per 100 million yuan)	Output value by comprehensive utilization of the three types of industrial waste (unit: 10,000 yuan)
2006	13.95	9163162.20
2007	7.98	12430524.20
2008	4.85	14805573.00
2009	3.72	14186126.70
2010	1.64	16066049.70
2011	1.16	18399111.83
2012	0.80	20567760.92
2013	0.73	22371742.45

Data source: *China Statistical Yearbook on Environment*, and *China Statistical Yearbook*

of raw chemical materials and chemical products was the industry that discharged the second-largest amount of wastewater. In 2013, the annual operating costs of industrial wastewater treatment facilities of China's manufacturing sector sharply declined to 256.42 million yuan, only 0.38% of those in 2012. Overall, the investment in wastewater treatment was significantly reduced, discharge of wastewater was not controlled well, and a certain rebound occurred in environmental governance. Therefore, it is imperative that China attaches great importance to this environmental problem in the long term. The control of discharge amount of wastewater remains a key point in environmental governance.

The discharge amount of wastewater per unit of output value assumed a steady downward trend during 2006–2012, but rose sharply in 2013. In 2013, the discharge amount of wastewater by China's manufacturing sector increased significantly at a higher rate than the total output value of China's manufacturing sector, and thus, the discharge amount of wastewater per unit of output value in 2013 was restored to the level of 2008. As shown in Fig. 3.17, the discharge amount of

Table 3.7 Dimensionless data on the environmental indexes of China's manufacturing sector

Year	Discharge amount of wastewater (unit: 10,000 tons)	Discharge amount of wastewater per unit of output value (unit: 10,000 tons per 100 million yuan)	Discharge amount of waste gas (unit: 100 million standard cubic meters)	Discharge amount of waste gas per unit of output value (unit: 100 million standard cubic meters per 100 million yuan)	Discharge amount of waste residue (unit: tons)
2006	1	0	1	0	0
2007	0.9385	0.2423	0.8504	0.2298	0.3177
2008	0.9468	0.5128	0.7715	0.5581	0.5223
2009	0.9656	0.5612	0.7390	0.5581	0.6518
2010	0.9572	0.8699	0.5496	0.8207	0.8648
2011	0.9727	0.9464	0.0000	0.5252	0.9083
2012	0.9847	1	0.1729	0.8535	0.9749
2013	0	0.4092	0.1122	1	1

Year	Discharge amount of waste residue per unit of output value (unit: tons per 100 million yuan)	Output value by comprehensive utilization of the three types of industrial waste (unit: 10,000 yuan)
2006	0	0
2007	0.4516	0.2474
2008	0.6884	0.4272
2009	0.7738	0.3803
2010	0.9312	0.5226
2011	0.9675	0.6992
2012	0.9947	0.8634
2013	1	1

Data source Author's calculation

wastewater per unit of output value was not high. In summary, China's manufacturing sector needs to abate wastewater pollution further.

(1) Discharge of waste gas by China's manufacturing sector

The discharge amount of waste gas by China's manufacturing sector assumed an upward trend during 2006–2011, somewhat decreased in 2011 and 2012, and slightly increased in 2013. Figure 3.18 shows the change trend of the discharge amount of waste gas by China's manufacturing sector. Among all cost indexes, this index is the only one that assumes an upward trend on the whole. The discharge of waste gas was mainly from the following manufacturing industries: smelting and pressing of ferrous metals, manufacture of non-metallic mineral products, smelting and pressing of non-ferrous metals, and manufacture of raw chemical materials and chemical products. Industries with high energy consumption and that are highly polluting always rank at the top in the IGP, thereby causing a continued increase in

Table 3.8 Comparison and rank of environmental indexes of China's manufacturing sector

Year	Discharge amount of wastewater (unit: 10,000 tons)	Discharge amount of wastewater per unit of output value (unit: 10,000 tons per 100 million yuan)	Discharge amount of waste gas (unit: 100 million standard cubic meters)	Discharge amount of waste gas per unit of output value (unit: 100 million standard cubic meters per 100 million yuan)	Discharge amount of waste residue (unit: tons)
2006	1	0	1	0	0
2007	0.9385	0.2423	0.8504	0.2298	0.3177
2008	0.9468	0.5128	0.7715	0.5581	0.5223
2009	0.9656	0.5612	0.7390	0.5581	0.6518
2010	0.9572	0.8699	0.5496	0.8207	0.8648
2011	0.9727	0.9464	0.0000	0.5252	0.9083
2012	0.9847	1	0.1729	0.8535	0.9749
2013	0	0.4092	0.1122	1	1
Weight	0.0957	0.1563	0.1658	0.1440	0.1525

Year	Discharge amount of waste residue per unit of output value (unit: tons per 100 million yuan)	Output value by comprehensive utilization of the three types of industrial waste (unit: 10,000 yuan)	Comprehensive evaluation score	Rank
2006	0	0	0.2615	8
2007	0.4516	0.2474	0.4495	7
2008	0.6884	0.4272	0.6173	6
2009	0.7738	0.3803	0.6462	5
2010	0.9312	0.5226	0.7753	2
2011	0.9675	0.6992	0.6925	3
2012	0.9947	0.8634	0.8159	1
2013	1	1	0.6648	4
Weight	0.1401	0.1456	–	

Data source Author's calculation

the discharge amount of waste gas, and to a certain extent, offsetting the decline in the discharge amount of waste gas brought about by technological progress.

Overall, the discharge amount of waste gas by China's manufacturing sector tended to decline slowly. Both the total output value of China's manufacturing sector and discharge amount of waste gas by China's manufacturing sector assumed an upward trend during 2006–2013. As shown in Fig. 3.19, the discharge amount of waste gas per unit of output value tended to decline slowly; specifically, this index declined steadily during 2006–2008, fluctuated very slightly during 2009–2012, and declined to the minimum level in 2013. Figure 3.20 shows the change trend of the discharge amount of waste gas per unit of output value. In 2007, 2008,

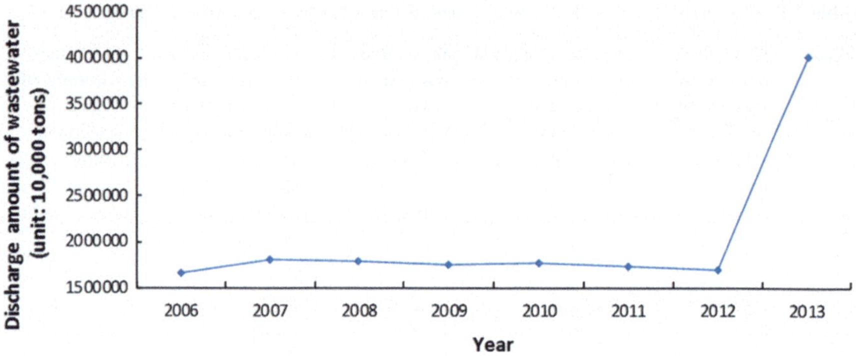

Fig. 3.16 Discharge amount of wastewater by China's manufacturing sector

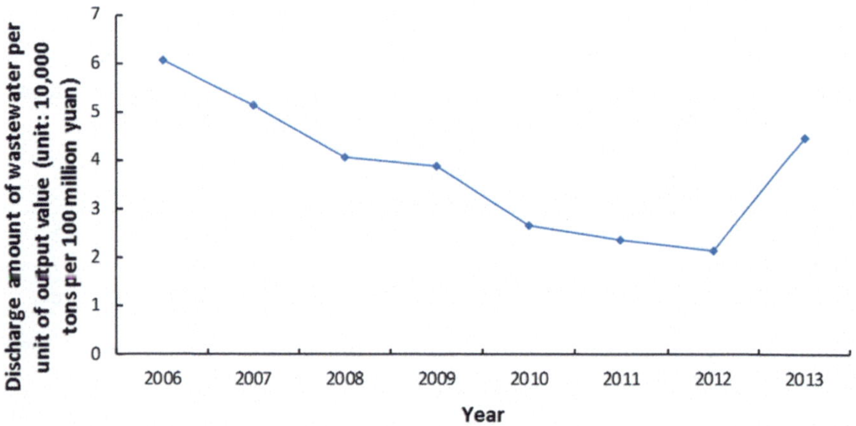

Fig. 3.17 Discharge amount of wastewater per unit output value by China's manufacturing sector

2010, 2012, and 2013, the growth rate of total output value of China's manufacturing sector was higher than that of the discharge amount of waste gas by China's manufacturing sector. Therefore, in 2006–2008, 2009–2010, and 2011–2013, the discharge amount of waste gas per unit of output value assumed a downward trend. The growth rate of total output value of China's manufacturing sector was approximately that of the discharge amount of waste gas by China's manufacturing sector in 2009, and thus, the discharge amount of waste gas per unit of output value varied slightly in 2008 and 2009. The growth rate of discharge amount of waste gas by China's manufacturing sector was higher than that of the total output value of China's manufacturing sector, and thus, the discharge amount of waste gas per unit of output value assumed an upward trend in 2010 and 2011. The discharge amount of waste gas per unit of output value fluctuated obviously during 2008–2012, indicating that China's manufacturing sector encountered a bottleneck in reducing

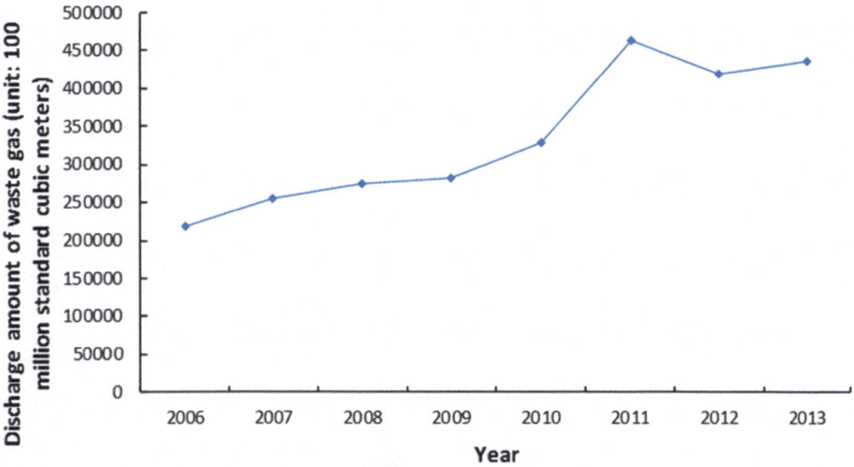

Fig. 3.18 Discharge amount of waste gas by China's manufacturing sector

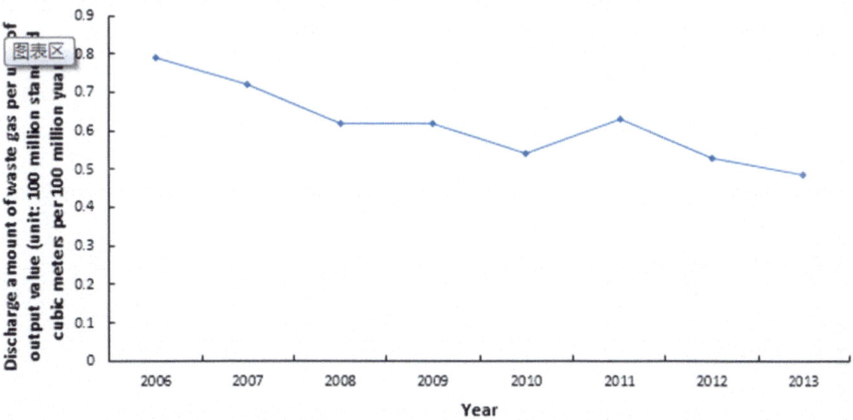

Fig. 3.19 Discharge amount of waste gas per unit output value by China's manufacturing sector

the discharge amount of waste gas discharge. In order to overcome the bottleneck, it is necessary to increase R&D input moderately.

(3) Discharge of waste residue by China's manufacturing sector

The discharge amount of waste residue by China's manufacturing sector continued to decline during 2006–2013, and the downward trend slowed during these years. Figure 3.21 shows the change trend of the discharge amount of waste residue by China's manufacturing sector during 2006–2013. Obviously, the discharge amount of waste residue decreased year by year. Judging by the slope of the straight lines formed every 2 years, the downward trend in the discharge amount of waste

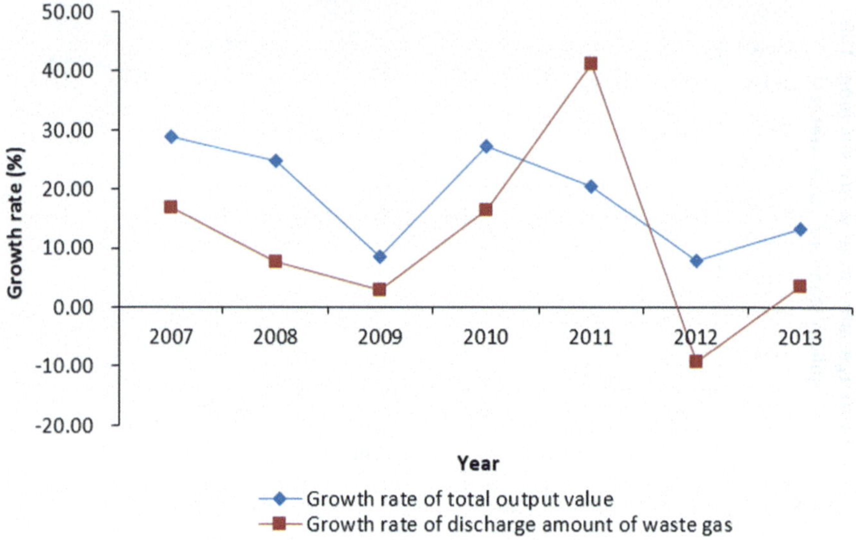

Fig. 3.20 Comparison between the growth rate of total output value and the growth rate of discharge amount of waste gas in China's manufacturing sector

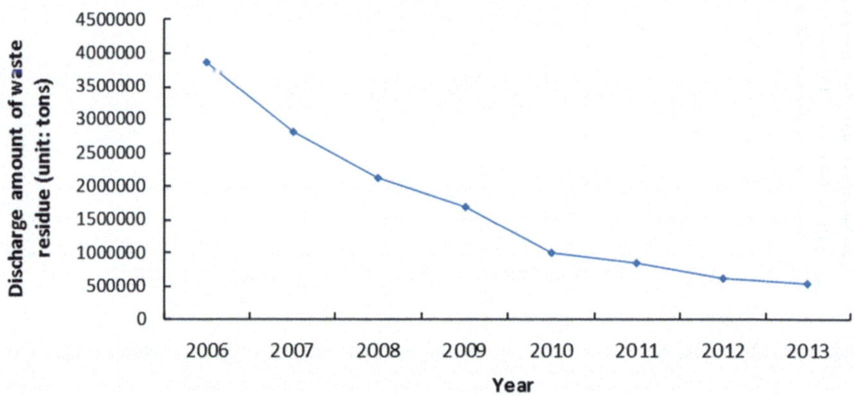

Fig. 3.21 Discharge amount of waste residue by China's manufacturing sector

residue slowed year by year. The *Law of the People's Republic of China on the Prevention and Control of Environmental Pollution Caused by Solid Waste,* effective from April 1, 1996, stipulates the following principles for the prevention and control of waste pollution: quantity reduction, harmlessness, cyclic utilization, and life-cycle administration. This law provides a powerful legal assurance for the treatment of waste residue. In addition, the abovementioned downward trend after 2004 can be ascribed to the following factors: (1) a large number of enterprises

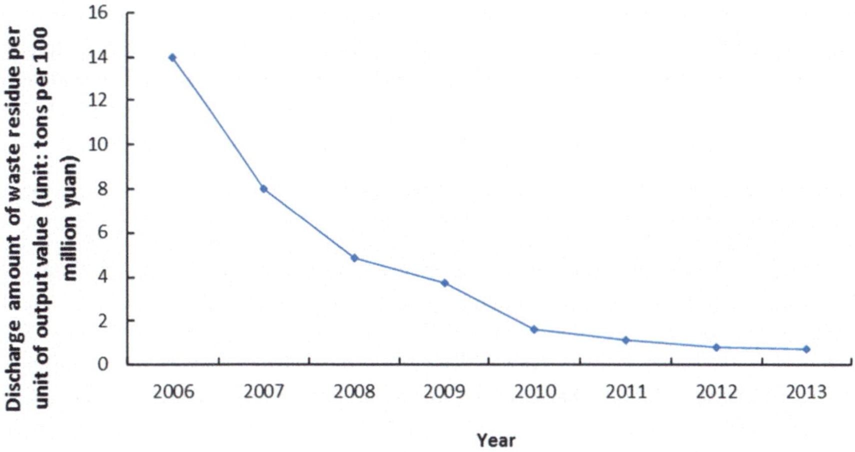

Fig. 3.22 Discharge amount of waste residue per unit output value in China's manufacturing sector

engaged in treatment of solid waste have emerged; (2) China has enhanced fundamental research in the treatment of solid waste; (3) some concentrated treatment facilities for hazardous waste have been constructed to improve the treatment of hazardous waste; (4) China has deepened its survey of the sources of solid waste; (5) the laws, regulations, and national standards for pollution control have been improved further; and (6) demonstration projects on the treatment of solid waste have provided guidance. The slowdown in the downward trend is because measures for the abatement and treatment of solid waste have been carried out effectively. Therefore, in order to reduce the discharge amount of solid waste, the key is to conduct technological innovations, which still remain to be explored further.

The discharge amount of waste residue by China's manufacturing sector decreased year by year, and there was a slowdown in the decline in the discharge amount of waste residue per unit of output value in the 8 years sample period. As shown in Fig. 3.22, the discharge amount of waste residue per unit of output value assumed a downward trend during 2006–2013 and reached a very low level in 2013, although there was a slowdown in the downward trend. The discharge amount of waste residue was mainly available from the following manufacturing industries: smelting and pressing of ferrous metals; manufacture of non-metallic mineral products; manufacture of raw chemical materials and chemical products; smelting and pressing of non-ferrous metals; and processing of petroleum, coking, processing of nuclear fuel. The manufacture of non-metallic mineral products contributed less to the total output value of China's manufacturing sector than did the discharge of waste residue by this industry; the manufacture of textiles contributed more to the total output value of China's manufacturing sector than did the discharge of waste residue by this industry; while the contribution rate of the 28 other industries was approximately in direct proportion to the contribution rate of their discharge of waste residue. From the perspective of China's manufacturing sector as a whole, the discharge amount of waste residue was reduced effectively. However, in order to reduce the discharge amount of waste residue continuously, it is necessary to closely monitor the major manufacturing industries that discharge waste residue, especially the manufacture of non-metallic mineral products.

(4) Comprehensive utilization of three types of industrial waste in China's manufacturing sector

Overall, the output value by comprehensive utilization of the three types of industrial waste in China's manufacturing sector increased. This index is a benefit index; therefore, the larger the index value is, the better it is. As shown in Fig. 3.23, this index increased during 2006–2008 and 2009–2013, and declined slightly during 2008–2009. This is because the discharge amount of wastewater and waste gas per unit of output value varied slightly during 2008–2009 (as shown in Figs. 3.17 and 3.19, respectively) and the discharge amount of waste residue per unit of output value decreased during 2008–2009 (as shown in Fig. 3.22). Therefore, it can be inferred that the comprehensive utilization amount of the three types of industrial waste decreased slightly during 2008–2009. Overall, there was an increase in this index. The *Suggestions Concerning the Further Promotion of Comprehensive Resources Utilization* promulgated in 1996 promoted comprehensive resource utilization in China. The *10th Five-year Plan on Resource Conservation and Comprehensive Utilization* promulgated in 2001 proposed long-term planning for the comprehensive utilization of the three types of industrial waste. The *Law on Promoting Clean Production*, effective from January 1, 2003, laid a legal basis for the comprehensive utilization of the three types of industrial waste. In the annual environmental protection plans for 2006–2012, projects on the comprehensive utilization of the three types of industrial waste always accounted for a large proportion of the total environmental protection plans. During the

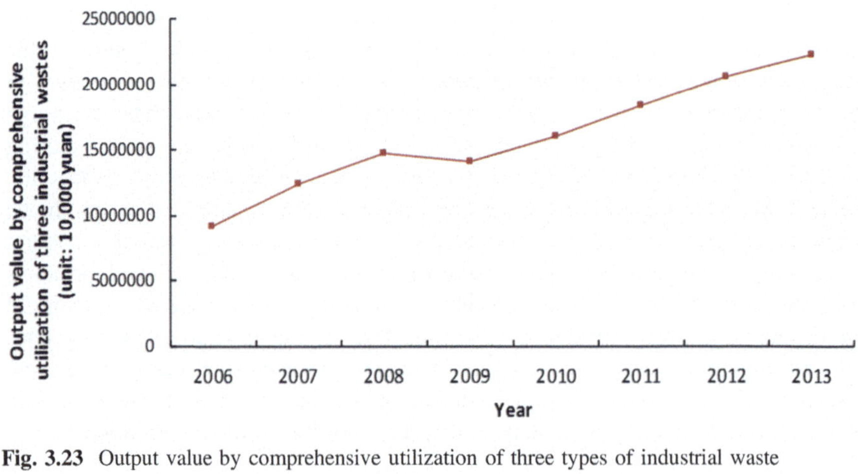

Fig. 3.23 Output value by comprehensive utilization of three types of industrial waste

8 years from 2006–2013, the output value by comprehensive utilization of the three types of industrial waste doubled, which was an exciting achievement. Continued improvement in the comprehensive utilization of the three types of industrial waste can be ascribed to constant technological progress, enhanced consciousness of cyclic utilization, and government incentives. Different levels of government and concerned authorities should pay attention to the main manufacturing industries and regions in which the three types of industrial waste are discharged; supervise the discharge of the three types of industrial waste in different manufacturing industries; and provide consultancy and information services for manufacturing enterprises. The intent is to improve the comprehensive utilization of the three types of industrial waste effectively, save resources, improve the environment, increase economic benefits, promote the transformation of the economic growth pattern, optimize resource allocation, and attain sustainable development.

3. Comprehensive analysis and evaluation of environmental efficiency in China's manufacturing sector

(1) Comprehensive analysis of environmental efficiency

Based on the comprehensive evaluation data on environmental efficiency for 2006–2013, this section analyzes China's environmental status and prospects. By using the deviation maximization method, this study calculates the weight of each environmental efficiency index of China's manufacturing sector, and thus, determines the extent to which these weights influence the comprehensive evaluation score of environmental efficiency. As shown in Fig. 3.24, the comprehensive evaluation score of environmental efficiency is mainly influenced by the following factors (in descending order): discharge amount of waste gas, discharge amount of wastewater per unit of output value, discharge amount of waste residue, total output

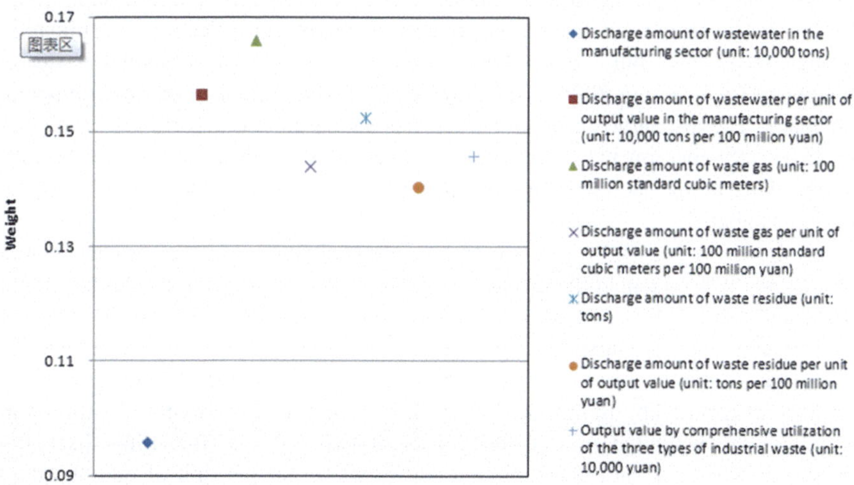

Fig. 3.24 Weight of each environmental efficiency index of China's manufacturing sector

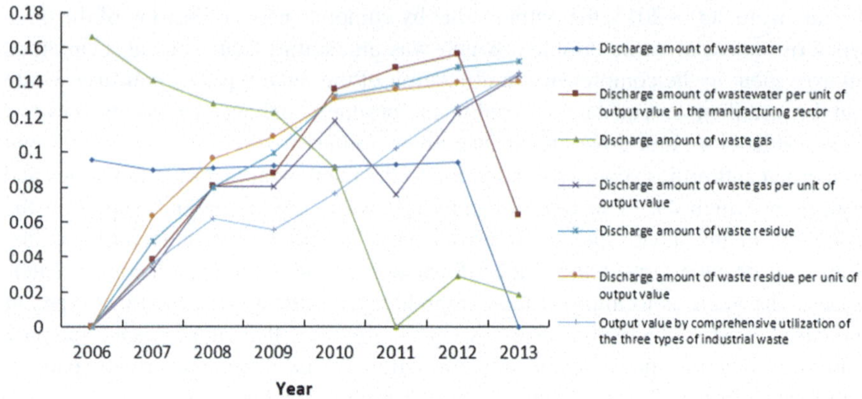

Fig. 3.25 Weighted normalized values of environmental efficiency indexes of China's manufacturing sector

value by comprehensive utilization of the three types of industrial waste, discharge amount of waste gas per unit of output value, discharge amount of waste residue per unit of output value, and discharge amount of wastewater.

Overall, the comprehensive evaluation score of environmental efficiency of China's manufacturing sector was on the increase in the 8 years sample period; in other words, environmental efficiency continued to be improved. An upward trend of the weighted normalized values of the environmental efficiency indexes indicates a positive influence on environmental efficiency, and a downward trend indicates a negative influence on environmental efficiency. As shown in Fig. 3.25, four environmental efficiency indexes (excluding the discharge amount of waste gas, discharge amount of wastewater, and discharge amount of wastewater per unit of output value) were on the increase, and their change trend was favorable and steady. The discharge amount of waste gas per unit of output value declined significantly during 2010–2011, but the discharge amount of waste gas assumed a steady downward trend during 2006–2011, having a negative influence on environmental efficiency. The discharge amount of waste gas increased during 2011–2012, having an active influence on the comprehensive evaluation score of environmental efficiency. This index declined slightly during 2012–2013. Overall, the pollution of waste gas abated.

As shown in Fig. 3.26, the comprehensive evaluation score of environmental efficiency of China's manufacturing sector assumed an upward trend during 2006–2010, and 2011–2012, and declined slightly during 2010–2011, and 2012–2013. Overall, this index assumed an upward trend, or rather, environmental efficiency improved constantly. In terms of environmental efficiency, the individual years of 2006–2013 ranked as follows: 8, 7, 6, 5, 2, 3, 1, and 4, respectively. As shown in Fig. 3.25, the weighted normalized values of two indexes (including discharge amount of waste gas and discharge amount of waste gas per unit of output value) decreased significantly during 2010–2011. As shown in Figs. 3.19 and 3.20, the

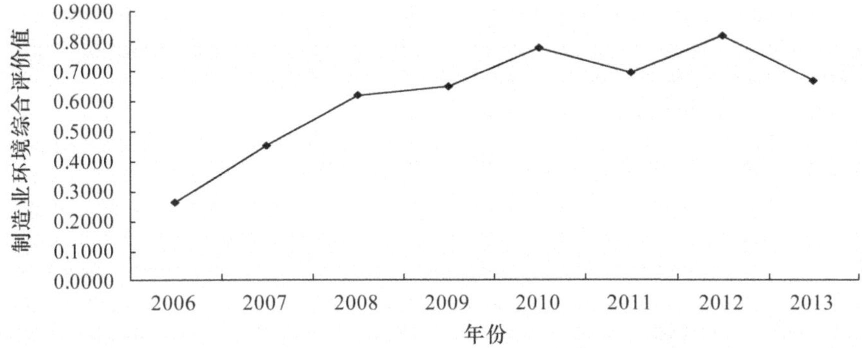

Fig. 3.26 Comprehensive evaluation score of environmental efficiency of China's manufacturing sector

discharge amount of waste gas increased remarkably during 2010–2011, increasing at a higher rate than the total output value of China's manufacturing sector; as a result, the discharge amount of waste gas per unit of output value increased negatively, offsetting the positive influence of the other indexes. This explains why environmental efficiency declined during 2010–2011.

(2) Comprehensive evaluation of environmental efficiency

Overall, the environmental efficiency of China's manufacturing sector improved constantly; the discharge amount of waste gas per unit of output value decreased in a wave-like manner, having a significant influence on environmental efficiency. Here, environmental efficiency refers to the ratio of total added value of China's manufacturing sector to the environmental impact of China's manufacturing sector. Research on environmental efficiency covers the discharge of the three types of industrial waste and comprehensive utilization of the three types of industrial waste. Environmental efficiency has a close bearing on sustainable development. Economic development will be negative if it is deficient in environmental efficiency. According to the above analyses, the environmental efficiency of China's manufacturing sector improved constantly in the 8 years sample period. Specifically, the discharge of waste residue and comprehensive utilization of the three types of industrial waste moved toward a favorable trend, and their change trend was very steady. The discharge amount of waste gas per unit of output value tended to decrease overall, but fluctuated obviously, having a remarkable influence on environmental efficiency. The discharge amount of waste gas should be taken seriously, because this index is not only the main factor that causes the fluctuation of the discharge amount of waste gas per unit of output value, but also the main factor that hinders the improvement in environmental efficiency. In order to reduce the discharge amount of waste gas, desulfurization technologies should be developed urgently and the carbon trading market system should be honed. The discharge of wastewater assumed a favorable trend during 2006–2012. The discharge of wastewater increased in 2013 due to the sharp increase in the discharge of wastewater in the smelting and pressing of ferrous metals, and manufacture of raw chemical materials and chemical products. Therefore, it is necessary to pay more attention to the discharge and treatment of wastewater in these two industries. In terms of the average discharge intensity of the three types of industrial waste during 2006 and 2013, the following industries were low-pollution industries: manufacture of general purpose machinery; manufacture of special purpose machinery; manufacture of railway, ship, aerospace and other transportation equipment, manufacture of electrical machinery and equipment; manufacture of computers, communication, and other electronic equipment; and manufacture of measuring instruments. The following industries were pollution-intensive industries: manufacture of non-metallic mineral products; smelting and pressing of ferrous metals; manufacture of textiles; manufacture of paper and paper products; printing, reproduction of recording media; processing of petroleum, coking, processing of nuclear fuel; and manufacture of raw chemical materials and chemical products. In preventing and controlling industrial pollution and improving environmental efficiency, special attention should be paid to these pollution-intensive sectors.

The environmental efficiency of China's manufacturing sector is moving toward a favorable trend, but China obviously lags behind other advanced countries in this aspect. China's investment in environmental treatment is increasing year by year, but lags far behind that of the US and Germany. In 2005, China's investment in the treatment of environmental pollution was just over 21% of such investment in the US. Since 2006, China has gradually caught up with and even surpassed the UK in terms of investment in pollution treatment. However, there has been no obvious rise in the proportion of China's investment in pollution treatment to total output value of its manufacturing sector. In summary, the treatment of environmental pollution still requires joint efforts by China's levels of governments and its enterprises.

(3) Suggestions for environmental efficiency of China's manufacturing sector

Based on the above analyses, this study offers the following suggestions for how to improve the environmental efficiency of China's manufacturing sector and to promote sustainable development of China's manufacturing sector:

First, China should carry out clean production. Clean production is a new approach to harmonizing environmental efficiency and economic development. Specifically, each step of the manufacturing process should be pollution-free, that is, industrial pollution should be abated or prevented from within the manufacturing enterprises.

Second, China should develop a circular economy. In order to do so, manufacturing enterprises should break the shackles of the traditional economic system, and attain technological innovations to an appropriate extent, thereby preventing any diseconomies.

Third, China should establish appropriate environmental regulations and coordinate the enforced implementation of environmental regulations with incentives to comply with environmental regulations. Enforced environmental regulations (including environmental standards, emissions standards and technical standards based on environmental standards, and regulations in other forms) are intended to force manufacturing enterprises to abide by the emissions standards and control the discharge amount of the three types of industrial waste. Incentives to comply with environmental regulations (including the emissions trading, emissions taxes, a subsidy and deposit refund system, and a voluntary negotiation system) are intended to stimulate green technology and management innovations.

3.5 Comprehensive Evaluation of China's New-Type Manufacturing Sector

To achieve sustainable development, China's manufacturing sector must embark on a new-type manufacturing sector path with Chinese characteristics in light of the scientific outlook on development. By evaluating the overall degree of newness of China's manufacturing sector during 2006–2003, this study analyzes the real development level of China's manufacturing sector over the past years and its development potential in the future. By using the deviation maximization method, this study determines a comprehensive index on the degree of newness of China's manufacturing sector during 2006–2013, as described in Table 3.9 and shown in Fig. 3.27.

During 2006–2013, the comprehensive index on the degree of newness of China's manufacturing sector continued to increase from 0.2088 in 2006 to 0.8052 in 2013. Evidently, China's manufacturing sector assumed a favorable development trend, and its degree of newness improved constantly.

As described in Table 3.9, the four measuring indexes enjoy different weights: the economic factor has the highest weight (0.3841), and plays a decisive role in evaluating the degree of newness of China's manufacturing sector; the technological factor has the second highest weight (0.2184), while the energy factor has the lowest weight (0.1887).

Table 3.9 Comprehensive index on the degree of newness of China's manufacturing sector

Weight	Economy (0.3841)	Technology (0.2184)	Energy (0.1887)	Environment (0.2088)	Comprehensive index
2006	0.0333	0.0728	0.3321	0.6517	0.2088
2007	0.2153	0.1335	0.3641	0.6436	0.3219
2008	0.2505	0.1745	0.3850	0.4611	0.2610
2009	0.4117	0.2696	0.4328	0.3730	0.3474
2010	0.7944	0.3752	0.4660	0.2712	0.4373
2011	0.8694	0.6258	0.5394	0.3471	0.6226
2012	0.8774	0.7730	0.5776	0.2014	0.6677
2013	0.8841	0.7908	0.6679	0.2316	0.8052

Data source Author's calculation

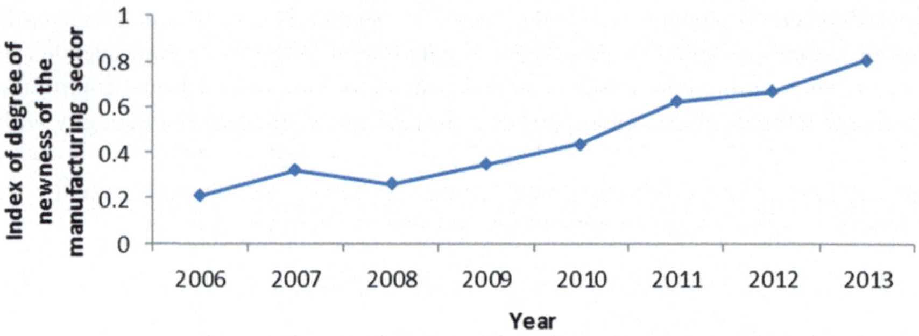

Fig. 3.27 Comprehensive index on the degree of newness of China's manufacturing sector

In fact, economic resourcefulness is an important index to measure the degree of newness of China's manufacturing sector. For a country still in the process of industrialization, economic resourcefulness is especially important. Only by creating significant economic benefits can China's manufacturing sector attain the motivation to achieve sustainable development and provide strong material support for developing technologies, improving production efficiency, offering more employment opportunities, enhancing taxation capacity, and protecting the environment.

Technological innovation capability is an important index to measure the degree of newness of China's manufacturing sector, and provides a powerful assurance for environmental protection, energy conservation, and economic resourcefulness. Only by taking full advantage of modern science and technologies and by resorting to technological innovations can China's manufacturing sector transform from extensive traditional manufacturing to a new pattern, from labor-intensive manufacturing to capital-intensive and technology-intensive manufacturing, and from an investment-driven growth pattern to an innovation-driven growth pattern, with a view to attaining sustainable development. Environmental and ecological protection is the prerequisite for sustainable social and economic development. Over the years, the extensive growth pattern of China's manufacturing sector has brought huge environmental damage and has impacted severely on the lives of Chinese people. Thus, it is reasonable to expect that the weight of the economic and technological indexes should be raised.

China's new-type manufacturing sector will inevitably undergo a tortuous road but embrace a bright future. Today, both the domestic environment and the international environment for China's manufacturing sector have undergone great changes. Domestically, the national economy is confronted with a downturn and economic slowdown. Internationally, the impact of the 2008 financial crisis remains. As a result, the development of China's manufacturing sector has slowed. According to our comprehensive evaluation of China's new-type manufacturing sector over 8 years, China's manufacturing sector is moving toward a favorable trend on the whole and its degree of newness has improved constantly. The abovementioned evaluation results provide a valuable basis for decision making when governments implement macro regulations for the manufacturing sector, formulate related economic policies, and appraise the operating performance of manufacturing enterprises.

3.6 Conclusions

This chapter sets forth the connotations of China's new-type manufacturing sector in the following four aspects: economic resourcefulness, technological innovation capability, energy conservation capability, and environmental protection capability. This chapter proposes an evaluation index system for China's new-type manufacturing sector, which comprises four main indexes and 31 subindexes. Moreover, this chapter evaluates the development status of China's manufacturing sector by using four indexes, namely, the economic index, technological index, energy index, and environmental index. The results show that in terms of economic resourcefulness, the economic index improved year by year, and the overall growth of China's manufacturing sector speeds up obviously. The employed population of China's manufacturing sector increases significantly year by year, indicating that China's manufacturing sector creates a large quantity of employment opportunities. Moreover, the gross profit of China's manufacturing sector increases significantly year by year. Therefore, the manufacturing sector promotes rapid development of China's national economy and has become the pillar sector of China's national economy. In terms of technological innovation capability, China's manufacturing sector is transforming from "Made in China" to "Created in China." In addition, the technological innovation index shows that China's manufacturing sector attached more importance to the improvement in technological innovation capability after China became the largest manufacturing country globally in 2010. In respect of energy consumption, China's manufacturing sector is increasingly energy-dependent. Therefore, governments or manufacturing enterprises should take this problem seriously, and governments should formulate appropriate policies to stimulate energy conservation and emissions reduction. Furthermore, the continued decline in the energy consumption per unit of output value shows that China's manufacturing sector has attached more importance to energy conservation and energy efficiency.

Currently, both domestic and international environments of China's manufacturing sector have undergone significant changes. Domestically, the national economy is confronted with a downturn and slowdown in economic growth. Internationally, the impact of the 2008 financial crisis remains. As a result, the development of China's manufacturing sector has slowed. Notwithstanding the negative impact, China's manufacturing sector is in a favorable trend on the whole and its degree of newness is improving constantly. In summary, China's manufacturing sector is certain to embrace a bright future.

References

National Bureau of Statistics. 2007. *China Statistical Yearbook* (2007)-VI. Beijing: China Statistics Press.

National Bureau of Statistics. 2008. *China Statistical Yearbook* (2008)-VI. Beijing: China Statistics Press.

National Bureau of Statistics. 2009. *China Statistical Yearbook* (2009)-VI. Beijing: China Statistics Press.

National Bureau of Statistics. 2010. *China Statistical Yearbook* (2010)-VI. Beijing: China Statistics Press.

National Bureau of Statistics. 2011. *China Statistical Yearbook* (2011)-VI. Beijing: China Statistics Press.

National Bureau of Statistics. 2012. *China Statistical Yearbook* (2012)-VI. Beijing: China Statistics Press.

National Bureau of Statistics. 2013. *China Statistical Yearbook* (2013)-VI. Beijing: China Statistics Press.

National Bureau of Statistics. 2014. *China Statistical Yearbook* (2014)-VI. Beijing: China Statistics Press.

Chapter 4
Development of China's Manufacturing Sector: Regional Research

Changping Xu and Minjie Wu

This chapter conducts an in-depth analysis of the development of China's manufacturing sector from a regional perspective. Regional manufacturing industries are an integral part of China's manufacturing sector. The development level of China's manufacturing sector varies among four major regions (including 31 provinces/municipalities). Based on the objective data on regional manufacturing industries, this chapter evaluates the degree of newness of regional manufacturing industries from four aspects: economic resourcefulness, technological innovations, resource utilization, and environmental protection. This chapter introduces regional manufacturing industries of China considering diverse aspects, such as, development level, achievements, contributions to economic and social development, pollutant emissions, and regional gaps. The regional research in this chapter is intended to establish the regional layout of China's manufacturing sector, sum up the development experience of advanced regions, reveal the development trends of regional manufacturing industries, and help different regions determine their development directions correctly and attain sustainable development (Data in this chapter are from References Part at the end of this chapter).

4.1 Overall Evaluation of Development of Regional Manufacturing Industries

With a consistent style and an appropriate evaluation index system, this chapter determines the top 10 manufacturing provinces, municipalities, and autonomous regions and the top 10 manufacturing cities in China's manufacturing sector,[1] with a view to uncovering the development status and trends of China's manufacturing sector.

[1]Unless otherwise noted, China herein does not include Hongkong, Macau, or Taiwan.

L. Li and Z. Du (eds.), *A Research Report on the Development of China's Manufacturing Sector (2016)*, Current Chinese Economic Report Series,
DOI 10.1007/978-981-10-4445-8_4

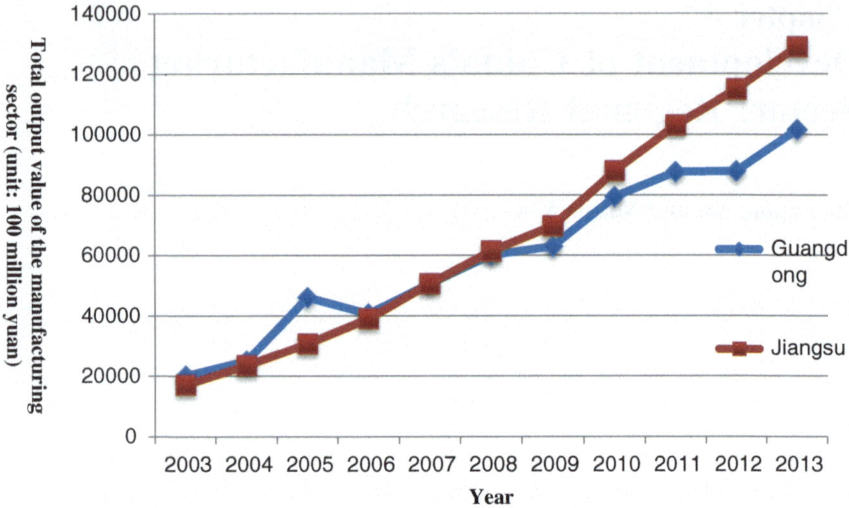

Fig. 4.1 Output value of manufacturing sector in Jiangsu and Guangdong from 2003 to 2013

4.1.1 Yangtze River Delta Leads Scale of the Manufacturing Sector

The Yangtze River Delta (including Jiangsu province, Zhejiang province, and Shanghai city) has the largest scale of manufacturing sector in China. In 2013, the output value of the Yangtze River Delta's manufacturing sector accounted for 24.02% of that of China's entire manufacturing sector,[2] while the output value of the manufacturing sector in the Pearl River Delta's (Guangdong province) accounted for 11.19% of China's entire manufacturing sector. Therefore, the scale of Yangtze River Delta's manufacturing sector was 2.15 times that of the Pearl River Delta's manufacturing sector. The scale of Jiangsu's manufacturing sector accounted for nearly 60% of the scale of the Yangtze River Delta's manufacturing sector. In 2013, the GDP of Jiangsu province was 5.916 trillion yuan, and the added value of Jiangsu's manufacturing sector was 2.423 trillion yuan. In this year, the GDP of Guangdong province was 6.216 trillion yuan, and the added value of Guangdong's manufacturing sector was 2.389 trillion yuan. Therefore, the GDP of Guangdong was slightly higher than the GDP of Jiangsu, while the scale of Guangdong's manufacturing sector was slightly smaller than the scale of Jiangsu's manufacturing sector. Figure 4.1 shows the change trend of output value of the manufacturing sector in Jiangsu and Guangdong during 2003–2013. In 2008,

[2]The data about the output value of the manufacturing sector of Xizang and Hainan are not available, and the scale of their manufacturing sectors is very small. Therefore, the data for the output value of China's manufacturing sector do not include the related data of the two provinces.

Jiangsu surpassed Guangdong in the output value of their manufacturing sectors, and the gap between them widened in subsequent years.

In summary, the Yangtze River Delta led the scale of the manufacturing sector, and the manufacturing sector in this region was differentiated; specifically, Jiangsu's manufacturing sector grew rapidly while the Shanghai's manufacturing sector slowed. Shanghai has gradually transformed from a manufacturing-centered city to a service-centered city, and gives priority to shipping, logistics, and financial industries; thus, the proportion of its manufacturing sector to the Yangtze River Delta's manufacturing sector continued to decline, from 7.8% in 2002 to 3.39% in 2013. In 2013, the year-on-year growth rate of Shanghai's manufacturing sector was only 2.32%, far below the average level of China's manufacturing sector. In future, the Yangtze River Delta's manufacturing sector needs to be integrated further; specifically, the different parts of this region should exert their respective advantages, complement each other, and develop hand in hand with each other.

4.1.2 *Pearl River Delta Actively Develops High-End Manufacturing Industries Amid Continuing Pressure of Low-End Manufacturing Industries*

The manufacturing sector of the Pearl River Delta possesses outstanding technological innovation capability. Since 2004, the technological innovation capability of Guangdong's manufacturing sector has ranked first among the four regions, which can be demonstrated by the industrial structure of Guangdong. In light of the technology intensity standards of the OECD, this section classifies Guangdong's manufacturing sectors and undertakes a comparison between Guangdong and Jiangsu from this perspective. As shown in Fig. 4.2, hi-tech manufacturing industries account for the largest proportion of Guangdong's manufacturing sector. In 2013, the high-tech proportion was 27.38%, which was nearly 10% points higher than that of Jiangsu. Some Guangdong-based hi-tech enterprises (e.g., Huawei and ZTE) have a leading status among similar enterprises, and are a major driving force for R&D in the international arena, making significant contributions to the improvement in the technological level of Guangdong's manufacturing sector.

It is noteworthy that low-tech manufacturing industries also accounted for a large proportion of Guangdong's manufacturing sector in 2013, which was 5% points higher than that of Jiangsu. Today, Guangdong's manufacturing sector is confronted with various disadvantageous factors, including disappearance of the demographic dividend, economic downturn, and weak demand in both domestic and international markets. In view of this, the Pearl River Delta needs to adopt the following measures: (1) speed up the adjustment of the industrial structure, and transfer medium and low-end manufacturing industries to the eastern, western, and northern parts of Guangdong and even China's inland regions; (2) raise labor productivity significantly, and increase the degree of mechanization, automation,

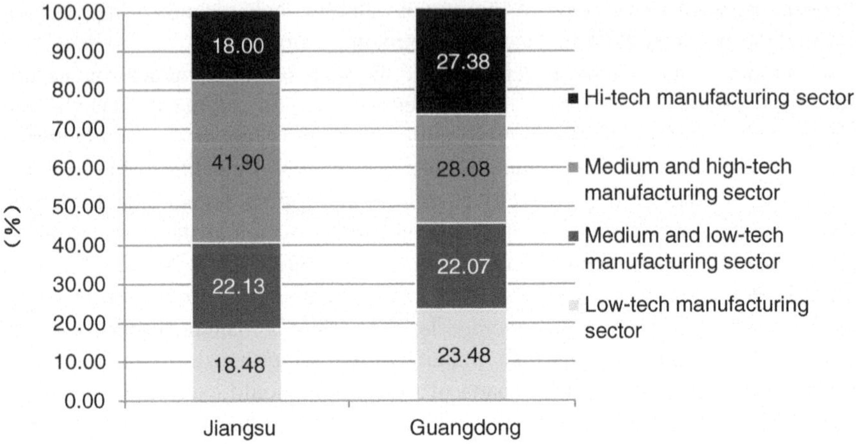

Fig. 4.2 Comparison between Jiangsu and Guangdong in the structure of the manufacturing sector in 2013

and artificial intelligence; (3) heighten the role and status of the Pearl River Delta in the global industrial chain, specifically, by developing toward the medium and high ends of the chain, attaching importance to the manufacture of key basic parts and components, and extending pure assembly industries to both the upstream and downstream of the manufacturing value chain, thereby building a complete industrial chain; and (4) attach importance to environmental protection and pollution treatment.

4.1.3 Northeast China's Manufacturing Sector's Slowdown

Northeast China's manufacturing sector's slowdown: setback or temporary adjustment?

In 2003, China's central government began to carry out the Northeast China Revitalizing strategy. During 2008–2012, the average economic growth rate of northeast China was as high as 12.4%, which was nearly 3% points higher than the average level of China. Since 2013, however, this region's economic growth rate has fallen markedly. For example, the year-on-year economic growth rate of Liaoning, Jilin, and Heilongjiang was 8.98, 8.73, and 5.05%, respectively, ranking at the bottom of the 31 provinces or municipalities of China. The economic growth rate of Heilongjiang was the lowest in China. The manufacturing sector accounted for about half of northeast China's economy; due to the slowdown in the growth of the region's manufacturing sector, the economic growth of this region has lost major impetus. Figure 4.3 shows the growth of output value of northeast China's

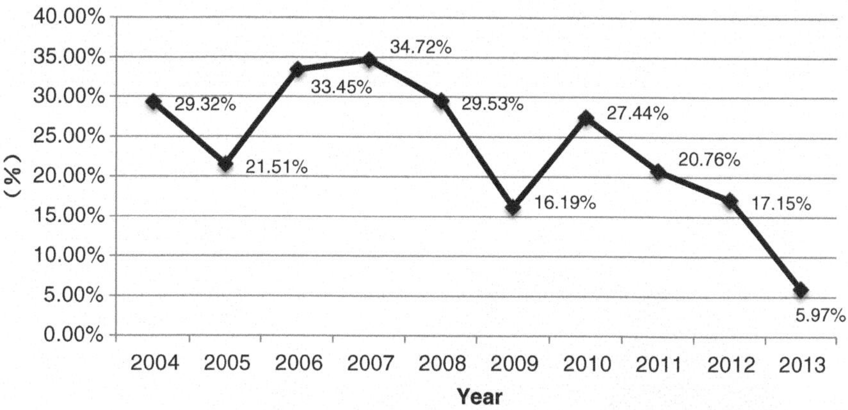

Fig. 4.3 Scale growth of northeast China's manufacturing sector from 2004 to 2013

manufacturing sector during Li and Zhanyuan 2004–2013. In the last decade, the growth rate of its manufacturing sector was clearly lowest in 2013.

Hengshan Fan, Deputy Secretary of the National Development and Reform Commission, pointed out that northeast China was an important industrial base for the equipment manufacturing, energy, and raw material industries. After 2013, these industries were severely influenced by sluggish countrywide demand for investment. For example, the equipment manufacturing industry was confronted with a decrease in market orders and enterprise profits; due to the price fall of raw materials (including coal, crude oil, and steel), a number of enterprises in the resources sector successively were faced with predicaments. After 2013 and the first quarter of 2014, investment growth in northeast China slowed markedly. Therefore, the driving force of investment on economic development weakened somewhat.

The scale growth of northeast China's manufacturing sector has slowed. In 2013, the profits of northeast China's manufacturing sector increased at a double-digit rate (16.8%), which was higher than that in the previous 2 years and higher than that in the Yangtze River Delta and the Beijing–Tianjin–Hebei region. In terms of economic resourcefulness, Liaoning province (especially the capital city, Shenyang) ranked second among similar regions of China, demonstrating the powerful industrial strength of northeast China. Northeast China's manufacturing sector is dominated by the equipment manufacturing industry and resource-oriented manufacturing industries. The two sectors fall under the production means of the manufacturing sector, which is in the upstream of the manufacturing sector chain and is highly susceptible to economic fluctuations. Resource-oriented manufacturing industries are characterized by high industrial pollution and low added value, and thus, are the major object for the upgrade and optimization of the industrial structure. The elimination and transfer of high-pollution manufacturing industries, which account for a large proportion of northeast China's manufacturing sector,

Table 4.1 R&D expenditure intensity of the four regions in 2013

Region	Beijing–Tianjin–Hebei Region	Northeast China	Yangtze River Delta	Pearl River Delta
R&D expenditure intensity	0.81	0.57	1.02	1.16

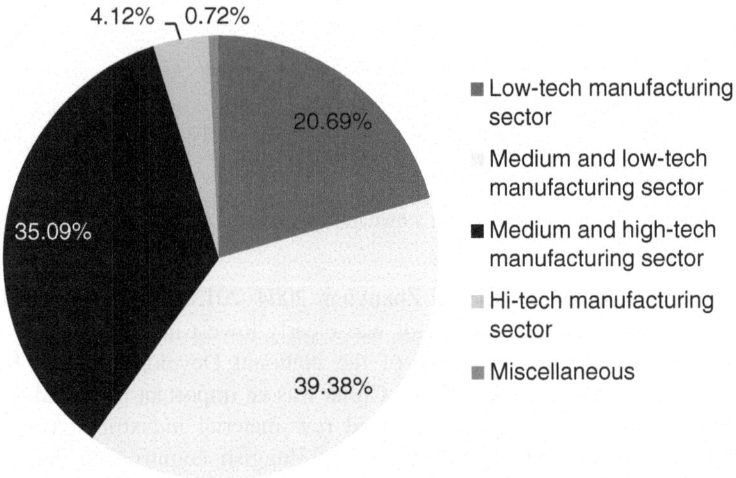

Fig. 4.4 Proportion of different types of manufacturing industries in Liaoning's manufacturing sector

have also made a significant impact on the scale growth of northeast China's manufacturing sector.

Regardless of whether northeast China has encountered a significant setback or a temporary adjustment, its manufacturing sector remains in a development stage characterized by extensive growth and an investment-driven pattern. Because of the deficiency in the region's R&D input (in 2013, its R&D input was the lowest among the four regions, and was only half that of the Pearl River Delta, as described in Table 4.1) and technological innovations, northeast China's manufacturing sector is dominated by medium and low-end manufacturing industries. Take Liaoning as an example. Figure 4.4 shows that the scale of Liaoning's hi-tech manufacturing sector accounts for only 4.12% of the scale of Liaoning's manufacturing sector (compared with 27.38% in the Pearl River Delta). In the context of resource shortages and economic downturn, Northeast China's manufacturing sector can hardly attain sustainable industrial competitiveness.

Northeast China has a strong industrial base in equipment manufacturing. In a working conference on revitalizing the old industrial bases of Northeast China,

Keqiang Li, Premier of the State Council, pointed out that northeast China should exert its own advantages, strengthen its equipment manufacturing industry, stimulate technological innovations and product innovations, and export equipment made by northeast China around the world. In order to revitalize the region, it is imperative that Northeast China learn lesson from previous experience, continuously deepen state-owned enterprise reform, increase input in technological innovations, speed up the transformation and upgrade of its manufacturing sector, and advance its equipment manufacturing industry to a higher level.

4.1.3.1 Beijing–Tianjin–Hebei Region's Manufacturing Sector's Key Problem: Eliminating Industrial Pollution

In 2013, the total output value of Beijing's manufacturing sector was 1.221 trillion yuan, a year-on-year increase of 10.8%; the total output value of Tianjin's manufacturing sector was 2.256 trillion yuan, a year-on-year increase of 14.19%; and the total output value of Hebei's manufacturing sector was 3.878 trillion yuan, a year-on-year increase of 8.58%.

Beijing and Tianjin far exceed Hebei in the growth of manufacturing sector, technological strength, and human resources, and thus can help Hebei to develop its manufacturing sector. On February 26, 2014, Jinping Xi, General Secretary of the CPC, expressly pointed out that Beijing, Tianjin, and Hebei should overcome a self-centered mindset and develop hand in hand with each other. Therefore, the Beijing–Tianjin–Hebei region should strengthen industrial division and cooperation, attain coordinated development of the region's manufacturing sector, and promote deep integration of advanced manufacturing industries.

It is noteworthy that this region currently is confronted with an embarrassing statistic: it contains half of China's top 10 cities with severe air pollution. On one hand, this is due to specific geographical and climatic conditions; on the other hand, massive emissions of industrial pollution aggravate the environmental pollution in this region. The manufacturing industries of Beijing and Tianjin possess excellent environmental protection capability while Hebei lags far behind them (for details, refer to Table 6.2.3), thereby affecting the overall performance of this region in this aspect. The scale of the manufacturing sector in this region is smaller than that in the other three regions, but the discharge amount of industrial waste gas and waste residue is larger than that in the other three regions (see Fig. 4.5). With the transfer of low-end and high-pollution manufacturing sectors in Beijing and Tianjin, Hebei will face a severe test because Hebei is already within the scope of radiation and spillover.

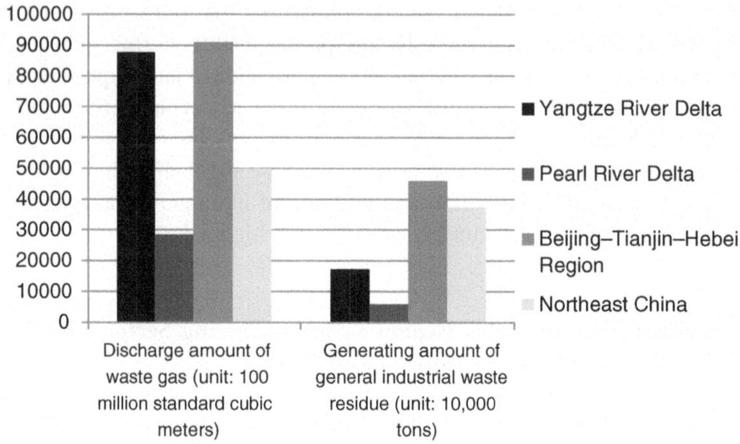

Fig. 4.5 Discharge amount of industrial waste gas and waste residue in the four regions

4.2 China's Manufacturing Sector: Top 10 Manufacturing Provinces

This study evaluates China's strong manufacturing provinces by numerical data, and attempts to ensure the authoritativeness, objectivity, scientificity, and completeness of the data. The data on the evaluation indexes are available from national-level and provincial-level statistical yearbooks and sector yearbooks. The provinces involved are comprehensively evaluated and ranked by the deviation maximization method. For details about the method, refer to Chap. 5.

4.2.1 Evaluation Indexes for Strong Manufacturing Provinces

The development status of the manufacturing sectors of different provinces is usually evaluated in terms of economic indexes, including single indexes and aggregate indexes, which mainly focus on industrial scale and economic benefits. To a certain extent, these evaluation indexes can reflect the development status of regional manufacturing industries, especially the contributions to national economic development and regional economic development. However, excessive energy consumption and increasingly severe environmental pollution have demonstrated that this evaluation method alone is not enough. We consider that the development status of regional manufacturing industries should be evaluated from the perspective of a new-type manufacturing sector, that is, it should be evaluated from the following four aspects: economic indexes, technological indexes, energy indexes, and environmental indexes. Specifically, economic indexes are used to measure the current contributions to the national economy; technological indexes are used to measure the future competitiveness of regional manufacturing sectors; energy indexes are used to measure the capability to utilize energy reasonably and intensively in the context of resource shortages; and environmental indexes are used to measure the environmental pollution caused by regional manufacturing sectors and the sustainable development capability of regional manufacturing sectors.

Based on a preliminary selection of 60 evaluation indexes, this section uses a combination of expert investigation and real data analysis to conduct analog calculation and evaluation for the development status of the manufacturing sectors of China's provinces, municipalities, and autonomous regions. Considering that the evaluation indexes should be consistent with the existing indexes used in the *China Statistical Yearbook* and should comply with the four principles of scientificity, comparability, systematicness, and operability, this study proposes an evaluation index system for China's strong manufacturing provinces, municipalities, and autonomous regions which comprises four major indexes and 32 subindexes (for details, see Table 4.2).

1. Economic resourcefulness

Economic resourcefulness is an important index to evaluate the development of regional manufacturing sectors. For countries in the process of industrialization, economic benefits are particularly important. Only by creating significant economic benefits can regional manufacturing sectors attain the impetus for sustainable development and provide powerful support for developing technologies, improving production efficiency, saving resources, and protecting the environment (for details, see Table 4.3).

A1 and A2 are output value subindexes, which are used to reflect the scale of the manufacturing sector (usually, the larger the scale of an enterprise scale is, the higher its output value is) and its contributions to the national economy. A3 and A4 are profit subindexes, which are used to reflect the earnings performance of the operating activities of manufacturing enterprises. A5 is an efficiency subindex, which is used to reflect the labor productivity of manufacturing enterprises. A6 is a market subindex, which is used to reflect the product sales performance of the manufacturing sector, as well as the degree to which the manufactured products cater to social needs. A7 and A8 are employment subindexes of the manufacturing sector; A7 is an aggregate employment subindex, which is used to reflect the capacity of the manufacturing sector to provide employment opportunities, and A8 is a relative subindex about employment, which is used to reflect the proportion of the employed population in the manufacturing sector to the total employed population. A9 and A10 are tax subindexes, which are used to reflect the tax contributions made by manufacturing enterprises. The following section describes the calculation methods for the above mentioned subindexes.

Table 4.2 Evaluation index system for China's strong manufacturing provinces, municipalities, and autonomous regions

General index	No.	Major index	No.	Subindex	
Evaluation index system for China's strong manufacturing provinces, municipalities, and autonomous regions	A	Economic resourcefulness	A1	Output value	Total output value of the manufacturing sector
			A2		Ratio of total output value of the manufacturing sector to total industrial output value
			A3	Profit	Total profit of manufacturing enterprises
			A4		Per-capita profit in the manufacturing sector
			A5	Efficiency	Labor productivity of the manufacturing sector
			A6	Market	Product sales rate of the manufacturing sector
			A7	Employment	Employed population in the manufacturing sector
			A8		Ratio of employed population in manufacturing sector to total employed population
			A9	Tax	Total taxes of manufacturing enterprises
			A10		Per-capita tax in the manufacturing sector
	B	Technological innovation capability	B1	R&D	R&D expenditure in the manufacturing sector
			B2		Full-time equivalent of R&D personnel in the manufacturing sector
			B3		R&D expenditure intensity in the manufacturing sector
			B4		Ratio of R&D personnel to total employed population in the manufacturing sector
			B5	Product development	Quantity of new product development projects in the manufacturing sector
			B6		New product development expenditures in the manufacturing sector
			B7	Patent	Quantity of patent applications in the manufacturing sector
			B8		Quantity of valid invention patents in the manufacturing sector

(continued)

Table 4.2 (continued)

General index	No.	Major index	No.	Subindex	
			B9	Technological achievement transformation	Sales income of new products in the manufacturing sector
			B10		Ratio of sales income of new products in the manufacturing sector
			B11		Input–output coefficient of technological innovations in the manufacturing sector
	C	Energy conservation capability	C1	Energy	Energy consumption in the manufacturing sector
			C2		Energy consumption per unit of output value in the manufacturing sector
			C3		Ratio of electricity consumption in the manufacturing sector
			C4		Ratio of coal consumption in the manufacturing sector
	D	Environmental protection capability	D1	Wastewater	Discharge amount of wastewater in the manufacturing sector
			D2		Discharge amount of wastewater per unit of output value in the manufacturing sector
			D3	Waste gas	Discharge amount of waste gas in the manufacturing sector
			D4		Discharge amount of waste gas per unit of output value in the manufacturing sector
			D5	Waste residue	Discharge amount of waste residue in the manufacturing sector
			D6		Discharge amount of waste residue per unit of output value in the manufacturing sector
			D7	Comprehensive utilization	Comprehensive utilization ratio of waste residue in the manufacturing sector

Table 4.3 Evaluation indexes for the economic resourcefulness of the manufacturing sector

No.	Economic index for strong manufacturing provinces		Unit
A1	Output value	Total output value of the manufacturing sector	100 million yuan
A2		Ratio of total output value of the manufacturing sector to total industrial output value	%
A3	Profit	Total profit of manufacturing enterprises	100 million yuan
A4		Per-capita profit in the manufacturing sector	10,000 yuan per capita
A5	Efficiency	Labor productivity of the manufacturing sector	10,000 yuan per capita
A6	Market	Product sales rate of the manufacturing sector	%
A7	Employment	Employed population in the manufacturing sector	10,000
A8		Ratio of employed population in the manufacturing sector to total employed population	%
A9	Tax	Total taxes of manufacturing enterprises	100,000,000 yuan
A10		Per-capita tax in the manufacturing sector	10,000 yuan per capita

$$\text{Total output value of manufacturing sector} = \sum_{j=1}^{m} TVP_j$$

$$\text{Ratio of total output value of the manufacturing sector to total industrial output value} = \frac{\sum_{j=1}^{m} TVP_j}{TP} \times 100\%$$

$$\text{Output value per manufacturing enterprise} = \frac{\sum_{j=1}^{m} TVP_j}{N}$$

where, TVP_j indicates the total output value of the j-th manufacturing sector, TP indicates the total industrial output value, and N indicates the quantity of state-owned manufacturing enterprises and manufacturing enterprises above a designated size.

$$\text{Per-capita profit in the manufacturing sector} = \frac{S}{L}$$

$$\text{Profit ratio of production in the manufacturing sector} = \frac{S}{\sum_{j=1}^{m} TVP_j} \times 100\%$$

$$\text{Labor productivity of the manufacturing sector} = \frac{\sum_{j=1}^{m} TVP_j}{L} \times 100\%$$

where, S indicates the total profit of manufacturing enterprises, and L indicates the total employed population of manufacturing enterprises.

$$\text{Product sales rate of the manufacturing sector} = \frac{\sum_{k=1}^{m} SR_k}{\sum_{j=1}^{m} TVP_j} \times 100\%$$

where, SR_k indicates the product sales income of the k-th manufacturing sector.

$$\text{Ratio of employed population in the manufacturing sector to total employed population} = \frac{L}{L_q} \times 100\%$$

$$\text{Per-capita profit in the manufacturing sector} = \frac{T}{L}$$

where, T indicates the total taxes of manufacturing enterprises, L indicates the employed population of manufacturing enterprises, and L_q indicates the total employed population of the region.

2. Technological innovation capability

Table 4.4 lists the indexes for evaluating the technological innovation capability of the manufacturing sector.

B1, B2, B3, and B4 are R&D subindexes of the manufacturing sector, which reflect the total R&D expenditure and R&D expenditure intensity of manufacturing enterprises. B5 and B6 are product development subindexes, which reflect the intensity of new product development by manufacturing enterprises to a certain extent. B7 and B8 are patent subindexes, which reflect the intensity of technological innovations made by manufacturing enterprises. B9, B10, and B11 are the subindexes on technological achievement transformation, which reflect the capability of manufacturing enterprises to transform and apply technological achievements. The above mentioned 11 subindexes reflect the technological strength and technological input/output status of the manufacturing sector in the following aspects: R&D expenditure/personnel input, technological output, and technological progress. The following section describes the calculation methods for the above mentioned subindexes.[3]

---

[3]The statistical caliber is subject to change. As defined by the main data bulletin (No. 2) of the second national R&D resource survey, the R&D expenditure intensity in the manufacturing sector refers to the ratio of R&D expenditure to main business income instead of the ratio of R&D expenditure to GDP.

Table 4.4 Evaluation indexes for the technological innovation capability of the manufacturing sector

No.	Evaluation index for the technological innovation capability of the manufacturing sector		Unit
B1	R&D	R&D expenditure in the manufacturing sector	10,000 yuan
B2		Full-time equivalent of R&D personnel in the manufacturing sector	Person-year
B3		R&D expenditure intensity in the manufacturing sector	%
B4		Ratio of R&D personnel to total employed population in the manufacturing sector	%
B5	Product development	Quantity of new product development projects in the manufacturing sector	
B6		New product development expenditures in the manufacturing sector	10,000 yuan
B7	Patent	Quantity of patent applications in the manufacturing sector	
B8		Quantity of valid invention patents in the manufacturing sector	
B9	Technological achievement transformation	Sales income of new products in the manufacturing sector	10,000 yuan
B10		Ratio of sales income of new products in the manufacturing sector	%
B11		Input–output coefficient of technological innovations in the manufacturing sector	

$$\text{R\&D expenditure intensity in the manufacturing sector} \\ = (\text{R\&D expenditure})/(\text{main business income}) \times 100\%$$

$$\text{Ratio of R\&D personnel to total employed population in the manufacturing sector} \\ = \frac{L'}{L} \times 100\%$$

where L', indicates the population of R&D personnel in the manufacturing sector, and L indicates the total employed population in the manufacturing sector.

$$\text{Ratio of sales income of new products in the manufacturing sector} = \frac{NPV}{TVP} \times 100\%$$

where, *NPV* indicates the sales income of new products in the manufacturing sector (unit: 100 million yuan), and *TVP* indicates the total output value[4] of the manufacturing sector

$$\text{Input–output coefficient of technological innovations in the manufacturing sector} = \frac{NPV}{NPR}$$

where *NPR* indicates the expenditure of new product development in the manufacturing sector (unit: 10,000 yuan).

3. Energy conservation capability

Energy resources provide the material basis for the production activities of the manufacturing sector, and are characterized by limited availability, that is, energy resources are non-renewable. With the improvement in humankind's cognitive ability and progress in science and technologies, both the availability and utilization efficiency of energy resources are increasingly high. Unreasonable resource utilization leads to energy shortages and environmental deterioration. In the context of energy shortages and environmental pollution, it is imperative that manufacturing enterprises strengthen reasonable and intensive utilization of energy resources, so as to enhance their competitiveness and attain sustainable development.

The energy conservation capability is measured mainly by four indexes: C1 (total energy consumption), C2 (energy consumption per unit of output value), C3 (ratio of electricity consumption to total energy consumption), and C4 (ratio of coal consumption to total energy consumption) (for details, see Table 4.5).

4. Environmental protection capability

Environmental harmony and ecological protection are the prerequisite for sustainable economic and social development. China's traditional manufacturing sector has an extensive production mode characterized by high growth, high consumption, and high pollution. This has given rise to severe resource shortages and ecological deterioration. Therefore, the environmental protection capability is an important index for measuring the degree of greenness of China's manufacturing sector (for details, see Table 4.6).

D1 and D2 reflect the total discharge amount and discharge intensity of wastewater in manufacturing activities, respectively; D3 and D4 reflect the total discharge amount and discharge intensity of waste gas in manufacturing activities, respectively; D5 and D6 reflect the total discharge amount and discharge intensity of waste residue in manufacturing activities, respectively; and D7 reflects the capability of manufacturing enterprises to utilize industrial waste comprehensively.

---

[4]Previously, NPV is used to denote the output value of new products. Subsequent to the data being available from the statistical yearbook, no data on the output value of new products are available. Here, the sales income of new products is used.

Table 4.5 Evaluation indexes for the energy conservation capability of the manufacturing sector

No.	Evaluation index for the environmental resource status of the manufacturing sector	Unit
C1	Energy consumption in the manufacturing sector	10,000 tons of standard coal
C2	Energy consumption per unit of output value in the manufacturing sector	10,000 tons of standard coal per 100 million yuan
C3	Ratio of electricity consumption to total energy consumption in the manufacturing sector	%
C4	Ratio of coal consumption to total energy consumption in the manufacturing sector	%

Table 4.6 Index set for environmental protection capability of manufacturing sector

No.	Evaluation index for the environmental protection capability of the manufacturing sector		Unit
D1	Wastewater	Discharge amount of wastewater in the manufacturing sector	10,000 tons
D2		Discharge amount of wastewater per unit of output value in the manufacturing sector	10,000 tons per 100 million yuan
D3	Waste gas	Discharge amount of waste gas in the manufacturing sector	100 million standard cubic meters
D4		Discharge amount of waste gas per unit of output value in the manufacturing sector	100 million standard cubic meters per 100 million yuan
D5	Waste residue	Discharge amount of waste residue in the manufacturing sector	10,000 tons
D6		Discharge amount of waste residue per unit of output value in the manufacturing sector	Ton per 100 million yuan
D7	Comprehensive utilization	Comprehensive utilization ratio of waste residue in the manufacturing sector	%

4.2.2 *Rank of China's Provinces, Municipalities, and Autonomous Regions by Single Index*

1. Ranking by economic resourcefulness

   (1) Rank by total output value of the manufacturing sector

The total output value of the manufacturing sector is an important index for evaluating the scale and level of regional manufacturing sectors. The total output value of China's manufacturing sector increased from 80.692 trillion yuan in

2012–90.778 trillion yuan[5] in 2013, an increase of 12.4995%. The average output value of the manufacturing sectors of provinces, municipalities, and autonomous regions ranked in the top 28 increased from 2.871 trillion yuan in 2012–3.236 trillion yuan in 2013. The rank of the 28 provinces, municipalities, and autonomous regions in 2013 is essentially consistent with that in 2011 and 2012, except that the rank of one or two provinces changed slightly. By the output value of the manufacturing sector, Jiangsu, Shandong, Guangdong, and Zhejiang rank stably in the top four (collectively accounting for 45% of the total output value of China's manufacturing sector). For details, see Fig. 4.6 and Table 4.7.

(2) Ranking by the ratio of output value of the manufacturing sector to total industrial output value

The ratio of output value of the manufacturing sector to total industrial output value is an important index used to measure the contributions to regional industrial development made by regional manufacturing sectors; to a certain extent, this index reflects the status of the manufacturing industries in the sector. As indicated in Table 4.8, this index is higher than 90% in nearly 30% of the above mentioned provinces, municipalities, and autonomous regions. Compared with the 2012 index, this index increased somewhat in most provinces, municipalities, and autonomous regions in 2013; it decreased only in Shandong, Jilin, Yunnan, Beijing, and Shanxi in 2013. In particular, the index declined by 12% points in Jilin in 2013, to some extent reflecting a significant adjustment of Jilin's industrial structure.

(3) Ranking by per-capita output value

The per-capita output value of the manufacturing sector is used to measure the labor productivity of the regional manufacturing sector. In terms of this index, Shandong ranked first (2.728 million yuan per capita), and Liaoning, Hebei, Inner Mongolia, and Anhui ranked next (2.642 million per capita, 2.58 million per capita, 2.542 million per capita, and 2.453 million per capita, respectively). Compared with 2012, this index increased slightly in most provinces, municipalities, and autonomous regions in 2013. The index rankings for Fujian and Chongqing improved very obviously, while those for Henan, Sichuan, Jilin, and Jiangsu declined notably in 2013. In particular, Henan declined most significantly, from 9th place in 2012 to 19th place in 2013. For details, see Table 4.9.

(4) Ranking by output value per manufacturing enterprise

The output value per manufacturing enterprise reflects the average scale of manufacturing enterprise in a region. If the value of this index is very large, it indicates that the average scale of the manufacturing enterprise is large, the industrial concentration ratio is high, and it is easy to make a scale effect. If the value of this index is very small, it indicates that there is a large number of small and medium-sized manufacturing enterprises in the region, and the market

[5]The statistical data do not include data for Xizang and Hainan.

Table 4.7 Output value of manufacturing sector in China's provinces, municipalities, and autonomous regions

Rank	Region	Total output value of the manufacturing sector (unit: 100 million yuan)	Proportion of total output value of China's manufacturing sector (%)	Cumulative percentage
1	Jiangsu	129,079.0300	14.22	14.22
2	Shandong	119,276.3701	13.14	27.36
3	Guangdong	101,623.5700	11.19	38.55
4	Zhejiang	58,246.3400	6.42	44.97
5	Henan	49,285.9683	5.43	50.40
6	Liaoning	47,427.1800	5.22	55.62
7	Hebei	38,777.4800	4.27	59.89
8	Hubei	36,221.9600	3.99	63.88
9	Fujian	31,241.1871	3.44	67.33
10	Shanghai	30,736.8400	3.39	70.71
11	Sichuan	30,410.0300	3.35	74.06
12	Anhui	29,540.4200	3.25	77.32
13	Hunan	28,732.7100	3.17	80.48
14	Jiangxi	22,647.3682	2.49	82.98
15	Tianjin	22,558.4144	2.48	85.46
16	Jilin	16,813.7856	1.85	87.31
17	Guangxi	16,344.8731	1.80	89.11
18	Chongqing	14,545.4724	1.60	90.72
19	Shaanxi	13,021.6153	1.43	92.15
20	Beijing	12,210.2148	1.35	93.50
21	Inner Mongolia	12,074.9660	1.33	94.83
22	Heilongjiang	9667.9465	1.07	95.89
23	Shanxi	8178.9660	0.90	96.79
24	Yunnan	7837.2800	0.86	97.66
25	Gansu	6214.3550	0.68	98.34
26	Xinjiang	5968.1188	0.66	99.00
27	Guizhou	5129.4400	0.57	99.56
28	Ningxia	2349.9931	0.26	99.82
29	Qinghai	1622.7514	0.18	100.00
30	Hainan	–	–	–
30	Xizang	–	–	–

Data source: *China Statistical Yearbook* (2014), and statistical yearbooks (2014) of China's provinces, municipalities, and autonomous regions

competition is very severe. Table 4.10 shows that this index is the highest in Tianjin, Qinghai, and Gansu; specifically, this index equals 400 million yuan per manufacturing enterprise above a designated size. Among the provinces,

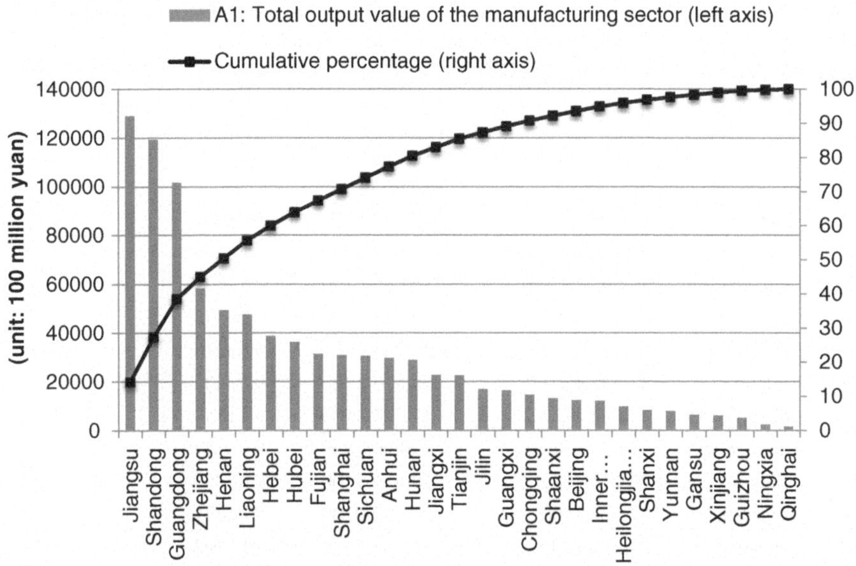

Fig. 4.6 Pareto distribution of output value of the manufacturing sector among China's provinces, municipalities, and autonomous regions in 2013

municipalities, and autonomous regions, the average value of this index is equal to 303 million yuan per manufacturing enterprise. In 2013, this index increased in most of the provinces, municipalities, and autonomous regions, and somewhat decreased in Qinghai, Shanxi, Ningxia, Jilin, and Guizhou (mainly distributed in the western regions of China), indicating more severe competition in the manufacturing sector in such regions.

(5) Ranking by total profit of manufacturing enterprises

The total profit of manufacturing enterprises is an important index used to measure the profitability of the regional manufacturing sector. According to the related statistical data, the average profit of China's provinces, municipalities, and autonomous regions was 155,847 billion yuan in 2012, and was 189.506 billion yuan in 2013, a year-on-year increase of 22%. In terms of this index, Jiangsu, Shandong, Guangdong, Henan, and Zhejiang ranked in the top five in 2013, when Shandong overtook Jiangsu to rank first. Gansu, Ningxia, Jiangxi, Chongqing, and Liaoning ranked in the top five in terms of profit growth (according to the *Statistical Yearbook of Gansu*, the industrial profit of Gansu was 76.845 billion yuan in 2013, whereas according to the *China Industry Statistical Yearbook*, the industrial profit of Gansu was 30.046 billion yuan in 2013; there is a great difference between the two statistical figures. Nevertheless, the total profit of Gansu's manufacturing sector had an abnormal increase in 2013, as in 2012, the total profit of Gansu's manufacturing sector was 6.32 billion yuan). In the meantime, Shanxi, Qinghai, Shaanxi, and Heilongjiang had negative growth in the profit of manufacturing enterprises (Table 4.11).

Table 4.8 Ratio of output value of the manufacturing sector to total industrial output value in China's provinces, municipalities, and autonomous regions in 2013

Rank	Region	Total output value of the manufacturing sector (unit: 100 million yuan)	Total industrial output value (unit: 100 million yuan)	Ratio of total output value of the manufacturing sector to total industrial output value (%)
1	Jiangsu	129,079.0300	134,648.9000	95.8634
2	Shanghai	30,736.8400	32,088.8800	95.7866
3	Guangdong	101,623.5700	109,673.1000	92.6605
4	Zhejiang	58,246.3400	62,980.2900	92.4834
5	Hubei	36,221.9600	39,208.9800	92.3818
6	Fujian	31,241.1871	33,853.3600	92.2839
7	Chongqing	14,545.4724	15,785.4100	92.1451
8	Shandong	119,276.3701	129,906.0000	91.8174
9	Jiangxi	22,647.3682	24,676.9100	91.7756
10	Liaoning	47,427.1800	52,892.0100	89.6679
11	Hunan	28,732.7100	32,157.7800	89.3492
12	Guangxi	16,344.8731	18,362.3800	89.0128
13	Anhui	29,540.4200	33,756.8200	87.5095
14	Sichuan	30,410.0300	35,328.5500	86.0778
15	Tianjin	22,558.4144	26,514.5100	85.0795
16	Henan	49,285.9683	58,779.9800	83.8482
17	Hebei	38,777.4800	46,316.6600	83.7225
18	Gansu	6214.3550	7996.5160	77.7133
19	Jilin	16,813.7856	22,061.3700	76.2137
20	Yunnan	7837.2800	10,289.0700	76.1709
21	Heilongjiang	9667.9465	13,719.3000	70.4697
22	Beijing	12,210.2148	17,370.9000	70.2912
23	Xinjiang	5968.1188	8679.6300	68.7601
24	Shaanxi	13,021.6153	18,982.4700	68.5981
25	Ningxia	2349.9931	3502.3150	67.0983
26	Qinghai	1622.7514	2537.1040	63.9608
27	Guizhou	5129.4400	8074.6000	63.5256
28	Inner Mongolia	12,074.9660	20,098.3500	60.0794
29	Shanxi	8178.9660	16,585.8000	49.3131
30	Hainan	–	–	–
30	Xizang	–	–	–

Data source: *China Statistical Yearbook* (2014), and statistical yearbooks (2014) of China's provinces, municipalities, and autonomous regions

Table 4.9 Per-capita output value of the manufacturing sector in China's provinces, municipalities, and autonomous regions in 2013

Rank	Region	Total output value of the manufacturing sector (unit: 100 million yuan)	Employed population in the manufacturing sector (unit: 10,000)	Per-capita output value in the manufacturing sector (unit: 10,000 yuan per capita)
1	Shandong	119,276.3701	437.2	272.8188
2	Liaoning	47,427.1800	179.5	264.2183
3	Hebei	38,777.4800	150.3	258.0005
4	Inner Mongolia	12,074.9660	47.5	254.2098
5	Anhui	29,540.4200	120.4	245.3523
6	Jiangsu	129,079.0300	555.4	232.4073
7	Hunan	28,732.7100	134.2	214.1037
8	Guangxi	16,344.8731	80.8	202.288
9	Jilin	16,813.7856	88.4	190.2012
10	Hubei	36,221.9600	190.9	189.7431
11	Guangdong	101,623.5700	540.9	187.8787
12	Ningxia	2349.9931	12.6	186.5074
13	Tianjin	22,558.4144	122.3	184.4515
14	Jiangxi	22,647.3682	125.8	180.0268
15	Xinjiang	5968.1188	33.7	177.0955
16	Chongqing	14,545.4724	86.4	168.3504
17	Zhejiang	58,246.3400	357.9	162.7447
18	Gansu	6214.3550	39.2	158.5295
19	Henan	49,285.9683	312.7	157.6142
20	Heilongjiang	9667.9465	65	148.7376
21	Sichuan	30,410.0300	204.6	148.6316
22	Shanghai	30,736.8400	211.8	145.122
23	Qinghai	1622.7514	11.8	137.5213
24	Fujian	31,241.1871	252.6	123.6785
25	Shaanxi	13,021.6153	107.9	120.6823
26	Beijing	12,210.2148	103.5	117.9731
27	Shanxi	8178.9660	73.1	111.8874
28	Guizhou	5129.4400	46.4	110.5483
29	Yunnan	7837.2800	74	105.9092
30	Hainan	–	–	–
30	Xizang	–	–	–

Data source: *China Statistical Yearbook* (2014), and statistical yearbooks (2014) of China's provinces, municipalities, and autonomous regions

Table 4.10 Output value per manufacturing enterprise in China's provinces, municipalities, and autonomous regions in 2013

Rank	Region	Quantity of state-owned manufacturing enterprises and manufacturing enterprises above designated size	Output value per manufacturing enterprise (unit: 100 million yuan per manufacturing enterprise)
1	Tianjin	5369	4.2016
2	Qinghai	387	4.1932
3	Gansu	1545	4.0222
4	Inner Mongolia	3057	3.9499
5	Shaanxi	3566	3.6516
6	Jilin	4718	3.5638
7	Shanxi	2327	3.5148
8	Beijing	3518	3.4708
9	Xinjiang	1747	3.4162
10	Guangxi	4925	3.3188
11	Jiangxi	6864	3.2994
12	Shanghai	9705	3.1671
13	Yunnan	2501	3.1337
14	Hebei	12,475	3.1084
15	Liaoning	15,346	3.0905
16	Shandong	38,940	3.0631
17	Chongqing	4910	2.9624
18	Jiangsu	45,811	2.8176
19	Sichuan	11,129	2.7325
20	Ningxia	881	2.6674
21	Hubei	13,596	2.6642
22	Henan	19,001	2.5939
23	Heilongjiang	3741	2.5843
24	Guangdong	40,261	2.5241
25	Hunan	11,762	2.4428
26	Guizhou	2300	2.2302
27	Fujian	14,872	2.1007
28	Anhui	15,463	1.9104
29	Zhejiang	39,050	1.4916
30	Hainan	–	–
30	Xizang	–	–

Data source: *China Statistical Yearbook* (2014), and statistical yearbooks (2014) of China's provinces, municipalities, and autonomous regions

Table 4.11 Total profit of manufacturing enterprises in 2013

Rank	Region	Total profit of manufacturing enterprises (unit: 100 million yuan)
1	Shandong	7683.8287
2	Jiangsu	7339.1000
3	Guangdong	5627.9000
4	Henan	4005.7300
5	Zhejiang	3226.8000
6	Liaoning	2597.8300
7	Shanghai	2313.4800
8	Hubei	2192.6400
9	Fujian	2019.0011
10	Hebei	1915.4400
11	Sichuan	1915.3100
12	Anhui	1881.1200
13	Hunan	1794.7900
14	Jiangxi	1613.7924
15	Tianjin	1499.1718
16	Jilin	1180.1304
17	Guangxi	882.1757
18	Beijing	880.1390
19	Chongqing	771.8792
20	Inner Mongolia	718.6548
21	Shaanxi	664.8198
22	Guizhou	484.1500
23	Gansu	472.7270
24	Yunnan	458.9200
25	Heilongjiang	387.2601
26	Xinjiang	212.6892
27	Shanxi	106.7828
28	Ningxia	70.1975
29	Qinghai	40.4222
30	Hainan	–
30	Xizang	–

Data source: *China Statistical Yearbook* (2014), and statistical yearbooks (2014) of China's provinces, municipalities, and autonomous regions

(6) Ranking by the ratio of employed population in the manufacturing sector to total employed population

The ratio of the employed population in the manufacturing sector to the total employed population reflects the capacity of the regional manufacturing sector to offer employment opportunities. According to Table 4.12, Tianjin, Fujian, Jiangsu,

Table 4.12 Ratio of employed population in the manufacturing sector to total employed population in China's provinces, municipalities, and autonomous regions in 2013

Rank	Region	Employed population in the manufacturing sector (unit: 10,000)	Ratio of employed population in the manufacturing sector to total employed population (%)
1	Tianjin	122.3	40.4431
2	Fujian	252.6	39.2236
3	Jiangsu	555.4	36.9454
4	Shanghai	211.8	34.2275
5	Shandong	437.2	33.8757
6	Zhejiang	357.9	33.3987
7	Henan	312.7	29.0613
8	Jiangxi	125.8	28.2697
9	Guangdong	540.9	27.4987
10	Hubei	190.9	27.4085
11	Jilin	88.4	26.1229
12	Liaoning	179.5	26.0485
13	Sichuan	204.6	24.1787
14	Anhui	120.4	23.1672
15	Hebei	150.3	23.0028
16	Hunan	134.2	22.3295
17	Chongqing	86.4	21.4925
18	Shaanxi	107.9	21.3537
19	Guangxi	80.8	20.0496
20	Qinghai	11.8	18.3801
21	Ningxia	12.6	17.4515
22	Yunnan	74	17.2857
23	Shanxi	73.1	15.7543
24	Guizhou	46.4	15.6387
25	Inner Mongolia	47.5	15.6353
26	Gansu	39.2	15.2767
27	Beijing	103.5	13.9431
28	Heilongjiang	65	13.8948
29	Xinjiang	33.7	10.8885
30	Hainan	10.6	10.7287
31	Xizang	1.1	3.5484

Data source: *China Statistical Yearbook* (2014), and statistical yearbooks (2014) of China's provinces, municipalities, and autonomous regions

Shanghai, and Shandong ranked in the top five in terms of this index (40.4431, 39.2236, 36.9454, 34.2275, and 33.8757%, respectively), indicating that the manufacturing sectors of these regions made significant contributions to employment. It is noteworthy that the index for these five places, all advanced areas in the eastern coastal regions of China, declined in 2013. In terms of this index, Xinjiang, Hainan, and Xizang still ranked at the bottom in 2013, but the index increased somewhat in the three areas, reflecting the evolution of industrial structure: in the process of economic development, labor resources transfer from the primary sector to the secondary sector first and then, transfer from the secondary sector to the tertiary sector continuously.

(7) Ranking by total taxes of manufacturing enterprises

The total tax of manufacturing enterprises reflects the capacity of the regional manufacturing sector to pay taxes, and is an important index to measure the social contribution ability of the regional manufacturing sector. According to statistical data on this index, Shandong, Jiangsu, Guangdong, Henan, and Zhejiang still ranked in the top five in 2013, remaining unchanged from 2012. In 2013, the ranking by this index changed slightly; the index increased in most of China's provinces, municipalities, and autonomous regions but declined in 22.58% of China's provinces, municipalities, and autonomous regions. In 2013, the index for Shanxi declined most significantly to 10.9 billion yuan). For details, see Table 4.13.

Table 4.13 Rank of China's provinces, municipalities, and autonomous regions by total tax of manufacturing enterprises in 2013

Rank	Region	Total tax of manufacturing enterprises (unit: 100 million yuan)
1	Shandong	12,206.4000
2	Jiangsu	12,103.4600
3	Guangdong	9703.7700
4	Henan	5840.3700
5	Zhejiang	5434.8200
6	Liaoning	4620.0800
7	Shanghai	4062.4700
8	Hubei	3957.7200
9	Hunan	3777.6900
10	Sichuan	3536.1900
11	Fujian	3245.5130
12	Anhui	3106.8100
13	Hebei	3030.1900
14	Tianjin	2605.7160
15	Jilin	2167.6850
16	Jiangxi	1867.3650
17	Guangxi	1681.4970
18	Chongqing	1568.2120
19	Yunnan	1537.4000

(continued)

Table 4.13 (continued)

Rank	Region	Total tax of manufacturing enterprises (unit: 100 million yuan)
20	Beijing	1499.8570
21	Shaanxi	1389.3250
22	Inner Mongolia	1195.1170
23	Guizhou	934.1100
24	Heilongjiang	893.4344
25	Xinjiang	567.1995
26	Gansu	354.8112
27	Shanxi	343.9823
28	Ningxia	182.4803
29	Qinghai	94.1151
30	Hainan	–
30	Xizang	–

Data source: *China Statistical Yearbook* (2014), and statistical yearbooks (2014) of China's provinces, municipalities, and autonomous regions

2. Technological innovation capability
   (1) Ranking by R&D expenditure in the manufacturing sector

R&D expenditure in the manufacturing sector directly reflects the expenditure status of the manufacturing sector on technological R&D, and indirectly reflects the growth potential and technological strength of the regional manufacturing sector. According to Table 4.14, Jiangsu, Guangdong, Shandong, Zhejiang, and Shanghai ranked in the top five in 2013, remaining unchanged from 2012. In 2013, Inner Mongolia moved up one place to enter the top 20; Heilongjiang moved down one

Table 4.14 R&D expenditure of manufacturing enterprises above designated size in 2013

Rank	Region	R&D expenditure in the manufacturing sector (unit: 10,000 yuan)	Percentage (%)
1	Jiangsu	12,395,745	14.9016
2	Guangdong	12,374,791	14.8764
3	Shandong	10,528,097	12.6564
4	Zhejiang	6,843,562	8.2270
5	Shanghai	4,047,800	4.8661
6	Liaoning	3,331,303	4.0047
7	Hubei	3,117,987	3.7483
8	Tianjin	3,000,377	3.6069
9	Henan	2,953,410	3.5505
10	Fujian	2,791,966	3.3564

(continued)

Table 4.14 (continued)

Rank	Region	R&D expenditure in the manufacturing sector (unit: 10,000 yuan)	Percentage (%)
11	Hunan	2,703,987	3.2506
12	Anhui	2,477,246	2.9780
13	Hebei	2,327,418	2.7979
14	Beijing	2,130,618	2.5613
15	Sichuan	1,688,902	2.0303
16	Shaanxi	1,401,480	1.6848
17	Chongqing	1,388,199	1.6688
18	Shanxi	1,237,698	1.4879
19	Jiangxi	1,106,443	1.3301
20	Inner Mongolia	1,004,406	1.2075
21	Heilongjiang	950,335	1.1424
22	Guangxi	817,063	0.9822
23	Jilin	698,136	0.8393
24	Yunnan	454,278	0.5461
25	Gansu	400,743	0.4818
26	Guizhou	342,541	0.4118
27	Xinjiang	314,257	0.3778
28	Ningxia	167,494	0.2014
29	Hainan	93,567	0.1125
30	Qinghai	89,540	0.1076
31	Xizang	4617	0.0056

Data source: *China Statistical Yearbook* (2014)

place, falling out of the top 20. Hainan ranked 29th, moving up one place; and Qinghai ranked 31st, moving down one place. Hubei and Henan in central China ranked in the top 10; while Sichuan in western China ranked 15th, the highest place in western China.

(2) Ranking by the full-time equivalent of R&D personnel in the manufacturing sector

The full-time equivalent of R&D personnel in manufacturing enterprises above a designated size reflects the status of technological labor input in the manufacturing sector. According to Table 4.15, Guangdong, Jiangsu, Zhejiang, Shandong, and Henan ranked in the top five. Compared with 2012, the ranking changed slightly in 2013; specifically, Anhui, Liaoning, Sichuan, Jiangxi, Inner Mongolia, and Gansu each moved up one place, Hubei moved down one place, and Beijing, Jilin, and Yunnan each moved down two places. Among the top 10 provinces, municipalities, and autonomous regions, 6 were in eastern China, 4 in central China, and none in western China. The technological labor input is a motivating force for the development of the regional manufacturing sector. A deficiency in this input will further widen the development gap between eastern and western China.

Table 4.15 Full-time equivalent of R&D personnel in manufacturing enterprises above designated size in China's provinces, municipalities, and autonomous regions in 2013

Rank	Region	Full-time equivalent of R&D personnel in the manufacturing sector (unit: person-year)	Percentage (%)
1	Guangdong	426,330	17.0945
2	Jiangsu	393,942	15.7959
3	Zhejiang	263,507	10.5658
4	Shandong	227,403	9.1182
5	Henan	125,091	5.0157
6	Fujian	100,200	4.0177
7	Shanghai	92,136	3.6944
8	Anhui	86,000	3.4483
9	Hubei	85,826	3.4414
10	Hunan	73,558	2.9494
11	Tianjin	68,175	2.7336
12	Hebei	65,049	2.6082
13	Liaoning	59,090	2.3693
14	Sichuan	58,148	2.3315
15	Beijing	58,036	2.3271
16	Shaanxi	45,809	1.8368
17	Heilongjiang	37,296	1.4954
18	Chongqing	36,605	1.4678
19	Shanxi	34,024	1.3642
20	Jiangxi	29,519	1.1836
21	Inner Mongolia	26,990	1.0822
22	Jilin	23,709	0.9507
23	Guangxi	20,700	0.8300
24	Guizhou	16,049	0.6435
25	Gansu	12,472	0.5001
26	Yunnan	11,811	0.4736
27	Xinjiang	6668	0.2674
28	Ningxia	4817	0.1931
29	Hainan	2882	0.1156
30	Qinghai	2039	0.0818
31	Xizang	81	0.0032

Data source: *China Statistical Yearbook* (2014)

(3) Ranking by the quantity of new product development projects in the manufacturing sector

The quantity of new product development projects in the manufacturing sector reflects the new product development status and technological strength. According to Table 4.16, Jiangsu, Zhejiang, Guangdong, Shandong, and Anhui ranked in the

Table 4.16 Quantity of new product development projects in the manufacturing sector in China's provinces, municipalities, and autonomous regions in 2013

Rank	Region	Quantity of new product development projects in the manufacturing sector	Percentage (%)
1	Jiangsu	58,353	16.2867
2	Zhejiang	47,778	13.3351
3	Guangdong	47,387	13.226
4	Shandong	31,100	8.6802
5	Anhui	17,320	4.8341
6	Shanghai	17,295	4.8271
7	Beijing	13,310	3.7149
8	Sichuan	12,681	3.5393
9	Tianjin	11,977	3.3429
10	Henan	11,150	3.112
11	Hubei	10,722	2.9926
12	Fujian	10,534	2.9401
13	Hunan	9089	2.5368
14	Liaoning	8568	2.3914
15	Hebei	7194	2.0079
16	Chongqing	6820	1.9035
17	Jilin	6516	1.8187
18	Shaanxi	6491	1.8117
19	Jiangxi	4381	1.2228
20	Heilongjiang	3438	0.9596
21	Guangxi	3332	0.93
22	Shanxi	2938	0.82
23	Guizhou	1908	0.5325
24	Yunnan	1903	0.5311
25	Gansu	1629	0.4547
26	Inner Mongolia	1581	0.4413
27	Xinjiang	1103	0.3079
28	Ningxia	966	0.2696
29	Hainan	704	0.1965
30	Qinghai	111	0.031
31	Xizang	8	0.0022

Data source: *China Statistical Yearbook* (2014)

top five in 2013, when Zhejiang replaced Guangdong in second place, and Anhui moved up one place, entering the top five. Table 4.16 shows that the new product development projects carried out by the east coastal regions of China account for a large proportion of total new product development projects in China's manufacturing sector. To a certain extent, this is because hi-tech manufacturing industries account for a large proportion, and economic strength and R&D input are very high in these regions. Compared with 2012, the ranking of the provinces, municipalities, and autonomous regions was relatively stable in 2013, except that Jilin moved up five places to 17th place.

(4) Ranking by the quantity of valid invention patents in the manufacturing sector

The quantity of valid invention patents in the manufacturing sector reflects the effect of technological innovation activities in the manufacturing sector, as well as the technological output of the manufacturing sector. According to Table 4.17, Guangdong, Jiangsu, Zhejiang, Shanghai, and Shandong ranked in the top five, with Guangdong retaining a stable ranking in first place. In 2013, the proportion of Guangdong's valid invention patents to China's total valid invention patents decreased slightly, but remained as high as 28.94, far higher than that of the four

Table 4.17 Quantity of valid invention patents in the manufacturing sector in China's provinces, municipalities, and autonomous regions in 2013

Rank	Region	Quantity of valid invention patents in the manufacturing sector	Percentage (%)
1	Guangdong	97,052	28.9361
2	Jiangsu	52,718	15.7179
3	Zhejiang	22,578	6.7316
4	Shanghai	20,140	6.0048
5	Shandong	18,340	5.4681
6	Beijing	16,402	4.8903
7	Anhui	13,582	4.0495
8	Hunan	10,512	3.1342
9	Tianjin	10,191	3.0385
10	Sichuan	9043	2.6962
11	Hubei	8745	2.6073
12	Fujian	7119	2.1225
13	Liaoning	6923	2.0641
14	Henan	6470	1.929
15	Shaanxi	5449	1.6246
16	Chongqing	4792	1.4287
17	Hebei	4049	1.2072
18	Shanxi	3008	0.8968
19	Jilin	2985	0.89
20	Heilongjiang	2342	0.6983
21	Jiangxi	2333	0.6956
22	Yunnan	2280	0.6798
23	Guizhou	1985	0.5918
24	Guangxi	1889	0.5632
25	Inner Mongolia	1444	0.4305

(continued)

Table 4.17 (continued)

Rank	Region	Quantity of valid invention patents in the manufacturing sector	Percentage (%)
26	Gansu	1028	0.3065
27	Xinjiang	695	0.2072
28	Hainan	683	0.2036
29	Ningxia	387	0.1154
30	Qinghai	205	0.0611
31	Xizang	32	0.0095

Data source: *China Statistical Yearbook on Science and Technology* (2014)

other provinces. This is mainly because Guangdong attaches much importance to technological talents and has an incentive policy for technological innovations. The valid invention patents of the top five provinces accounted for 62.86% of China's total valid invention patents, indicating that China's technological innovations were mainly conducted in several key regions. In 2013, Jiangxi moved up two places and Guangxi moved down two places to 24th place; the ranks of the other provinces, municipalities, and autonomous regions remained stable.

(5) Ranking by R&D expenditure intensity in the manufacturing sector

The R&D expenditure intensity in the manufacturing sector is measured by the ratio of R&D expenditure to main business income in the regional manufacturing sector. Although eastern China enjoys an overwhelming advantage in other technological indexes, both central and western China have some provinces, municipalities, and autonomous regions that ranked in the top 10 of this index. Specifically, Chongqing in western China ranked 7th, and Hunan and Hubei in central China ranked 8th and 10th, respectively. Among the top 15, 6 were distributed in central and western China, indicating that certain regions in central and western China attach importance to science and technologies, and enhance R&D expenditure intensity. In addition, the base numbers for main business income are very small, explicitly affecting the rankings. It is noteworthy that small base numbers for main business income did not change the lowest ranks of Jilin, Xinjiang, and Sichuan. This indicates that such regions are deficient in the technological force supporting the development of their regional manufacturing sectors. Therefore, these regions should guard against the possibility that their manufacturing sectors may fall into difficulties in the future (Table 4.18).

(6) Ranking by sales income of new products in the manufacturing sector

The sales income of new products in the manufacturing sector reflects the performance of new product development and technological achievement transformation. According to Table 4.19, Jiangsu, Guangdong, Zhejiang, Shandong, and Shanghai ranked in the top five of this index. In addition, the five provinces or municipalities were also ranked at the top by R&D expenditure and technological

Table 4.18 R&D expenditure intensity in the manufacturing sector in China's provinces, municipalities, and autonomous regions in 2013

Rank	Region	R&D expenditure intensity in the manufacturing sector (%)
1	Guangdong	1.1938
2	Shanghai	1.1721
3	Beijing	1.1440
4	Tianjin	1.1108
5	Zhejiang	1.1080
6	Jiangsu	0.9372
7	Chongqing	0.9004
8	Hunan	0.8552
9	Fujian	0.8500
10	Hubei	0.8235
11	Shandong	0.7957
12	Shaanxi	0.7890
13	Anhui	0.7489
14	Heilongjiang	0.7003
15	Shanxi	0.6725
16	Liaoning	0.6388
17	Hainan	0.5703
18	Inner Mongolia	0.5137
19	Hebei	0.5085
20	Guizhou	0.4980
21	Henan	0.4967
22	Ningxia	0.4964
23	Xizang	0.4945
24	Guangxi	0.4885
25	Sichuan	0.4791
26	Gansu	0.4746
27	Yunnan	0.4648
28	Qinghai	0.4378
29	Jiangxi	0.4144
30	Xinjiang	0.3651
31	Jilin	0.3180

Data source: *China Statistical Yearbook* (2014), and statistical yearbooks (2014) of China's provinces, municipalities, and autonomous regions

labor input. Compared with 2012, the ranks of Henan and Jilin changed significantly in 2013; specifically, Henan moved up five places to 8th place, and Jilin moved down five places to 21st place. The ranks of other provinces, municipalities, and autonomous regions were relatively stable.

Table 4.19 Sales income of new products in the manufacturing sector in China's provinces, municipalities, and autonomous regions in 2013

Rank	Region	Sales income of new products (unit: 10,000 yuan)
1	Jiangsu	197,142,112
2	Guangdong	180,137,410
3	Zhejiang	148,820,993
4	Shandong	142,841,782
5	Shanghai	76,883,835
6	Hunan	57,246,324
7	Tianjin	55,696,886
8	Henan	47,914,474
9	Hubei	46,544,784
10	Anhui	43,790,809
11	Liaoning	40,931,774
12	Beijing	36,727,656
13	Fujian	34,400,997
14	Hebei	29,160,256
15	Chongqing	26,961,130
16	Sichuan	24,758,761
17	Jiangxi	16,829,309
18	Guangxi	15,866,038
19	Shanxi	10,272,735
20	Shaanxi	10,154,791
21	Jilin	7,031,878
22	Inner Mongolia	6,285,040
23	Gansu	6,185,275
24	Heilongjiang	5,825,023
25	Yunnan	4,433,810
26	Guizhou	3,683,200
27	Xinjiang	3,533,318
28	Ningxia	2,796,416
29	Hainan	1,601,202
30	Qinghai	125,430
31	Xizang	23,454

Data source: *China Statistical Yearbook on Science and Technology* (2014)

(7) Ranking by ratio of sales income of new products to main business income in the manufacturing sector

The ratio of sales income of new products to main business income in the manufacturing sector can reflect the performance of technological achievement transformation and the degree of technological innovations. Table 4.20 lists the

Table 4.20 Ratio of sales income of new products to main business income in China's provinces, municipalities, and autonomous regions in 2013

Rank	Region	Ratio of sales income of new products to main business income (%)
1	Zhejiang	24.0945
2	Shanghai	22.2635
3	Tianjin	20.62
4	Beijing	19.7197
5	Hunan	18.1064
6	Chongqing	17.4878
7	Guangdong	17.3786
8	Jiangsu	14.9045
9	Anhui	13.2381
10	Hubei	12.2924
11	Shandong	10.7953
12	Fujian	10.4731
13	Hainan	9.7593
14	Guangxi	9.4859
15	Ningxia	8.2869
16	Henan	8.059
17	Liaoning	7.8488
18	Gansu	7.3254
19	Sichuan	7.0234
20	Hebei	6.3716
21	Jiangxi	6.3031
22	Shaanxi	5.7168
23	Shanxi	5.5816
24	Guizhou	5.3547
25	Yunnan	4.5367
26	Heilongjiang	4.2926
27	Xinjiang	4.1047
28	Inner Mongolia	3.2147
29	Jilin	3.2035
30	Xizang	2.5119
31	Qinghai	0.6132

Data source: *China Statistical Yearbook on Science and Technology* (2014)

statistical data for this index in 2013. Compared with the last year, Zhejiang ranked first in 2013, moving up two places. The rankings of this index fluctuated very obviously in 2013. Specifically, Ningxia moved up five places, Henan moved up 10 places, Guizhou moved downwards 5 places, and Jilin moved down 16 places. The ranks of provinces, municipalities, and autonomous regions in eastern, central, and western China all fluctuated significantly, indicating contingency in technological achievement transformation by manufacturing enterprises in the short term.

(8) Ranking by new product development expenditure of manufacturing enterprises

The new product development expenditure of manufacturing enterprises reflects the expenditure of manufacturing enterprises on new product development. According to Table 4.21, the provinces and municipalities in the top five were all distributed in central China. This can be ascribed to two reasons: the economic foundation of eastern China is good; and eastern China attaches great importance to technological innovations and new product development. Hubei, Anhui, and Hunan in central China entered the top 10, indicating that central China also attaches importance to new product development and maintains stable expenditure input in new product development. Compared with 2012, the ranking in 2013 was essentially stable, except for slight changes of one or two places.

3. Energy conservation capability

(1) Energy consumption of the manufacturing sector

China is a large energy-consuming country, and petroleum, electricity, and coal are the common energy resources consumed by China. Some are directly available from nature as primary sources, and others are secondary energy. Conversion efficiency varies among the different energy resources. If energy consumption were measured in terms of material consumption of only one specific energy resource, two problems would arise: (1) this method is not comprehensive enough; (2) if material consumption of different energy resources simply were added up, it would be very difficult to unify the calculation units. Therefore, energy consumption is uniformly measured in terms of standard coal. According to Table 4.22, Hebei ranked first, and Jiangsu, Shandong, Guangdong, and Henan following in descending order. For this index, the top provinces consumed 35.36% of total energy consumption of China's manufacturing sector. Xizang, Hainan, Beijing, Qinghai, and Chongqing ranked in the bottom five, collectively consuming 3.34% of the total energy consumption of China's manufacturing sector. In order to reduce energy consumption and pollutant emissions significantly, the key is to conserve energy in a few highly energy-consuming regions of China.

Note:

(1) The statistical data on energy consumption are available from the statistical yearbooks of China's provinces, municipalities, and autonomous regions, and mainly refer to data on industrial energy consumption. The statistical yearbooks (2014) for Beijing, Zhejiang, Chongqing, and Xinjiang cite industrial energy consumption data for 2012 because their related records on industrial energy consumption are not available.

Table 4.21 New product development expenditure in China's provinces, municipalities, and autonomous regions in 2013

Rank	Region	New product development expenditure (unit: 10,000 yuan)
1	Jiangsu	16,693,195
2	Guangdong	14,065,712
3	Shandong	10,206,343
4	Zhejiang	8,216,556
5	Shanghai	5,282,586
6	Liaoning	3,360,539
7	Hubei	3,317,175
8	Anhui	3,244,687
9	Hunan	2,959,845
10	Beijing	2,931,908
11	Henan	2,660,106
12	Fujian	2,656,091
13	Tianjin	2,459,585
14	Sichuan	2,135,771
15	Hebei	2,025,041
16	Shaanxi	1,799,803
17	Chongqing	1,438,649
18	Shanxi	991,958
19	Jiangxi	977,849
20	Guangxi	849,395
21	Heilongjiang	782,854
22	Jilin	740,849
23	Inner Mongolia	619,217
24	Yunnan	496,845
25	Gansu	403,460
26	Guizhou	403,004
27	Xinjiang	394,450
28	Ningxia	149,924
29	Hainan	114,916
30	Qinghai	87,949
31	Xizang	1177

Data source: *China Statistical Yearbook on Science and Technology* (2014)

(2) Ranking by energy consumption per unit output value in the manufacturing Sector

The energy consumption per unit of output value in the manufacturing sector reflects the efficiency of energy utilization in the regional manufacturing sector. The

Table 4.22 Energy consumption in manufacturing sectors in China's provinces, municipalities, and autonomous regions in 2013

Rank	Region	Total energy consumption in the manufacturing sector (unit: 10,000 tons of standard coal)
1	Hebei	25,104.3113
2	Jiangsu	21,825.1000
3	Shandong	20,441.0000
4	Guangdong	17,895.7800
5	Henan	17,335.8800
6	Sichuan	14,287.0000
7	Liaoning	13,188.4277
8	Hubei	12,380.0000
9	Shanxi	12,345.6100
10	Inner Mongolia	12,299.9200
11	Zhejiang	11,868.8552
12	Hunan	10,318.7700
13	Xinjiang	8742.1433
14	Anhui	8119.4800
15	Fujian	7928.6300
16	Yunnan	7621.5600
17	Guangxi	6869.7900
18	Shaanxi	6765.0500
19	Jilin	6760.8200
20	Tianjin	6059.6310
21	Guizhou	6006.2600
22	Shanghai	5998.5900
23	Heilongjiang	5933.5000
24	Gansu	5308.4200
25	Jiangxi	5176.3400
26	Ningxia	3899.9671
27	Chongqing	3513.5270
28	Qinghai	3015.3100
29	Beijing	2275.7000
30	Hainan	879.4000
31	Xizang	–

Data source: Statistical yearbooks (2014) of China's provinces, municipalities, and autonomous regions

smaller the value of this index is, the higher the efficiency of energy utilization is. According to Table 4.23, the index of Beijing was the lowest (specifically, at 1459 tons of standard coal per 100 million yuan in 2013, compared to 1418 tons of standard coal per 100 million yuan in 2012). Overall, this index declined to a varying extent in most of China's provinces, municipalities, and autonomous

Table 4.23 Energy consumption per unit output value in manufacturing sector in China's provinces, municipalities, and autonomous regions in 2013

Rank	Region	Energy consumption per unit of output value in the manufacturing sector (unit: 10,000 tons of standard coal per 100 million yuan)
1	Qinghai	1.1885
2	Xinjiang	1.1606
3	Ningxia	1.1135
4	Shanxi	0.7443
5	Guizhou	0.7438
6	Yunnan	0.7407
7	Gansu	0.6638
8	Inner Mongolia	0.612
9	Hebei	0.542
10	Hainan	0.4943
11	Heilongjiang	0.4325
12	Sichuan	0.4044
13	Guangxi	0.3741
14	Shaanxi	0.3564
15	Hunan	0.3209
16	Hubei	0.3157
17	Jilin	0.3065
18	Henan	0.2949
19	Chongqing	0.2683
20	Liaoning	0.2493
21	Anhui	0.2405
22	Fujian	0.2342
23	Tianjin	0.2285
24	Jiangxi	0.2098
25	Zhejiang	0.2007
26	Shanghai	0.1869
27	Guangdong	0.1632
28	Jiangsu	0.1621
29	Shandong	0.1574
30	Beijing	0.1459
31	Xizang	–

Data source: Statistical yearbooks (2014) of China's provinces, municipalities, and autonomous regions

regions. In terms of the value of this index, Qinghai, Xinjiang, Ningxia, Shanxi, and Guizhou ranked in the top five, indicating low efficiency of energy utilization in their manufacturing sectors. This can be ascribed to the low economic and technological development levels in these provinces and autonomous regions. By contrast, Beijing, Jiangsu, Shandong, Guangdong, and Shanghai had the lowest energy consumption per unit of output value. These five provinces and municipalities are all in developed manufacturing regions in eastern China. Their outstanding advantage in the output value of the manufacturing sector offset the impacts of the high total energy consumption of their manufacturing sectors.

4. Environmental protection capability[6]

(1) Ranking by discharge amount of wastewater in the manufacturing sector

The discharge amount of wastewater in the manufacturing sector reflects the degree of pollution to the water environment caused by the regional manufacturing sector. According to Table 4.24, Jiangsu ranked first, and Shandong, Guangdong, Zhejiang, and Henan following in descending order. Overall, the discharge amount of wastewater in regions of eastern China with developed manufacturing industries was noticeably larger than that in other regions of China; in other words, rapid economic growth has increased pollutant emissions. Guangxi continued to be an exception in this respect; specifically, Guangxi had a large discharge amount of wastewater in the manufacturing sector, but it is underdeveloped both economically and technologically. Therefore, China's local governments should take measures to prevent further aggravation of this situation; enhance the administration and supervision of high-pollution manufacturing industries; treat industrial wastewater effectively; encourage the adjustment of industrial structure; and reduce the ratio of high-pollution low-efficiency manufacturing industries across all regional manufacturing sectors.

(2) Ranking by discharge amount of wastewater per unit output value in the manufacturing sector

The discharge amount of wastewater per unit of output value in the manufacturing sector reflects the water environment cost incurred by the regional manufacturing sector. According to Table 4.25, Beijing, Tianjin, Shandong, Shanghai, and Liaoning ranked in the bottom five, indicating a small discharge amount of wastewater per unit of output value in the manufacturing sector. The five provinces and municipalities are all distributed in eastern China, and are characterized by a small discharge amount of wastewater per unit of output value but huge total discharge amount of wastewater in the manufacturing sector. By contrast, Guangxi, Ningxia, Xizang, Yunnan, and Xinjiang ranked in the top five, indicating a large discharge amount of wastewater in the manufacturing sector. Therefore, the governments of these regions should optimize their industrial structure, control the quantity of high-pollution and high-emission manufacturing enterprises, and take appropriate measures to abate water pollution. Compared with the rank in 2012, the rank of certain provinces and autonomous regions changed significantly in 2013;

[6]The existing official statistics lack detailed data about pollutant emissions in the manufacturing sectors of different provinces, municipalities, and autonomous regions. Therefore, this section uses data about industrial pollutant emissions instead.

Table 4.24 Discharge amount of wastewater in manufacturing sector in China's provinces, municipalities, and autonomous regions in 2013

Rank	Region	Discharge amount of wastewater in the manufacturing sector (unit: 10,000 tons))
1	Jiangsu	220,559
2	Shandong	181,179
3	Guangdong	170,463
4	Zhejiang	163,674
5	Henan	130,789
6	Hebei	109,876
7	Fujian	104,658
8	Hunan	92,311
9	Guangxi	89,508
10	Hubei	84,993
11	Liaoning	78,286
12	Anhui	70,972
13	Jiangxi	68,230
14	Sichuan	64,864
15	Heilongjiang	47,796
16	Shanxi	47,795
17	Shanghai	45,426
18	Jilin	42,656
19	Yunnan	41,844
20	Inner Mongolia	36,986
21	Shaanxi	34,871
22	Xinjiang	34,718
23	Chongqing	33,451
24	Guizhou	22,898
25	Gansu	20,171
26	Tianjin	18,692
27	Ningxia	15,708
28	Beijing	9486
29	Qinghai	8395
30	Hainan	6744
31	Xizang	400

Data source: *China Statistical Yearbook on Environment* (2014)

specifically, Xizang moved up four places to 3rd place, Shanxi moved up four places to 10th place, Inner Mongolia moved up four places to 22nd place, and Heilongjiang moved down five places to 8th place).

(3) Ranking by discharge amount of waste gas in the manufacturing sector

Table 4.25 Discharge amount of wastewater per unit output value in the manufacturing sector in China's provinces, municipalities, and autonomous regions in 2013

Rank	Region	Discharge amount of wastewater per unit of output value in the manufacturing sector (unit: 10,000 tons per 100 million yuan)
1	Guangxi	4.8745
2	Ningxia	4.4850
3	Xizang	4.1199
4	Yunnan	4.0668
5	Xinjiang	3.9999
6	Hainan	3.7904
7	Qinghai	3.6368
8	Heilongjiang	3.4839
9	Fujian	3.0915
10	Shanxi	2.8817
11	Hunan	2.8706
12	Guizhou	2.8358
13	Jiangxi	2.7649
14	Zhejiang	2.5988
15	Gansu	2.5225
16	Hebei	2.3723
17	Henan	2.2251
18	Hubei	2.1677
19	Chongqing	2.1191
20	Anhui	2.1024
21	Jilin	1.9335
22	Inner Mongolia	1.8403
23	Shaanxi	1.837
24	Sichuan	1.8360
25	Jiangsu	1.6380
26	Guangdong	1.5543
27	Liaoning	1.4801
28	Shanghai	1.4156
29	Shandong	1.3947
30	Tianjin	0.7050
31	Beijing	0.5461

Data source: *China Statistical Yearbook* (2013) and *China Statistical Yearbook on Environment* (2013)

The discharge amount of waste gas in the manufacturing sector reflects the degree of air pollution caused by the regional manufacturing sector. According to Table 4.26, Hebei, Jiangsu, Shandong, Shanxi, and Henan ranked in the top five. For these provinces (excluding Shanxi), economic resourcefulness was very high

Table 4.26 Discharge amount of waste gas in the manufacturing sector in China's provinces, municipalities, and autonomous regions in 2013

Rank	Region	Discharge amount of waste gas in the manufacturing sector (unit: 100 million standard cubic meters)
1	Hebei	79,121.3
2	Jiangsu	49,797.3
3	Shandong	47,159.8
4	Shanxi	41,276
5	Henan	37,665.3
6	Inner Mongolia	31,128.4
7	Liaoning	29,443.5
8	Guangdong	28,433.7
9	Anhui	28,335.4
10	Zhejiang	24,564.8
11	Guizhou	24,466.5
12	Guangxi	21,369.4
13	Hubei	19,986.9
14	Sichuan	19,760.6
15	Xinjiang	18,464.5
16	Hunan	17,276.4
17	Shaanxi	16,279.5
18	Fujian	16,183.2
19	Yunnan	15,958.1
20	Jiangxi	15,573.8
21	Shanghai	13,344.1
22	Gansu	12,676.7
23	Heilongjiang	10,622
24	Jilin	9803.6
25	Chongqing	9532.4
26	Ningxia	8909.2
27	Tianjin	8080
28	Qinghai	5620.6
29	Hainan	4721.1
30	Beijing	3692.2
31	Xizang	114.7

Data source: *China Statistical Yearbook on Environment* (2013)

and the discharge amount of waste gas in the manufacturing sector was very large. By contrast, Tianjin and Beijing did not have a large discharge amount of waste gas in 2013 although they have highly developed manufacturing industries. Through reasonable guidance by governments and effective measures by manufacturing enterprises, the contradiction between the development of the manufacturing sector and environmental protection could be mitigated to a certain extent. Compared with the rank in 2012, Guizhou moved up nine places in 2013 to 11th place.

(4) Ranking by discharge amount of waste gas per unit output value in the manufacturing sector

Table 4.27 Discharge amount of waste gas per unit output value in the manufacturing sector in China's provinces, municipalities, and autonomous regions in 2013

Rank	Region	Discharge amount of waste gas per unit of output value in the manufacturing sector (unit: 100 million standard cubic meters per 100 million yuan)
1	Guizhou	3.0301
2	Hainan	2.6535
3	Ningxia	2.5438
4	Shanxi	2.4886
5	Qinghai	2.4349
6	Xinjiang	2.1273
7	Hebei	1.7083
8	Gansu	1.5853
9	Yunnan	1.551
10	Inner Mongolia	1.5488
11	Xizang	1.1814
12	Guangxi	1.1638
13	Shaanxi	0.8576
14	Anhui	0.8394
15	Heilongjiang	0.7742
16	Henan	0.6408
17	Jiangxi	0.6311
18	Chongqing	0.6039
19	Sichuan	0.5593
20	Liaoning	0.5567
21	Hunan	0.5372
22	Hubei	0.5098
23	Fujian	0.478
24	Jilin	0.4444
25	Shanghai	0.4158
26	Zhejiang	0.39
27	Jiangsu	0.3698
28	Shandong	0.363
29	Tianjin	0.3047
30	Guangdong	0.2593
31	Beijing	0.2126

Data source: *China Statistical Yearbook* (2014), *China Statistical Yearbook on Environment* (2013)

The discharge amount of waste gas per unit of output value in the manufacturing sector reflects the atmospheric environment cost incurred by the regional manufacturing sector. According to Table 4.27, Beijing, Guangdong, Tianjin, Shandong, and Jiangsu ranked in the bottom five, indicating a small discharge amount of waste gas per unit of output value in the manufacturing sector. By contrast, Guizhou, Hainan, Shanxi, Ningxia, and Qinghai ranked in the top five, indicating a large discharge amount of waste gas per unit of output value in the manufacturing sector. For China's 31 provinces, municipalities, and autonomous regions, the index did not assume any downward trend. Obviously, China's local governments do not attach due importance to the discharge of industrial waste gas; alternatively, the existing measures taken by local governments are merely short-acting measures. Compared with the rank in 2012, the ranks of Hainan and Guangxi changed significantly in 2013; specifically, Hainan moved up 10 places to 2nd place, and Guangxi moved down five places to 12th place. The ranks of other provinces, municipalities, and autonomous regions remained stable in 2013.

(5) Ranking by generating amount of industrial waste residue

The discharge amount of waste residue in the manufacturing sector reflects the degree of pollution to the space environment caused by the regional manufacturing sector. Due to data availability limitations, this section uses the generating amount of industrial waste residue of China's provinces, municipalities, and autonomous regions instead, thereby reflecting the discharge amount of waste residue in the manufacturing sector of China's provinces, municipalities, and autonomous regions to a certain extent. According to Table 4.28, Henan, Shanxi, Liaoning, Inner Mongolia, and Shandong ranked in the top five, while Xizang, Hainan, Beijing, Tianjin, and Shanghai ranked in the bottom five.

(6) Ranking by generating amount of industrial waste residue per unit of industrial output value

The generating amount of industrial waste residue per unit of industrial output value reflects the space environment cost incurred by industrial production. According to Table 4.29, Qinghai, Xizang, Shanxi, Yunnan, and Xinjiang ranked in the top five, indicating a large generating amount of industrial waste residue per unit of industrial output value. Guangdong, Tianjin, Beijing, Shanghai, and Zhejiang ranked in the bottom five, indicating a small generating amount of industrial waste residue per unit of industrial output value; the five provinces and municipalities are economically developed regions, and have obvious effects in the prevention and abatement of solid waste pollution.

(7) Ranking by comprehensive utilization ratio of waste residue in the manufacturing sector

The comprehensive utilization ratio of waste residue in the manufacturing sector refers to the ratio of the amount of waste residue that can be converted into utilizable resources by means of recycling, processing, cycling, and exchange into the

Table 4.28 Generating amount of general industrial waste residue in China's provinces, municipalities, and autonomous regions in 2013

Rank	Region	Generating amount of general industrial waste residue (unit: 10,000 tons)
1	Hebei	43,289
2	Shanxi	30,520
3	Liaoning	26,759
4	Inner Mongolia	20,081
5	Shandong	18,172
6	Henan	16,270
7	Yunnan	16,040
8	Sichuan	14,007
9	Qinghai	12,377
10	Anhui	11,937
11	Jiangxi	11,518
12	Jiangsu	10,856
13	Xinjiang	9283
14	Fujian	8535
15	Guizhou	8194
16	Hubei	8181
17	Hunan	7806
18	Guangxi	7676
19	Shaanxi	7491
20	Heilongjiang	6094
21	Guangdong	5912
22	Gansu	5907
23	Jilin	4591
24	Zhejiang	4300
25	Ningxia	3277
26	Chongqing	3162
27	Shanghai	2054
28	Tianjin	1592
29	Beijing	1044
30	Hainan	415
31	Xizang	362

Data source: *China Statistical Yearbook* (2014), *China Statistical Yearbook on Environment* (2013)

total amount of generated waste residue. This index reflects the capability to utilize waste residue circularly in the manufacturing sector. According to Table 4.30, Tianjin, Qinghai, Shanghai, Hainan, and Zhejiang ranked in the top five, and strong manufacturing provinces had outstanding performance in this aspect. Xizang ranked last (its index was only 1.86%), lagging far behind other provinces, municipalities, and autonomous regions. It is noteworthy that the index for Tianjin

Table 4.29 Generating amount of general industrial waste residue per unit industrial output value in China's provinces, municipalities, and autonomous regions in 2013

Rank	Region	Generating amount of general industrial waste residue per unit of output value (unit: 10,000 tons per 100 million yuan)
1	Qinghai	5.3619
2	Xizang	3.7285
3	Shanxi	1.8401
4	Yunnan	1.5589
5	Xinjiang	1.0695
6	Guizhou	1.0148
7	Inner Mongolia	0.9991
8	Ningxia	0.9357
9	Hebei	0.9346
10	Gansu	0.7387
11	Liaoning	0.5059
12	Jiangxi	0.4668
13	Heilongjiang	0.4442
14	Guangxi	0.4180
15	Sichuan	0.3965
16	Shaanxi	0.3946
17	Anhui	0.3536
18	Henan	0.2768
19	Fujian	0.2521
20	Hunan	0.2427
21	Hainan	0.2332
22	Hubei	0.2087
23	Jilin	0.2081
24	Chongqing	0.2003
25	Shandong	0.1399
26	Jiangsu	0.0806
27	Zhejiang	0.0683
28	Shanghai	0.0640
29	Beijing	0.0601
30	Tianjin	0.0600
31	Guangdong	0.0539

Data source: *China Statistical Yearbook* (2014), and *China Statistical Yearbook on Environment* (2013)

was 99.62%, and for Qinghai was 97.63%. A higher comprehensive utilization capability for waste residue serves to improve the cyclic utilization of resources and reduce pollutant emissions to the environment.

Table 4.30 Comprehensive utilization amount and ratio of general industrial waste residue in China's provinces, municipalities, and autonomous regions in 2012

Rank	Region	Comprehensive utilization amount of general industrial waste residue (unit: 10,000 tons)	Comprehensive utilization rate of waste residue (%)
1	Tianjin	1582	99.6200
2	Qinghai	6798	97.6300
3	Shanghai	1995	97.3400
4	Hainan	271	96.3333
5	Zhejiang	4091	95.0309
6	Xinjiang	4814	94.2350
7	Jiangsu	10,502	93.5692
8	Sichuan	5780	92.5567
9	Guangdong	5024	91.9053
10	Fujian	7544	91.4900
11	Shandong	17,134	88.7735
12	Anhui	10,462	87.4869
13	Henan	12,466	85.3747
14	Heilongjiang	4145	83.3455
15	Hubei	6196	82.5842
16	Jiangxi	6431	81.7727
17	Chongqing	2695	81.6500
18	Hunan	5011	81.4900
19	Ningxia	2398	81.3940
20	Jilin	3712	80.5738
21	Beijing	904	78.9600
22	Shaanxi	4758	76.1230
23	Guangxi	5425	75.5414
24	Guizhou	4160	70.8133
25	Shanxi	19,815	70.3209
26	Gansu	3300	69.2492
27	Liaoning	11,742	60.4738
28	Yunnan	8414	56.3775
29	Inner Mongolia	9984	53.2144
30	Hebei	18,356	41.0450
31	Xizang	5	1.8600

Data source: *China Statistical Yearbook* (2013), and *China Statistical Yearbook on Environment* (2013)

4.2.3 Rank of China's Provinces, Municipalities, and Autonomous Regions by Cluster Indexes

This research study evaluates related indexes by using the deviation maximization method rather than the principal component analysis method, thereby preventing the loss of data information arising from the latter method. This study evaluates the new-type manufacturing sector in terms of four main indexes: economic resourcefulness, technological innovation capability, energy conservation capability, and environmental protection capability. In order to ensure data availability and simplify the evaluation index system, this study eventually uses 32 subindexes to evaluate and rank strong manufacturing provinces, municipalities, and autonomous regions. This comprehensive evaluation index system is consistent with the evaluation index system for China's strong manufacturing provinces, municipalities, and autonomous regions. This system uses a combination of aggregate indexes and unit indexes, thereby preventing the one-sidedness arising from aggregate indexes or unit indexes alone, as well as ensuring consistency.

1. Economic resourcefulness

A1–A10 are basic subindexes used to evaluate the economic resourcefulness of each regional manufacturing sector comprehensively. By using the deviation maximization method, this study assigns a weight to each of the 10 subindexes, and obtains a comprehensive score for the economic resourcefulness of each regional manufacturing sector for 2013 according to the normalized value of each subindex. According to Table 4.31, Shandong, Jiangsu, and Guangdong ranked in the top three in 2013 (the same as in 2012), except that Shandong replaced Jiangsu in first place. Among other provinces, municipalities, and autonomous regions, Zhejiang, Gansu, Fujian, and Tianjin moved up in the ranking most significantly (by at least five places). By contrast, some provinces, municipalities, and autonomous regions moved down in the ranking very significantly. For example, Jilin moved down eight places. In addition, Jilin and Heilongjiang in northeast China moved down in the ranking significantly.

2. Technological innovation capability

B1–B11 are basic subindexes used to evaluate the technological innovation capability of each regional manufacturing sector comprehensively. By using the deviation maximization method, this study assigns a weight to each of the 11 subindexes, and obtains a comprehensive score for the technological innovation capability of each regional manufacturing sector for 2013 according to the normalized value of each subindex. According to Table 4.32, the ranks of certain provinces, municipalities, and autonomous regions changed significantly in 2013 from 2012. Specifically, Jiangxi and Ningxia moved up six and five places, respectively; Guizhou moved down seven places, and Heilongjiang and Jilin each moved down six places. The ranks of other provinces, municipalities, and autonomous regions changed slightly. In terms of technological innovation capability, Guangdong, Jiangsu,

Table 4.31 Comprehensive evaluation of economic resourcefulness of China's regional manufacturing sectors

Rank	Region	Comprehensive evaluation score
1	Shandong	1.4423
2	Jiangsu	1.3799
3	Guangdong	1.091
4	Liaoning	0.9533
5	Zhejiang	0.8483
6	Henan	0.841
7	Anhui	0.8407
8	Hunan	0.8095
9	Hebei	0.7903
10	Hubei	0.7881
11	Tianjin	0.7791
12	Shanghai	0.7784
13	Fujian	0.6907
14	Jiangxi	0.6747
15	Jilin	0.6649
16	Sichuan	0.6321
17	Guangxi	0.622
18	Inner Mongolia	0.5929
19	Chongqing	0.5591
20	Shaanxi	0.4818
21	Gansu	0.4673
22	Ningxia	0.4394
23	Xinjiang	0.3905
24	Yunnan	0.3657
25	Beijing	0.3369
26	Guizhou	0.3295
27	Heilongjiang	0.3099
28	Qinghai	0.3078
29	Shanxi	0.0715
29	Xizang	–
29	Hainan	–

Data source: *China Statistical Yearbook* (2014), and statistical yearbooks (2014) of China's provinces, municipalities, and autonomous regions

Zhejiang, Shandong, and Shanghai still ranked in the top five in 2013. Central China has an obvious advantage in technological innovation capability.

3. Energy conservation capability

C1–C4 are basic subindexes used to evaluate the energy conservation capability of each regional manufacturing sector comprehensively. By using the deviation maximization method, this study assigns a weight to each of the four subindexes,

Table 4.32 Comprehensive evaluation of technological innovation capability of China's regional manufacturing sectors

Rank	Region	Comprehensive evaluation score
1	Guangdong	0.8850
2	Jiangsu	0.8283
3	Zhejiang	0.7491
4	Shandong	0.5560
5	Shanghai	0.5107
6	Tianjin	0.4756
7	Beijing	0.4280
8	Hunan	0.4008
9	Anhui	0.3720
10	Hubei	0.3413
11	Chongqing	0.3389
12	Fujian	0.3227
13	Henan	0.2855
14	Liaoning	0.2548
15	Hebei	0.2247
16	Sichuan	0.2075
17	Shaanxi	0.1943
18	Guangxi	0.1918
19	Jiangxi	0.1728
20	Shanxi	0.1711
21	Ningxia	0.1656
22	Heilongjiang	0.1653
23	Hainan	0.1582
24	Gansu	0.1552
25	Inner Mongolia	0.1411
26	Guizhou	0.1174
27	Xizang	0.1123
28	Jilin	0.1007
29	Yunnan	0.0981
30	Xinjiang	0.0775
31	Qinghai	0.0299

Data source: *China Statistical Yearbook* (2014), and statistical yearbooks (2014) of China's provinces, municipalities, and autonomous regions

and obtains a comprehensive score for the energy conservation capability of each regional manufacturing sector for 2013 according to the normalized value of each subindex. Based on the comprehensive scores, this research study ranks China's provinces, municipalities, and autonomous regions. According to Table 4.33, Beijing, Zhejiang, Shanghai, Hainan, and Guangdong ranked in the top five, and Hebei, Shanxi, Xinjiang, Hubei, and Guizhou ranked in the bottom five (Xizang was excluded from the ranking). Beijing, Zhejiang, Shanghai, and Guangdong are

Table 4.33 Comprehensive evaluation of energy conservation capability of China's regional manufacturing sectors

Rank	Region	Comprehensive evaluation score
1	Beijing	1.1643
2	Zhejiang	1.0015
3	Shanghai	0.9461
4	Hainan	0.9298
5	Guangdong	0.905
6	Fujian	0.8521
7	Qinghai	0.8502
8	Tianjin	0.8124
9	Gansu	0.7617
10	Chongqing	0.7506
11	Jiangsu	0.7326
12	Jiangxi	0.7115
13	Heilongjiang	0.7114
14	Shaanxi	0.6788
15	Hunan	0.666
16	Anhui	0.6647
17	Guangxi	0.6646
18	Jilin	0.654
19	Inner Mongolia	0.6468
20	Ningxia	0.6404
21	Henan	0.6372
22	Guizhou	0.6285
23	Liaoning	0.6234
24	Hubei	0.6034
25	Shandong	0.5965
26	Sichuan	0.5708
27	Xinjiang	0.5374
28	Yunnan	0.5153
29	Shanxi	0.4092
30	Hebei	0.2648
31	Xizang	–

Data source: *China Statistical Yearbook* (2014), and statistical yearbooks (2014) of China's provinces, municipalities, and autonomous regions

economically developed provinces and municipalities in eastern China. In these subindexes, a high energy-conservation capability represents a high efficiency of energy utilization and an optimum structure of energy utilization. The manufacturing sectors of Hebei and Hubei had remarkable economic resourcefulness (ranking 9th and 10th, respectively) but inferior energy conservation capability (ranking 30th and 27th, respectively), indicating an extensive development pattern (for details, see Table 4.33).

4. Environmental protection capability

D1–D7 are basic subindexes used to evaluate the environmental protection capability of each regional manufacturing sector comprehensively. By using the deviation maximization method, this research study assigns a weight to each of the seven subindexes, and obtains a comprehensive score for the environmental protection capability of each regional manufacturing sector for 2013 according to the normalized value of each subindex. Based on the comprehensive scores, this study ranks China's provinces, municipalities, and autonomous regions. According to Table 4.34, Beijing,

Table 4.34 Comprehensive evaluation of environmental protection capability of China's regional manufacturing sectors

Rank	Region	Comprehensive evaluation score
1	Beijing	0.9604
2	Tianjin	0.9577
3	Shanghai	0.8905
4	Jilin	0.8451
5	Chongqing	0.8405
6	Shaanxi	0.7985
7	Sichuan	0.7903
8	Hubei	0.7756
9	Guangdong	0.7608
10	Heilongjiang	0.7519
11	Jiangxi	0.7469
12	Hunan	0.7446
13	Anhui	0.7446
14	Fujian	0.7423
15	Zhejiang	0.7315
16	Gansu	0.7295
17	Liaoning	0.6984
18	Hainan	0.6938
19	Henan	0.6797
20	Shandong	0.679
21	Jiangsu	0.6654
22	Inner Mongolia	0.6476
23	Xinjiang	0.6292
24	Ningxia	0.6183
25	Guangxi	0.6109
26	Qinghai	0.5987
27	Xizang	0.5975
28	Guizhou	0.59
29	Yunnan	0.5858
30	Shanxi	0.491
31	Hebei	0.4017

Data source: *China Statistical Yearbook* (2013)

Tianjin, Shanghai, Jilin, and Chongqing ranked in the top five, while Hebei, Shanxi, Yunnan, Guizhou and Xizang ranked in the bottom five. Hebei's manufacturing sector was more developed than the other four other provinces and autonomous regions in the bottom five, but Hebei ranked top in terms of pollution-related subindexes. This indicates that Hebei's manufacturing sector incurred a high environmental cost, thereby calling for an improvement in the development pattern. Jiangsu and Guangdong are large manufacturing provinces, which keep far ahead of other provinces, municipalities, and autonomous regions by quantity. Because of the large scale of the manufacturing sector, the amount of pollutant discharge by the two provinces is very large. Therefore, the comprehensive scores for their environmental protection capability are not high (for details, see Table 4.34).

4.2.4 Rank of China's Top 10 Manufacturing Provinces, Municipalities, and Autonomous Regions

The above mentioned 32 subindexes are used to evaluate the degree of newness of each of China's regional manufacturing sector comprehensively. By using the deviation maximization method, this study assigns a weight to each of the 32 subindexes, and obtains a comprehensive score for each regional manufacturing sector of China for 2013 according to the normalized value of each subindex. According to the comprehensive scores, this study ranks China's provinces, municipalities, and autonomous regions (for details, see Table 4.35).

Based on the analyses of the 32 subindexes in four aspects, namely, economic resourcefulness, technological innovation capability, energy conservation capability, and environmental protection capability, this study uses the deviation maximization method to determine China's top 10 manufacturing provinces, municipalities, and autonomous regions for 2013, as follows: Guangdong, Jiangsu, Tianjin, Shanghai, Zhejiang, Shandong, Beijing, Hunan, Anhui, and Chongqing. The following section provides a comprehensive evaluation of the top 10. In the following, we highlight three key points of the overall ranking.

(1) In terms of comprehensive strength, Guangdong ranked first instead of Jiangsu in second place. Guangdong and Jiangsu are strong manufacturing provinces, which frequently rank in the top two by their comprehensive strength. Specifically, Guangdong ranked first in 2008, 2009, 2010, and 2013, and Jiangsu ranked first in 2005, 2007, 2011, and 2012. In recent years, Jiangsu has kept ahead of Guangdong in terms of economic resourcefulness, and energy conservation capability, while Guangdong has kept slightly ahead of Jiangsu in terms of environmental protection capability. In recent years, Jiangsu has ranked first in terms of R&D expenditure input, but has lagged behind Guangdong in terms of technological output. In terms of technological innovation capability, Guangdong has kept ahead of Jiangsu. Overall, Guangdong and Jiangsu are of equal comprehensive strength in the current stage.

Table 4.35 Rank of China's provinces, municipalities, and autonomous regions by degree of newness of manufacturing sector

Rank	Region	Comprehensive evaluation score
1	Guangdong	0.7489
2	Jiangsu	0.7319
3	Tianjin	0.7254
4	Shanghai	0.7193
5	Zhejiang	0.7113
6	Shandong	0.6908
7	Beijing	0.6621
8	Hunan	0.6336
9	Anhui	0.6197
10	Chongqing	0.6175
11	Hubei	0.6
12	Fujian	0.5931
13	Jilin	0.5741
14	Liaoning	0.5693
15	Jiangxi	0.5663
16	Henan	0.5623
17	Sichuan	0.5361
18	Guangxi	0.5093
19	Gansu	0.4916
20	Shaanxi	0.4909
21	Heilongjiang	0.4831
22	Inner Mongolia	0.4677
23	Ningxia	0.4418
24	Guizhou	0.4087
25	Xinjiang	0.4004
26	Yunnan	0.3965
27	Hebei	0.3931
28	Qinghai	0.3388
29	Shanxi	0.3109
30	Xizang	–
31	Hainan	–

Data source: *China Statistical Yearbook* (2014)

(2) In terms of comprehensive development capability, Hebei and Shaanxi have been on a downward trend. At one stage, the two provinces ranked in China's top 10 manufacturing provinces, municipalities, and autonomous regions every year. Specifically, Shaanxi ranked 9th (its highest place) in 2003, but then moved down rapidly in subsequent years, ranking 20th in 2013. Meanwhile, Hebei ranked in China's top 10 manufacturing provinces, municipalities, and autonomous regions in 2005 and 2006, ranked 19th in 2012, and then 27th in

2013, moving down significantly over the years. In spite of the difference in geological locations, these two provinces share the same feature: they are both resource-oriented manufacturing provinces. For example, Hebei's Tangshan is well known for its petroleum industry, Shaanxi's Yulin is well known for its petroleum and coal industries, and Shaanxi's Yan'an is well known for its natural gas industry. In future development strategies for their manufacturing sectors, the two provinces should adjust their industrial structures and extricate themselves from excessive reliance on natural resources.

(3) Jilin's manufacturing sector is sluggish. Thanks to the Northeast China Revitalizing strategy, in 2008, Jilin began to move up the ranking rapidly. Specifically, Jilin ranked 18th in 2008, 12th in 2009, 13th in 2010, and was in the top 10 in 2011, although it fell to the 13th place in 2012, indicating an obvious regression in economic resourcefulness. Fundamentally, this can be ascribed to deficiency in R&D expenditure and new product development inputs. Therefore, the deficiency in technological innovations is a major constraint to the development of Jilin's manufacturing sector.

4.3 The Top 10 Cities of China's Manufacturing Sector

At present, there are more than 600 small and big cities in China. We continued to adopt the concept of "central city" in order to select the sample cities from multiple cities. A central city is in a core position to play a leading role in the entire city system. We defined central cities as provincial capitals, sub-provincial cities, and a few larger cities (excluding municipalities). Based on the two criteria of higher manufacturing development level and ease of data acquisition, we selected 33 cities in China as the sample cities, namely, Changchun, Changsha, Chengdu, Dalian, Dongguan, Fuzhou, Guangzhou, Guiyang, Haikou, Hangzhou, Harbin, Hefei, Hohhot, Jinan, Kunming, Lanzhou, Nanchang, Nanjing, Nanning, Ningbo, Qingdao, Shenyang, Shenzhen, Shijiazhuang, Suzhou, Taiyuan, Urumqi, Wuhan, Wuxi, Xiamen, Xi'an, Yinchuan, and Zhengzhou. Beijing, Tianjin, Shanghai, and Chongqing fall within China's provincial classification, and therefore, do not qualify for the ranking of strong manufacturing cities despite their high overall manufacturing development capabilities. On the other hand, since the Dalian Statistical Yearbook 2014 and the Shijiazhuang Statistical Yearbook 2014 had not been published at the time of the research, and some of the manufacturing statistics are missing in the Dongguan Statistical Yearbook 2014, only 30 cities in the above mentioned list were included in the final sample due to availability of related data. The three cities excluded from the sample were Dongguan, Dalian, and Shijiazhuang.

4.3.1 Individual Index Ranking of Manufacturing Cities

1. Economic resourcefulness ranking

(1) Total manufacturing output ranking

The total manufacturing output reflects the total scale and total level of manufacturing production of a region in a specific period of time. The total overall manufacturing output of all major cities in China in 2013 grew in comparison to 2012. The average manufacturing output jumped from 756.098 billion yuan in 2012 to 803.429 billion yuan in 2013, which accounted for an increase of 6.26% (see Table 4.36). Among the 30 sample cities, the average manufacturing output of 13 cities in the sample was higher than the average output of the entire manufacturing sector whereas that of the other 17 cities was lower than the average output of the entire manufacturing sector. The contribution of the manufacturing sector exhibited a Pareto distribution (see Fig. 4.7) in which the total cumulative manufacturing output of the first 7 cities accounted for more than 50% of the total output of the entire industry whereas the total cumulative manufacturing output of the latter 13 accounted for less than 15% of the total output of the entire sector.

(2) Ratio ranking of total manufacturing output to total industrial output

The ratio of total manufacturing output to total industrial output reflects the status of manufacturing among all the industries in the region and serves as a measure of the contribution that the manufacturing sector has made to the total regional industrial development. As shown in Table 4.37, the total manufacturing output of 8 cities accounted for more than 95% of their total industrial output and the total manufacturing output of 22 cities accounted for more than 90% of their total industrial output. These observations clearly indicate that the manufacturing sector is the main sector in China's cities. Of the sample cities, the top five cities with the highest ratio of total manufacturing output to total industrial output were Wuxi, Suzhou, Nanjing, Shenyang, and Qingdao. Suzhou was ranked in the top five based on three different criteria, while the placement of Jiangsu province cities in the top three confirmed its status as a typical big manufacturing province (Table 4.37).

(3) Ranking by manufacturing labor productivity rate

The output value of the manufacturing sector reflects the manufacturing output in a region while the ratio of total manufacturing output to the number of manufacturing employees reflects the labor productivity level of manufacturing in a region. As shown in Table 4.38, the average output per worker was highest in Shenyang's manufacturing sector, at 0.3691 million yuan/person; followed after a big gap by Hohhot, Urumqi, Yinchuan, Changchun at 0.2578 million yuan/person,

Table 4.36 Total manufacturing output of each major city in 2013

Rank	City	Total output (billion yuan)	% to Total manufacturing (%)	Cumulative percentage (%)
1	Suzhou	2972.4014	12.3322	12.3322
2	Shenzhen	2178.8348	9.0397	21.3719
3	Guangzhou	1575.5098	6.5366	27.9085
4	Qingdao	1502.1386	6.2322	34.1407
5	Wuxi	1463.6835	6.0727	40.2133
6	Shenyang	1334.0934	5.5350	45.7483
7	Nanjing	1226.2727	5.0877	50.8360
8	Ningbo	1202.0339	4.9871	55.8231
9	Hangzhou	1150.8620	4.7748	60.5979
10	Zhengzhou	1027.5599	4.2632	64.8611
11	Wuhan	940.6530	3.9027	68.7638
12	Chengdu	882.1390	3.6599	72.4237
13	Changchun	865.7543	3.5919	76.0156
14	Changsha	796.3728	3.3041	79.3197
15	Hefei	686.4809	2.8481	82.1678
16	Fuzhou	619.4780	2.5701	84.7379
17	Xiamen	446.3791	1.8520	86.5899
18	Jinan	443.2906	1.8392	88.4291
19	Xi'an	417.4207	1.7318	90.1609
20	Nanchang	393.9306	1.6344	91.7953
21	Kunming	334.4228	1.3875	93.1828
22	Harbin	324.0741	1.3445	94.5273
23	Nanning	238.5245	0.9896	95.5169
24	Urumqi	208.5437	0.8652	96.3821
25	Lanzhou	207.4902	0.8609	97.2430
26	Taiyuan	199.8366	0.8291	98.0721
27	Guiyang	187.2016	0.7767	98.8488
28	Hohhot	116.2664	0.4824	99.3311
29	Yinchuan	115.0351	0.4773	99.8084
30	Haikou	46.1795	0.1916	100.0000
31	Shijiazhuang	–	–	–
32	Dongguan	–	–	–
33	Dalian	–	–	–

Data source: China City Statistical Yearbook 2014 and China Statistical Yearbook 2014 of each province

0.2465 million yuan/person,0.2308 million yuan/person, and 0.2151 million yuan/person, respectively. Compared to the labor productivity rates of other regions, Shenyang's leading advantage was significant.

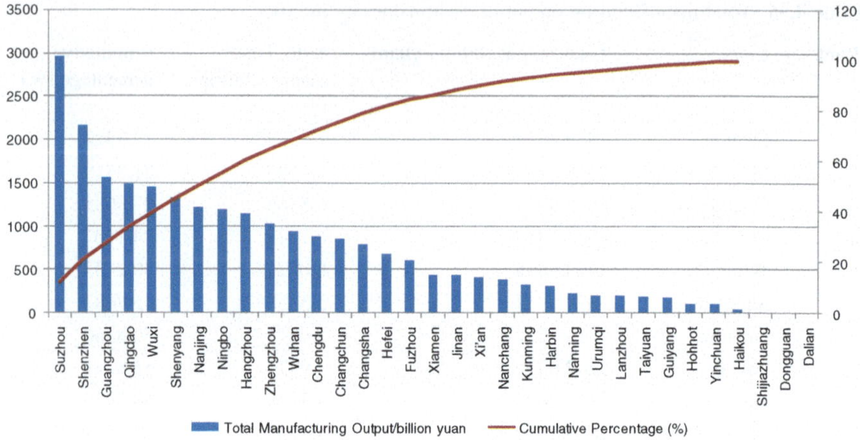

Fig. 4.7 Pareto distribution of city manufacturing output in 2013

(4) Ranking of total manufacturing profits

The total profit of manufacturing enterprises reflects the regional profitability of its manufacturing sector, and is a crucial index that measures economic resourcefulness of the manufacturing sector. As seen from Table 4.39, the average total manufacturing profits of all major cities was 44.9959 billion yuan. There were 14 cities that earned above this average, of which Suzhou, Guangzhou, Shenzhen, Nanjing, and Zhengzhou were the top five. The total manufacturing profits of Suzhou were 125.48 billion yuan, far surpassing the profits of Guangzhou and Shenzhen, which were ranked second and third, respectively. Although Lanzhou was ranked last, its manufacturing profits turned positive after a loss of 2.88 billion yuan in the previous year. Overall, city manufacturing has improved significantly.

(5) Ratio ranking of manufacturing workers to urban workers

The ratio of the number of manufacturing workers to the number of urban workers reflects how attractive a manufacturing sector is in a city in terms of numbers of workers employed. As shown in Table 4.40, which ranks cities by this ratio, Suzhou, Wuxi, Qingdao, Xiamen, and Ningbo, were the top five with ratios of 66.03, 53.78, 52.81, 49.07, and 46.93%, respectively. These five countries showed a higher contribution to provision of employment in the manufacturing sector, whereas Haikou, Yinchuan, and Hohhot were ranked in the bottom three cities. Among them, the ratio of Haikou was 11.3%, reflecting that its manufacturing sector contributed a smaller proportion to the overall industrial structure of the city than the manufacturing sectors of the other cities did.

Table 4.37 Ratio of total manufacturing output to total industrial output of major cities in 2013

Rank	City	Total manufacturing output (billion yuan)	Total industrial output (billion yuan)	Total manufacturing output to total industrial output ratio (%)
1	Wuxi	1463.6835	1487.6331	98.3901
2	Suzhou	2972.4014	3027.6294	98.1759
3	Nanjing	1226.2727	1256.3093	97.6091
4	Shenyang	1334.0934	1373.5178	97.1297
5	Qingdao	1502.1386	1551.2681	96.8330
6	Chengdu	882.1390	917.1160	96.1862
7	Changsha	796.3728	828.9133	96.0743
8	Harbin	324.0741	339.9263	95.3366
9	Xiamen	446.3791	471.6210	94.6478
10	Shenzhen	2178.8348	2309.5209	94.3414
11	Changchun	865.7543	922.8043	93.8178
12	Nanning	238.5245	255.7135	93.2780
13	Guiyang	187.2016	201.4314	92.9357
14	Xi'an	417.4207	449.7620	92.8092
15	Jinan	443.2906	477.7474	92.7877
16	Hangzhou	1150.8620	1241.8004	92.6769
17	Haikou	46.1795	49.9551	92.4420
18	Ningbo	1202.0339	1301.0089	92.3924
19	Guangzhou	1575.5098	1719.8718	91.6062
20	Fuzhou	619.4780	678.6330	91.2832
21	Hefei	686.4809	752.6575	91.2076
22	Wuhan	940.6530	1039.4070	90.4990
23	Nanchang	393.9306	443.7520	88.7727
24	Lanzhou	207.4902	241.6199	85.8746
25	Urumqi	208.5437	242.8853	85.8610
26	Kunming	334.4228	391.7845	85.3589
27	Zhengzhou	1027.5599	1215.3523	84.5483
28	Hohhot	116.2664	149.5376	77.7506
29	Taiyuan	199.8366	264.8840	75.4431
30	Yinchuan	115.0351	192.8030	59.6646
31	Dongguan	–	1191.8171	–
32	Dalian	–	–	–
33	Shijiazhuang	–	–	–

Data source: China City Statistical Yearbook 2014 and China Statistical Yearbook 2014 of each city

(6) Ranking of total taxes and profits of manufacturing enterprises

Total taxes and profits of manufacturing enterprises reflect the ability of the manufacturing sector to contribute to generating income tax for the country and respective city, and is an important index that measures the contribution made by the manufacturing sector to society. According to an analysis conducted on the total taxes and profits of manufacturing enterprises in major cities, the top five cities were Suzhou, Shenzhen, Guangzhou, Qingdao, and Nanjing at 172.15 billion yuan,

Table 4.38 Labor productivity rate of manufacturing workers in major cities in 2013

Rank	City	Total manufacturing output (hundred million yuan)	Total number of manufacturing workers (person)	Labor productivity rate per manufacturing worker (million yuan/person)
1	Shenyang	13,340.9344	361,432	0.3691
2	Hohhot	1162.664	45,100	0.2578
3	Urumqi	2085.4373	84,600	0.2465
4	Yinchuan	1150.3513	49,833	0.2308
5	Changchun	8657.5433	402,468	0.2151
6	Qingdao	15,021.3864	714,400	0.2103
7	Nanjing	12,262.72,719	585,501	0.2094
8	Changsha	7963.7281	380,497	0.2093
9	Hefei	6864.8089	339,898	0.2020
10	Wuxi	14,636.8354	733,700	0.1995
11	Wuhan	9406.53	528,600	0.1780
12	Lanzhou	2074.9018	123,600	0.1679
13	Nanning	2385.2448	147,762	0.1614
14	Guangzhou	15,755.098	985,000	0.1600
15	Hangzhou	11,508.6196	726,790	0.1583
16	Ningbo	12,020.3392	787,900	0.1526
17	Zhengzhou	10,275.5992	681,300	0.1508
18	Suzhou	29,724.0143	1,991,676	0.1492
19	Kunming	3344.2283	226,816	0.1474
20	Fuzhou	6194.78	435,660	0.1422
21	Jinan	4432.9060	325,105	0.1364
22	Nanchang	3939.3062	299,459	0.1315
23	Harbin	3240.7409	292,400	0.1108
24	Guiyang	1872.0162	177,200	0.1056
25	Taiyuan	1998.3658	208,100	0.0960
26	Xi'an	4174.2072	449,800	0.0928
27	Haikou	461.795	53,100	0.0870
28	Shenzhen	21,788.3475	2,558,749	0.0852
29	Chengdu	8821.3900	1,045,500	0.0844
30	Xiamen	4463.7907	540,800	0.0825
31	Dongguan	–	1,991,100	–
32	Dalian	–	547,800	–
33	Shijiazhuang	–	225,200	–

Data source: China City Statistical Yearbook 2014 and China Statistical Yearbook 2014 of each city. Above-scale industries were used as the statistical criterion for total manufacturing output and the total number of manufacturing workers in the city was used as the statistical criterion for the number of manufacturing workers

Table 4.39 Total profits of manufacturing enterprises in 2013

Rank	City	Total profits of manufacturing enterprises (hundred million yuan)
1	Suzhou	1254.8333
2	Guangzhou	982.6348
3	Shenzhen	982.2355
4	Nanjing	940.0228
5	Zhengzhou	908.3125
6	Qingdao	878.4306
7	Hangzhou	814.4319
8	Wuxi	782.9699
9	Changchun	757.4212
10	Shenyang	732.0854
11	Ningbo	588.057
12	Changsha	562.5081
13	Chengdu	544.3548
14	Hefei	472.7624
15	Wuhan	440.68
16	Fuzhou	372.5682
17	Dongguan	276.8933
18	Jinan	243.107
19	Nanchang	237.1998
20	Xiamen	212.0381
21	Xi'an	169.5842
22	Nanning	159.2115
23	Kunming	140.7076
24	Harbin	121.7636
25	Guiyang	111.8003
26	Hohhot	89.1836
27	Urumqi	48.2303
28	Yinchuan	42.1519
29	Haikou	33.0222
30	Taiyuan	32.5849
31	Lanzhou	16.94
32	Dalian	–
33	Shijiazhuang	–

Data source: China City Statistical Yearbook 2014 and China Statistical Yearbook 2014 of each city

153.07 billion yuan, 146.72 billion yuan, 139.47 billion yuan, and 132.82 billion yuan, respectively. The average total taxes and profits of the manufacturing enterprises of the 32 cities in 2013 was 73.93 billion yuan; the total taxes and profits of the manufacturing sectors of 16 cities exceeded the average (see Table 4.41).

Table 4.40 Ratio of manufacturing workers to urban workers in the major cities in 2013

Rank	City	Total number of manufacturing workers (person)	Ratio of number of manufacturing workers to number of urban workers (%)
1	Suzhou	882,800	66.0284
2	Wuxi	473,800	53.7798
3	Qingdao	681,200	52.8062
4	Xiamen	579,500	49.0686
5	Ningbo	819,900	46.9319
6	Shenzhen	1,263,800	45.1357
7	Dalian	460,500	41.3004
8	Nanjing	535,500	36.3297
9	Guangzhou	1,171,100	35.8244
10	Zhengzhou	553,100	34.1420
11	Fuzhou	462,600	32.1921
12	Changsha	367,400	29.6290
13	Nanchang	281,600	29.2723
14	Hangzhou	804,700	28.5456
15	Changchun	282,900	28.2900
16	Xi'an	465,420	28.1051
17	Chengdu	589,100	27.7485
18	Wuhan	517,200	26.9515
19	Shenyang	325,500	26.6803
20	Jinan	344,500	24.3463
21	Hefei	280,300	24.2474
22	Taiyuan	247,500	23.9246
23	Shijiazhuang	212,309	23.4119
24	Lanzhou	123,600	22.8466
25	Guiyang	179,600	21.7170
26	Nanning	172,900	21.5050
27	Harbin	274,900	19.8771
28	Hohhot	61,500	17.9300
29	Kunming	214,500	17.9048
30	Urumqi	88,900	14.1111
31	Yinchuan	44,300	13.2239
32	Haikou	51,000	11.3333
33	Dongguan	–	–

Data source: China City Statistical Yearbook 2014 with statistical criteria set as "entire city"

(7) Ranking of average income tax rate of manufacturing workers

The average income tax rate of manufacturing workers reflects the ability of an individual to contribute to generating income tax for the country, and is an

Table 4.41 Total taxes and profits of manufacturing enterprises of the major cities in 2013

Rank	City	Total taxes and profits of manufacturing enterprises
1	Suzhou	1721.5352
2	Shenzhen	1530.7331
3	Guangzhou	1467.2063
4	Qingdao	1394.7245
5	Nanjing	1328.2639
6	Changsha	1316.2313
7	Zhengzhou	1283.3612
8	Wuxi	1214.4877
9	Chengdu	1123.4214
10	Changchun	1121.1553
11	Dalian	1100.6098
12	Shenyang	1083.7655
13	Ningbo	1003.133
14	Wuhan	972.57
15	Shijiazhuang	779.4742
16	Hefei	744.0054
17	Fuzhou	549.8924
18	Hangzhou	538.5915
19	Jinan	422.4378
20	Kunming	389.2314
21	Xiamen	382.5929
22	Nanchang	373.3612
23	Harbin	303.1799
24	Xi'an	295.3278
25	Guiyang	261.9567
26	Nanning	188.6509
27	Taiyuan	174.4791
28	Lanzhou	168.9
29	Hohhot	144.7999
30	Urumqi	107.8347
31	Yinchuan	104.3957
32	Haikou	68.8813
33	Dongguan	–

Data source: Statistical Yearbooks of each city

important index that measures the contribution made to society. According to an analysis conducted on the average income tax rate of manufacturing workers in the major cities (see Table 4.42), the top five cities were Changchun, Shijiazhuang, Changsha, Shenyang, and Hefei at 396,300 yuan/person, 367,100 yuan/person,

Table 4.42 Average income tax rate of manufacturing workers of the major cities in 2013

Rank	City	Average income tax rate of manufacturing workers (ten thousand yuan/person)
1	Changchun	396,300
2	Shijiazhuang	367,100
3	Changsha	358,300
4	Shenyang	333,000
5	Hefei	265,400
6	Wuxi	256,300
7	Nanjing	248,000
8	Dalian	239,000
9	Yinchuan	235,700
10	Hohhot	235,400
11	Zhengzhou	232,000
12	Qingdao	204,700
13	Suzhou	195,000
14	Chengdu	190,700
15	Wuhan	188,000
16	Kunming	181,500
17	Guiyang	145,900
18	Lanzhou	136,700
19	Haikou	135,100
20	Nanchang	132,600
21	Guangzhou	125,300
22	Jinan	122,600
23	Ningbo	122,300
24	Urumqi	121,300
25	Shenzhen	121,100
26	Fuzhou	118,900
27	Harbin	110,300
28	Nanning	109,100
29	Taiyuan	70,500
30	Hangzhou	66,900
31	Xiamen	66,000
32	Xi'an	63,500

Data source: Statistical Yearbook 2014 of each city

358,300 yuan/person, 333,000 yuan/person, and 265,400 yuan/person. It is noteworthy that all of these five cities are located in the central region and that Changchun and Shenyang form the key industrial base of northeastern China.

2. Ranking of environmental resource protection capability

(1) Ranking of industrial wastewater emissions

Since the cities did not have statistical data targeted specifically at wastewater emissions of the manufacturing sector, we employed statistical data of wastewater emissions as the criteria to analyze the water pollution level created by industries in each city. To a certain extent, this reflects the water pollution level created by the manufacturing sector in each city. As shown in Table 4.43, Suzhou had the most industrial wastewater emissions in China in 2013, followed by Hangzhou, Shijiazhuang, Xiamen, and Dalian. Overall, wastewater emissions of the strong industrial regions located in the northern part of China were significantly higher than those of other regions and exhibited an upward trend in the overall share of the total wastewater emissions. Compared to industrial wastewater emissions in 2012, 20 cities achieved reductions in industrial wastewater emissions. Of these, Wuxi had the most notable reduction of industrial water emissions of 35.58%. On the other hand, the industrial wastewater emissions of another 13 cities showed an upward trend; Xi'an, in particular, had a rapid increase of 181.73% in industrial wastewater emissions.

(2) Ranking of wastewater emissions per industrial output unit

The index of wastewater emissions per industrial output unit could be employed to analyze the cost to the water environment of a city's industrial production, which could reflect, to a certain extent, the cost of manufacturing to the water environment. In other words, the lower wastewater emissions are per unit of output, the lower is the cost of the manufacturing unit to the water environment. As observed from Table 4.44, the five cities in the sample with the highest wastewater emissions per industrial output unit were Xiamen, Nanning, Shijiazhuang, Yinchuan, and Hangzhou, indicating that the cost incurred to the water environment by manufacturing production in these cities was higher than in other cities. On the other hand, the bottom ranked five cities with the lowest wastewater emissions per industrial output unit were Changsha, Shenzhen, Changchun, Shenyang, and Qingdao, indicating that the cost incurred to the water environment by manufacturing production in these cities was lower than in other cities.

(3) Ranking of industrial sulfur dioxide emissions

Manufacturing emissions reflect the air pollution level caused by manufacturing in the region. Due to limited availability of urban emissions data, we adopted industrial sulfur dioxide emissions to reflect the air pollution level caused by the industries in each city. To a certain extent, these emissions reflect the air pollution level caused by the manufacturing sector in each city. As shown in Table 4.45, Shijiazhuang was the city with the highest sulfur dioxide emissions, followed by Suzhou, Ningbo, Shenyang, and Dongguan. We observe that while the manufacturing sector in Suzhou, Shenyang, and Dongguan had relatively strong economic output, they also had higher manufacturing emissions. On the contrary, emissions in

Table 4.43 Industrial wastewater emissions of the major cities in 2013

Rank	City	Industrial wastewater emissions of 2012 (thousand tons)	Industrial wastewater emissions of 2013 (thousand tons)	Growth rate of industrial wastewater emissions (%)
1	Suzhou	70,754.00	66,916.00	−5.42
2	Hangzhou	42,724.00	39,186.00	−8.28
3	Shijiazhuang	31,058.00	27,753.00	−10.64
4	Xiamen	26,948.00	27,256.00	1.14
5	Dalian	30,795.00	26,154.00	−15.07
6	Nanjing	24,223.00	25,291.00	4.41
7	Dongguan	26,909.00	23,463.00	−12.81
8	Wuxi	35,846.00	23,093.00	−35.58
9	Guangzhou	22,677.00	22,558.00	−0.52
10	Ningbo	20,125.00	19,666.00	−2.28
11	Wuhan	20,851.00	14,700.00	−29.50
12	Shenzhen	13,831.00	12,012.00	−13.15
13	Zhengzhou	12,538.00	11,835.00	−5.61
14	Qingdao	11,146.00	10,641.00	−4.53
15	Nanchang	10,929.00	10,602.00	−2.99
16	Chengdu	11,780.00	10,524.00	−10.66
17	Nanning	12,496.00	9752.00	−21.96
18	Xi'an	3185.00	8973.00	181.73
19	Jinan	6653.00	8596.00	29.20
20	Shenyang	7705.00	8533.00	10.75
21	Yinchuan	5963.00	6194.00	3.87
22	Hefei	5971.00	6018.00	0.79
23	Changchun	5309.00	5482.00	3.26
24	Lanzhou	4652.00	4910.00	5.55
25	Urumqi	5980.00	4889.00	−18.24
26	Kunming	5211.00	4808.00	−7.73
27	Fuzhou	5333.00	4682.00	−12.21
28	Harbin	6497.00	4487.00	−30.94
29	Taiyuan	3161.00	4085.00	29.23
30	Changsha	3777.00	4049.00	7.20
31	Guiyang	2008.00	2262.00	12.65
32	Hohhot	2187.00	2082.00	−4.80
33	Haikou	879.00	824.00	−6.26

Data source: China City Statistical Yearbook 2014

Shenzhen were not high even though it is a strong manufacturing region, which reflects that although there is conflict between manufacturing development and environmental protection, the conflict could be alleviated to some extent provided there is reasonable government guidance and effective corporate controls. Overall, the industrial sulfur dioxide emissions of the majority of the cities in 2013 were

Table 4.44 Industrial wastewater emissions per industrial output unit of the major cities in 2013

Rank	City	Industrial wastewater emissions per industrial output unit (thousand tons/billion yuan)
1	Xiamen	577.92
2	Nanning	381.36
3	Shijiazhuang	363.99
4	Yinchuan	321.26
5	Hangzhou	315.56
6	Dalian	227
7	Nanchang	221.92
8	Suzhou	221.02
9	Lanzhou	203.21
10	Nanjing	201.31
11	Urumqi	201.29
12	Xi'an	199.51
13	Dongguan	196.87
14	Jinan	193.71
15	Haikou	164.95
16	Wuxi	155.23
17	Taiyuan	154.22
18	Ningbo	151.16
19	Wuhan	141.43
20	Hohhot	139.23
21	Harbin	132
22	Guangzhou	131.16
23	Kunming	122.72
24	Chengdu	114.75
25	Guiyang	112.3
26	Zhengzhou	97.38
27	Hefei	79.96
28	Fuzhou	68.99
29	Qingdao	68.6
30	Shenyang	62.13
31	Changchun	59.41
32	Shenzhen	52.01
33	Changsha	48.85

Data source: China City Statistical Yearbook 2014 and Statistical Yearbook 2014 of each city

significantly reduced from 2012. However, five cities maintained an upward trend; in particular, Shenyang showed a significant increase of 35.05%. More attention should be devoted to monitor air pollution caused by the development of the manufacturing sector in Shenyang.

Table 4.45 Industrial emissions (sulfur dioxide) of the major cities in 2013

Rank	City	Industrial sulfur dioxide emissions of 2012 (tons)	Industrial sulfur dioxide emissions of 2013 (tons)	Growth rate of industrial sulfur dioxide emissions (%)
1	Shijiazhuang	179,942	181,532	0.88
2	Suzhou	183,401	164,969	−10.05
3	Ningbo	144,356	134,630	−6.74
4	Shenyang	96,756	130,672	35.05
5	Dongguan	118,663	112,132	−5.50
6	Nanjing	119,155	110,665	−7.13
7	Zhengzhou	110,056	106,123	−3.57
8	Dalian	114,589	102,938	−10.17
9	Kunming	113,277	101,669	−10.25
10	Wuhan	100,072	97,600	−2.47
11	Hohhot	99,375	96,190	−3.21
12	Yinchuan	105,743	92,369	−12.65
13	Taiyuan	101,780	88,900	−12.65
14	Wuxi	99,857	83,213	−16.67
15	Hangzhou	86,181	82,021	−4.83
16	Jinan	103,187	81,118	−21.39
17	Fuzhou	76,255	76,043	−0.28
18	Urumqi	118,059	74,216	−37.14
19	Lanzhou	68,654	72,148	5.09
20	Guiyang	65,259	70,602	8.19
21	Qingdao	72,563	69,337	−4.45
22	Harbin	80,740	65,987	−18.27
23	Xi'an	83,063	64,664	−22.15
24	Guangzhou	68,379	63,331	−7.38
25	Changchun	69,046	57,246	−17.09
26	Chengdu	56,730	52,040	−8.27
27	Hefei	45,571.8	41,483	−8.97
28	Nanchang	43,470	40,756	−6.24
29	Nanning	30,626	33,045	7.90
30	Changsha	21,210	21,173	−0.17
31	Xiamen	19,276	18,772	−2.61
32	Shenzhen	9847	8193	−16.80
33	Haikou	1731	1798	3.87

Data source: China City Statistical Yearbook 2014 of each city

(4) Ranking of sulfur dioxide emissions per industrial output unit

Waste gas emissions per manufacturing output unit is the ratio of manufacturing emissions to total manufacturing output, which reflects the costs incurred to the environment by the manufacturing sector in the region. Due to limited data availability, we adopted the ratio ranking of industrial sulfur dioxide emissions to total industrial output since the ratios could reflect, to a certain extent, the costs incurred to the environment by the manufacturing sector in each city. According to Table 4.46, Shenzhen produced the lowest sulfur dioxide emissions per manufacturing output unit, followed by Changsha, Haikou, Guangzhou, and Xiamen. On the other hand, the top five emitting cities of sulfur dioxide per manufacturing output unit were Hohhot, Yinchuan, Guiyang, Taiyuan, and Urumqi, as shown in Table 4.46. Although Shenzhen and Guangzhou are the top manufacturing cities, their ratios of sulfur

Table 4.46 Industrial sulfur dioxide emissions per industrial output unit of the major cities in 2013

Rank	City	Industrial sulfur dioxide emissions per industrial output unit (tons/hundred million yuan)
1	Hohhot	643.25
2	Yinchuan	479.09
3	Guiyang	350.50
4	Taiyuan	335.62
5	Urumqi	305.56
6	Lanzhou	298.60
7	Kunming	259.50
8	Shijiazhuang	238.08
9	Harbin	194.12
10	Jinan	182.80
11	Xi'an	143.77
12	Nanning	129.23
13	Fuzhou	112.05
14	Ningbo	103.48
15	Shenyang	95.14
16	Dongguan	94.09
17	Wuhan	93.90
18	Dalian	89.34
19	Nanjing	88.09
20	Zhengzhou	87.32
21	Nanchang	85.31
22	Hangzhou	66.05
23	Changchun	62.04
24	Chengdu	56.74

(continued)

Table 4.46 (continued)

Rank	City	Industrial sulfur dioxide emissions per industrial output unit (tons/hundred million yuan)
25	Wuxi	55.94
26	Hefei	55.12
27	Suzhou	54.49
28	Qingdao	44.70
29	Xiamen	39.80
30	Guangzhou	36.82
31	Haikou	35.99
32	Changsha	25.54
33	Shenzhen	3.55

Data source: China City Statistical Yearbook 2014

dioxide emissions per manufacturing output unit were lower than other top manufacturing cities, which indicates that the local governments have paid more attention to pollution controls for industrial emissions and their environmental regulatory measures implemented were more effective. For such cities as Hohhot, Yinchuan, and Urumqi, urban industrial development lagged relative to other cities while their sulfur dioxide emissions per industrial output unit were higher, which should serve to alert them to devote more attention to resolving the pollution issue.

(5) Industrial solid waste incineration ranking

Manufacturing solid waste incineration reflects the level of air pollution caused by manufacturing in the region. Similarly, owingto limited data availability, we compiled a ranking of industrial solid waste incineration of the major cities, which, to a certain extent, reflects solid waste incineration generated by manufacturing in each major city. As shown in Table 4.47, Kunming was the city with the highest industrial solid waste incineration, followed by Taiyuan, Suzhou, Shijiazhuang, and Zhengzhou. On the other hand, Haikou, Changsha, Xiamen, Shenzhen, and Nanchanghad the lowest solid waste incineration. Compared to 2012, more than half of the cities reduced their urban industrial solid waste incineration in 2013, showing a good overall trend. Simultaneously, we noticed that industrial solid waste incineration in some cities increased, especially in Shijiazhuang (105.18%).

(6) Industrial solid waste incineration per industrial output unit ranking

The amount of industrial solid waste incineration per industrial output unit is the ratio of industrial solid waste incineration to industrial output, which reflects the environmental cost incurred by industrial production to the air. As observed from Table 4.48, Taiyuan had the highest industrial solid waste incineration per manufacturing output unit, followed by Kunming, Hohhot, Guiyang, and Urumqi, primarily due to the location of these provinces in the Midwest region, which is rich in mineral resources. In particular, black metal smelting and rolling processing,

Table 4.47 Industrial solid waste incineration of the major cities in 2013

Rank	City	Industrial solid waste incineration of 2012 (ten thousand tons)	Industrial solid waste incineration of 2013 (ten thousand tons)	Growth rate of industrial solid waste incineration (%)
1	Kunming	30,026.80	33,191.20	10.5386
2	Taiyuan	27,873.90	26,321.30	−5.5701
3	Suzhou	22,140.00	24,860.00	12.2855
4	Shijiazhuang	7594.90	15,583.20	105.1798
5	Zhengzhou	15,002.30	15,487.20	3.2322
6	Wuhan	13,813.00	13,841.40	0.2056
7	Urumqi	12,997.00	12,997.00	0.0000
8	Nanjing	16,159.50	12,530.40	−22.4580
9	Ningbo	12,466.30	12,466.30	0.0000
10	Guiyang	11,223.80	11,043.70	−1.6046
11	Wuxi	9220.00	10,518.90	14.0879
12	Hefei	10,768.50	10,241.00	−4.8985
13	Jinan	10,122.10	9323.90	−7.8857
14	Hohhot	11,219.60	8756.50	−21.9535
15	Qingdao	8534.50	8218.30	−3.7050
16	Fuzhou	7284.00	8096.30	11.1518
17	Shenyang	7037.50	7909.30	12.3879
18	Hangzhou	7068.40	6874.80	−2.7390
19	Yinchuan	7287.20	6856.90	−5.9049
20	Lanzhou	6278.80	6245.70	−0.5272
21	Changchun	4695.50	6060.20	29.0640
22	Harbin	5711.80	5745.30	0.5865
23	Guangzhou	6149.60	5555.60	−9.6592
24	Chengdu	5852.70	5333.50	−8.8711
25	Dalian	7066.90	5179.80	−26.7034
26	Nanning	3564.90	3963.20	11.1728
27	Xi'an	2582.20	2548.50	−1.3051
28	Nanchang	1863.20	2242.30	20.3467
29	Shenzhen	1052.30	1113.10	5.7778
30	Xiamen	1144.70	1028.80	−10.1249
31	Changsha	1035.10	1005.60	−2.8500
32	Haikou	61.50	64.10	4.2276
33	Dongguan	–	–	–

Data source: China Statistical Yearbook 2013 and 2014 of each city

Table 4.48 Industrial solid waste incineration per manufacturing unit output of the major cities in 2013

Rank	City	Industrial solid waste incineration per manufacturing output unit (tons/hundred million yuan)
1	Taiyuan	9.937
2	Kunming	8.472
3	Hohhot	5.856
4	Guiyang	5.483
5	Urumqi	5.351
6	Yinchuan	3.556
7	Dalian	3.404
8	Lanzhou	2.585
9	Jinan	2.101
10	Shijiazhuang	2.044
11	Harbin	1.69
12	Nanning	1.55
13	Hefei	1.361
14	Wuhan	1.332
15	Zhengzhou	1.274
16	Fuzhou	1.193
17	Nanjing	0.997
18	Ningbo	0.958
19	Suzhou	0.821
20	Wuxi	0.707
21	Changchun	0.657
22	Chengdu	0.582
23	Shenyang	0.576
24	Xi'an	0.567
25	Hangzhou	0.554
26	Qingdao	0.53
27	Nanchang	0.469
28	Dongguan	0.453
29	Guangzhou	0.323
30	Xiamen	0.218
31	Haikou	0.128
32	Changsha	0.121
33	Shenzhen	0.048

Data source: China Statistical Yearbook 2014 of each city

petroleum processing and coking, and chemical materials and products make up a large proportion of the overall industrial structure, which tends to generate serious industrial solid waste incineration pollutants.

(7) Comprehensive utilization rate of industrial solid waste incineration ranking

The comprehensive utilization rate of industrial solid waste incineration refers to the percentage of solid waste from which resources, energy, and other raw materials were extracted or transformed through recycling, processing, and exchange to the total amount of solid waste. It reflects the capabilities of manufacturing industries in solid waste recycling and utilization. According to Table 4.49, Changchun had the highest comprehensive utilization rate, followed by Chengdu, Jinan, Shijiazhuang, and Suzhou. While Suzhou is the highest ranked city by total manufacturing output, it nevertheless had an outstanding performance with regard to its comprehensive utilization of solid waste.

4.3.2 Methods and Indexes for the Evaluation of Strong Manufacturing Cities

1. Evaluation methods and data sources

The evaluation method adopted in this chapter for the analysis and ranking of the manufacturing industries in top cities is the same as that in Chap. 5, namely, the maximizing deviation method. At the same time, in order to ensure accurate evaluation results, the sample data were adopted from the Statistical Yearbook 2014 of each city and the China City Statistical Yearbook 2014.

2. Evaluation index system for China's 10 strongest manufacturing cities

The principle of index design in the evaluation of the top manufacturing cities was used in the principle of index establishment in the evaluation of manufacturing provinces. Both principles are consistent; the only difference lies in the selection of indexes. At the same time, since the statistical indexes and data of technological innovation and energy consumption were significantly incomplete at the city level, the assessment of the overall development capabilities of urban manufacturing was conducted via two areas: economic resourcefulness and environmental protection capability (Table 4.50).

(1) Economic resourcefulness

Economic resourcefulness is an integral part of the overall development capabilities of the urban manufacturing sector. A city can maintain sustainable development that supports technological innovation, efficiency improvement, and environmental protection only if economic resourcefulness brings economic benefits. Table 4.51 shows the 10 indexes that reflect the economic resourcefulness of a city.

A1 and A2 are production indexes that reflect the overall and relative size, respectively, of manufacturing output to total industrial output; A3 and A4 are

Table 4.49 Comprehensive utilization rate of manufacturing solid waste of the major cities in 2013

Rank	City	Comprehensive utilization rate of industrial solid waste (%)
1	Changchun	99.7900
2	Chengdu	99.0000
3	Jinan	98.7200
4	Shijiazhuang	98.6100
5	Suzhou	97.9000
6	Nanchang	97.8000
7	Lanzhou	97.4000
8	Xi'an	95.4300
9	Guangzhou	95.1700
10	Wuhan	95.0000
11	Qingdao	94.8700
12	Nanning	94.6400
13	Fuzhou	94.3200
14	Xiamen	94.1800
15	Hangzhou	94.0000
16	Harbin	93.8500
17	Haikou	93.7600
18	Hefei	93.2700
19	Shenyang	92.6900
20	Nanjing	91.2000
21	Wuxi	91.0000
22	Dalian	90.3300
23	Ningbo	90.0600
24	Urumqi	87.6400
25	Changsha	85.6700
26	Yinchuan	84.7800
27	Dongguan	78.9400
28	Shenzhen	78.6900
29	Zhengzhou	73.5500
30	Guiyang	60.7500
31	Taiyuan	54.5100
32	Kunming	40.9000
33	Hohhot	35.7400

Data source: China Statistical Yearbook 2014 of each city

profit indexes that reflect the total profit of the manufacturing sector and the profit margin per manufacturing worker, respectively; A5 is an efficiency index that reflects the efficiency level of labor; A6 is a market index that reflects the extent of needs met by the manufacturing sector in society; A7 and A8 are employment indexes that reflect the number of people employed in manufacturing and the ratio

Table 4.50 Evaluation index system for overall development capabilities of urban manufacturing industries

Overall index	Serial no.	Main index	Serial no.	Sub index	
Index system of strong manufacturing cities	A		A1	Output	Total manufacturing output
			A2		Ratio of total manufacturing output to total industrial output
			A3	Profit	Total profits of manufacturing enterprises
			A4		Profit margin per manufacturing worker
			A5	Efficiency	Labor productivity rate per manufacturing worker
		Economic resourcefulness	A6	Market	Manufacturing product sales rate
			A7	Employment	Number of manufacturing workers
			A8		Ratio of number of manufacturing workers to total number of urban workers
			A9	Taxes	Total taxes and profits of manufacturing enterprises
			A10		Average income tax rate of manufacturing worker
	B	Environmental protection capability	B1	Wastewater	Wastewater emissions from manufacturing industries
			B2		Wastewater emissions per

(continued)

Table 4.50 (continued)

Overall index	Serial no.	Main index	Serial no.	Sub index	
					manufacturing output unit
			B3	Waste gas	Sulfur dioxide emissions from manufacturing industries
			B4		Sulfur dioxide emissions per manufacturing output unit
			B5	Solid	Solid waste incineration from manufacturing industries
			B6		Solid waste incineration per manufacturing
			B7	Overall	Comprehensive utilization rate of manufacturing solid waste

Table 4.51 List of indexes for economic resourcefulness of urban manufacturing industries

Serial No.	List of indexes for economic resourcefulness	
A1	Output	Total manufacturing output (hundred million yuan)
A2		Ratio of total manufacturing output to total industrial output (%)
A3	Profit	Total profits of manufacturing enterprises (hundred million yuan)
A4		Profit margin per manufacturing worker(ten thousand yuan/person)
A5	Efficiency	Labor productivity rate per manufacturing worker (ten thousand yuan/person)
A6	Market	Manufacturing product sales rate (%)
A7	Employment	Number of manufacturing workers (people)
A8		Ratio of number of manufacturing workers to number of urban workers (%)
A9	Taxes	Total taxes and profits of manufacturing enterprises (hundred million yuan)
A10		Average income tax rate of manufacturing worker (ten thousand yuan/person)

of this number to the total number of people employed in the city, respectively; and A9 and A10 are tax indexes that reflect the total taxes and profits of urban manufacturing enterprises and the average taxes paid by manufacturing workers.

Other ratios available are as follows.

Profit margin per worker = Total profits of manufacturing/Total employed in manufacturing
Labor productivity rate = Total manufacturing output/Total number of manufacturing workers
Manufacturing product sales rate = Total manufacturing sales/Total manufacturing output

(1) Environmental protection capability

The environmental protection capability index primarily reflects the abilities of urban manufacturing in managing pollution emissions and controlling pollution during the manufacturing process. Table 4.52 lists the seven evaluation indexes for assessing the environmental protection capability of the manufacturing sector. B1–B6 are the emissions indexes of the three types of waste. Among them, B1 and B2 reflect the status of wastewater emissions of manufacturing enterprises; B3 and B4 reflect the status of waste gas emissions of manufacturing enterprises; and B5 and B6 reflect the status of solid waste incineration of manufacturing enterprises. B7 is the index of pollution control that reflects the status of comprehensive utilization of solid waste.

Table 4.52 List of indexes for environmental protection capabilities of urban manufacturing industries

Serial No.	List of indexes for environmental protection	
B1	Wastewater	Manufacturing wastewater emissions (ten thousand tons)
B2		Wastewater emissions per manufacturing output unit (ten thousand tons/hundred million yuan)
B3	Waste gas	Manufacturing sulfur dioxide emissions (tons)
B4		Sulfur dioxide emissions per manufacturing output unit (ton/hundred million yuan)
B5	Solid waste	Manufacturing solid waste incineration (ten thousand tons)
B6		Wastewater emissions per manufacturing output unit (ten thousand ton/hundred million yuan)
B7	Overall	Comprehensive utilization rate of manufacturing solid waste

4.3.3 Ranking of Evaluation Indexes for China's Strong Manufacturing Cities

Different aspects were employed for conducting the individual evaluation of the economic resourcefulness and environmental protection capability of the 30 sample cities.

1. Economic resourcefulness ranking

A1–A10 were used as base indexes to perform a comprehensive evaluation of the economic resourcefulness of urban manufacturing industries. The maximizing deviation method was employed; weights were assigned to the 10 indexes for economic resourcefulness of manufacturing industries and in conjunction with the standardized value of each index, the 2013 comprehensive evaluation values of economic resourcefulness of China's manufacturing cities were generated (see Table 4.53).

As observed from Table 4.53, the top 10 manufacturing cities of economic resourcefulness were Suzhou, Shenyang, Nanjing, Qingdao, Changchun, Shenzhen, Guangzhou, Changsha, Zhengzhou, and Hangzhou. Compared to the ranking in previous years, the following characteristics are shown in the 2013 ranking.

First, by regional classification, 6 of the 2013 Top 10 manufacturing cities in terms of economic resourcefulness were located in the eastern region, two in the central region, two in the northeastern region,[7] and none in the western region. This indicates that there are still significant regional differences in economic resourcefulness among China's manufacturing industries.

Second, both Shenyang and Changchun were ranked in the top five in terms of economic resourcefulness. Among the two, Shenyang was ranked second on the list. Although Jilin regressed in the ranking of provinces, the ranking of its capital, Changchun climbed to fifth place, which explains why regional manufacturing development is generally focused on central cities. In addition, due to the relatively smaller labor base in manufacturing in Shenyang and Changchun compared with other cities, the profit margin per manufacturing worker and labor productivity rate of manufacturing workers in these two cities had a significant efficiency advantage.

Third, Jiangsu province maintained the economic resourcefulness advantage in its manufacturing sector with two of its cities (Suzhou and Nanjing) ranked in the top three. Suzhou was ranked 1st and Nanjing 3rd while Wuxi was ranked 11th. When considering the scale of manufacturing in the three cities, we observe that the

[7]According to the latest classification of China's National Bureau of Statistics, the eastern region includes the following 10 provinces: Beijing, Tianjin, Hebei, Shanghai, Jiangsu, Zhejiang, Fujian, Shandong, Guangdong, and Hainan; the central region includes the following 6 provinces: Shanxi, Anhui, Jiangxi, Henan, Hubei, and Hunan; the western region includes the following 12 provinces: Inner Mongolia, Guangxi, Chongqing, Sichuan, Guizhou, Yunnan, Tibet, Shaanxi, Gansu, Qinghai, Ningxia, and Xinjiang; and the northeastern region includes the following 3 provinces: Liaoning, Jilin, and Heilongjiang.

Table 4.53 2013 Comprehensive evaluation of economic resourcefulness of urban manufacturing industries

Rank	City	Score
1	Suzhou	0.7112
2	Shenyang	0.6379
3	Nanjing	0.6354
4	Qingdao	0.6232
5	Changchun	0.594
6	Shenzhen	0.5904
7	Guangzhou	0.5804
8	Changsha	0.5518
9	Zhengzhou	0.5242
10	Hangzhou	0.5097
11	Wuxi	0.5000
12	Ningbo	0.4579
13	Hefei	0.4329
14	Wuhan	0.4312
15	Hohhot	0.3714
16	Chengdu	0.3619
17	Fuzhou	0.3563
18	Nanchang	0.2859
19	Jinan	0.2796
20	Nanning	0.2759
21	Xiamen	0.2483
22	Harbin	0.2188
23	Urumqi	0.2157
24	Kunming	0.2017
25	Guiyang	0.2015
26	Yinchuan	0.1963
27	Xi'an	0.1816
28	Lanzhou	0.1793
29	Haikou	0.1613
30	Taiyuan	0.1336

total manufacturing output has accounted for more than 97% of the total industrial output whereas the total manufacturing output in Wuxi reached 98.4% of its total industrial output. The industrial systems of these three cities are typically focused on manufacturing. Among them, Suzhou's total manufacturing profits was ranked the highest among the sample cities, showing strong profitability (Fig. 4.8).

1. Ranking of environmental protection capability

B1–B7 were employed as base indexes to perform a comprehensive evaluation of environmental protection capability of urban manufacturing industries. The maximizing deviation method was employed; weights were assigned to the seven indexes for environmental protection capability of manufacturing industries; and in

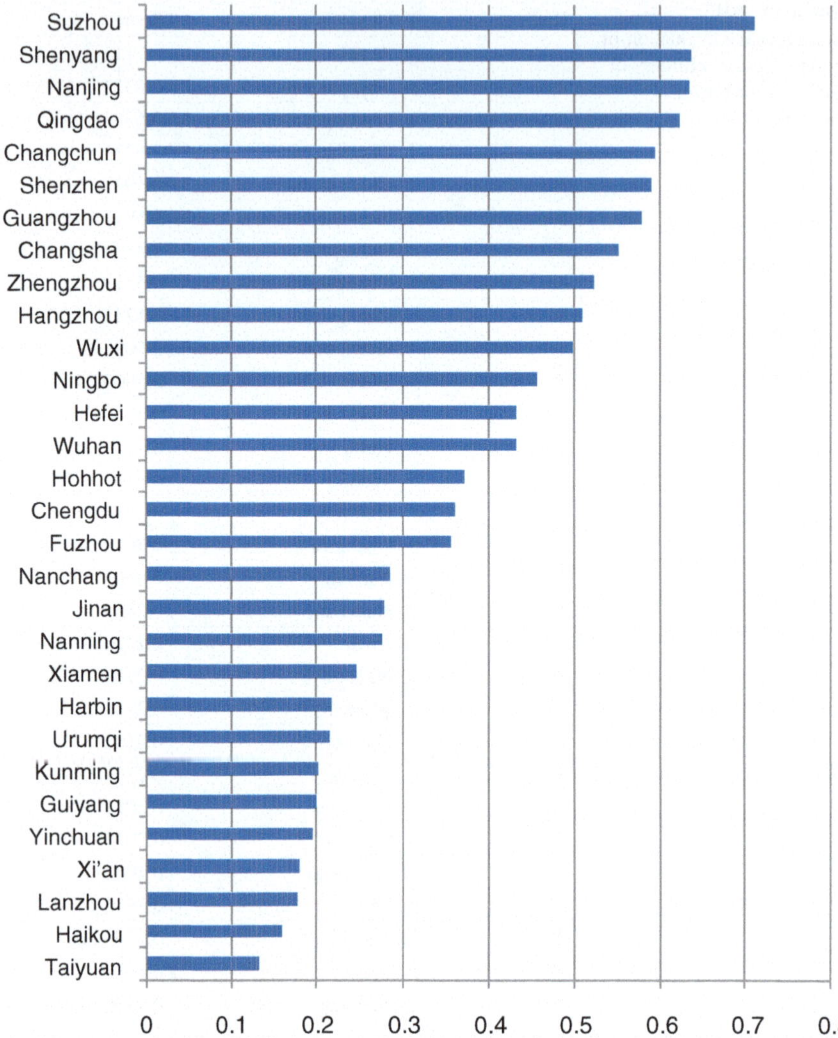

Fig. 4.8 Comprehensive evaluation of economic resourcefulness of urban manufacturing industries

conjunction with the standardized value of each index, the 2013 comprehensive evaluation values on environmental protection capability of manufacturing cities in China were generated (see Table 4.54).

As shown in Table 4.54, the top 10 manufacturing cities in terms of environmental protection capability were Shenzhen, Changsha, Haikou, Changchun, Hefei, Chengdu, Qingdao, Guangzhou, Fuzhou, and Xi'an. The ranking of the 2013 environmental protection capability of manufacturing cities reveals the following.

Table 4.54 2013 Comprehensive evaluation on environmental protection capability of urban manufacturing industries

Rank	City	Score
1	Shenzhen	0.8693
2	Changsha	0.8522
3	Haikou	0.8284
4	Changchun	0.7442
5	Hefei	0.7396
6	Chengdu	0.7323
7	Qingdao	0.7289
8	Guangzhou	0.7114
9	Fuzhou	0.7113
10	Xi'an	0.7035
11	Harbin	0.6903
12	Zhengzhou	0.6832
13	Nanchang	0.6779
14	Shenyang	0.6748
15	Dongguan	0.6728
16	Jinan	0.6707
17	Nanning	0.667
18	Wuxi	0.6644
19	Guiyang	0.6578
20	Xiamen	0.6489
21	Wuhan	0.6297
22	Hohhot	0.6263
23	Lanzhou	0.6191
24	Ningbo	0.6058
25	Nanjing	0.6033
26	Hangzhou	0.6027
27	Dalian	0.6002
28	Urumqi	0.568
29	Kunming	0.5467
30	Yinchuan	0.5404
31	Taiyuan	0.5118
32	Shijiazhuang	0.4148
33	Suzhou	0.4122

First, among the top 10 manufacturing cities by environmental protection capability, 5 were in the eastern region, 2 each in the central and western regions, and 1 in the northeastern region. Overall, due to the relatively larger scale of the manufacturing sector in the eastern region, the total emissions of the three types of waste were higher. However, the emissions intensity of the three types of waste was relatively low.

Second, similar to 2012, the environmental protection capabilities of three cities in Jiangsu were still ranked near the bottom of the list. Among them, Wuxi was ranked 18th, Nanjing 25th, and Suzhou last, which strongly contrasts their economic resourcefulness in manufacturing (Suzhou was ranked 2nd, Nanjing 3rd, and Wuxi 6th). Jiangsu's environmental pollution as the price incurred by manufacturing development remains unresolved. With traditionally big manufacturing cities, Jiangsu province continues to face enormous pressure from the structural adjustment upgrade of its manufacturing sector and industrial pollution control.

Finally, Suzhou and Shenzhen, both cities with traditionally strong economies in China, performed significantly differently in terms of environmental protection. Shenzhen ranked first in environmental protection capability whereas Suzhou ranked last. Comparing the various environmental protection indexes, it is observed that Suzhou faces enormous pressure in both wastewater emissions and total industrial sulfur dioxide emissions. Suzhou ranked first in wastewater emissions whereas Shenzhen performed well in total emissions of the three types of waste and the emissions intensity of the three types of waste. Although both cities are strong manufacturing cities, Suzhou indeed has paid a much higher environmental price than Shenzhen (see Fig. 4.9).

4.3.4 Ranking of the Top 10 Manufacturing Cities in China

Owing to the lack of a technological innovation index and energy-related indexes, the list of top 10 manufacturing cities in 2013 is an overall ranking based on comprehensive considerations of economic resourcefulness and environmental protection capability of the 30 cities.

As observed from Table 4.55, the top 10 manufacturing cities with the strongest overall development capabilities were Shenzhen, Qingdao, Changchun, Changsha, Shenyang, Guangzhou, Nanjing, Suzhou, Wuxi, and Hefei.

First, like in 2012, cities in the eastern region continued to maintain their traditional manufacturing advantage whereas cities in the central and western regions still lagged in urban manufacturing development. Among the top 10 cities of overall urban manufacturing development capabilities, 6 were in the eastern region, 2 each were in the central and northeastern regions, and none were in the western region. Significant differences exist in overall urban manufacturing development capabilities across regions

Second, among the top 10 cities, 3 were in the Yangtze River Delta (Nanjing, Suzhou, and Wuxi), 2 were in the Pearl River Delta (Shenzhen and Guangzhou), and 2 were in the northeastern region (Changchun and Shenyang). Comparing the traditional advantages of the Yangtze River Delta and Pearl River Delta, the economic resourcefulness of urban manufacturing in the northeastern region was more effective, showing a significantly higher advantage in profit margin per manufacturing worker and labor productivity rate. On the other hand, the technological innovation capability in the northeastern region was weaker. Due to unavailability of technology-related indexes, technological considerations were not included in

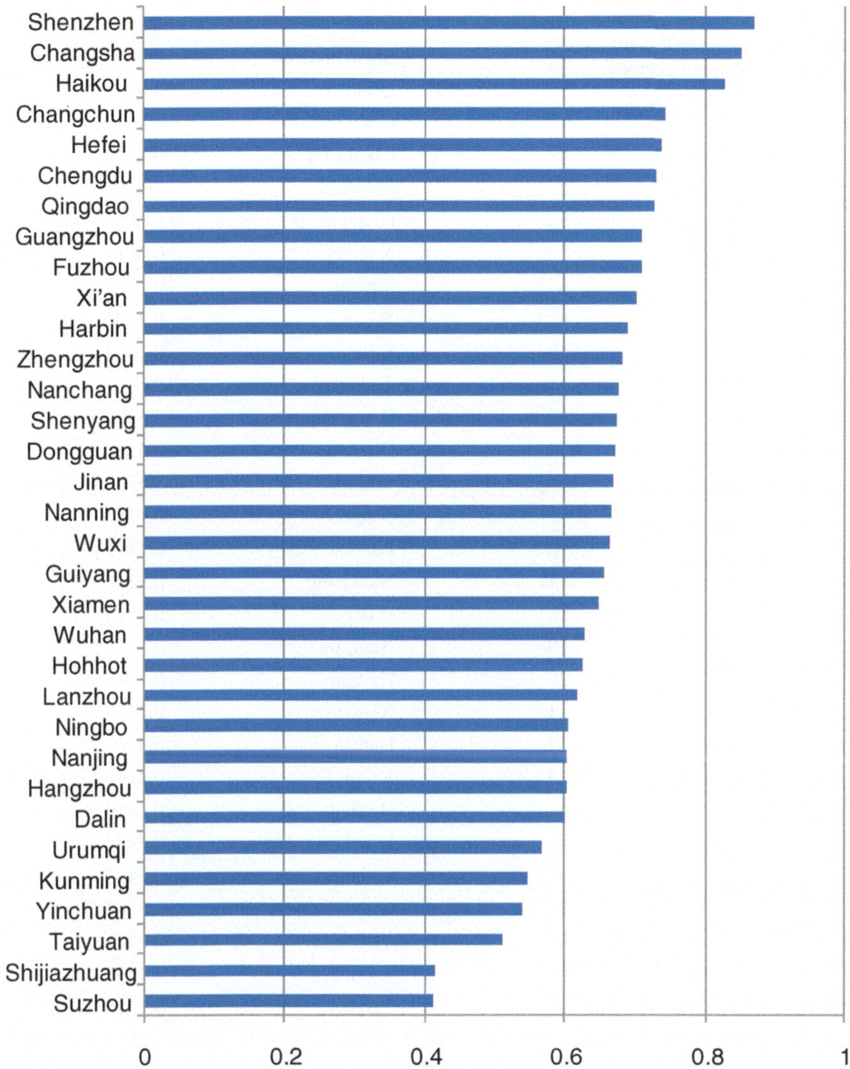

Fig. 4.9 Comprehensive evaluation of environmental protection capability of urban manufacturing industries

the overall ranking, and therefore, the overall ranking of the northeastern region was raised to a certain extent.

Third, all of the three sample cities in Jiangsu province (Nanjing, Suzhou, and Wuxi) were in the top 10 cities of overall manufacturing development capabilities, ranking 7th, 8th, and 9th, respectively. This indicates that the overall development

Table 4.55 Comprehensive evaluation of capabilities of urban manufacturing sector

Rank	City	Score
1	Shenzhen	0.7153
2	Qingdao	0.7083
3	Changchun	0.7049
4	Changsha	0.6964
5	Shenyang	0.6906
6	Guangzhou	0.6767
7	Nanjing	0.6611
8	Suzhou	0.6469
9	Wuxi	0.6009
10	Hefei	0.5923
11	Zhengzhou	0.5905
12	Hangzhou	0.5901
13	Chengdu	0.5558
14	Wuhan	0.5509
15	Ningbo	0.5487
16	Fuzhou	0.5345
17	Nanchang	0.4836
18	Jinan	0.4825
19	Nanning	0.4713
20	Haikou	0.4605
21	Xiamen	0.4489
22	Harbin	0.4409
23	Xi'an	0.4265
24	Hohhot	0.4059
25	Lanzhou	0.3952
26	Urumqi	0.3807
27	Guiyang	0.3582
28	Yinchuan	0.3529
29	Kunming	0.2795
30	Taiyuan	0.2484

capabilities of urban manufacturing in Jiangsu were relatively strong. At the same time, however, the environmental pollution problems caused by urban manufacturing in Jiangsu were more serious than in other cities in the sample because it is a traditionally big manufacturing province. If Jiangsu province aims to achieve further urban manufacturing development, it must learn from the experiences of such cities as Shenzhen about how to handle environmental pollution problems arising from urban manufacturing development (see Fig. 4.10).

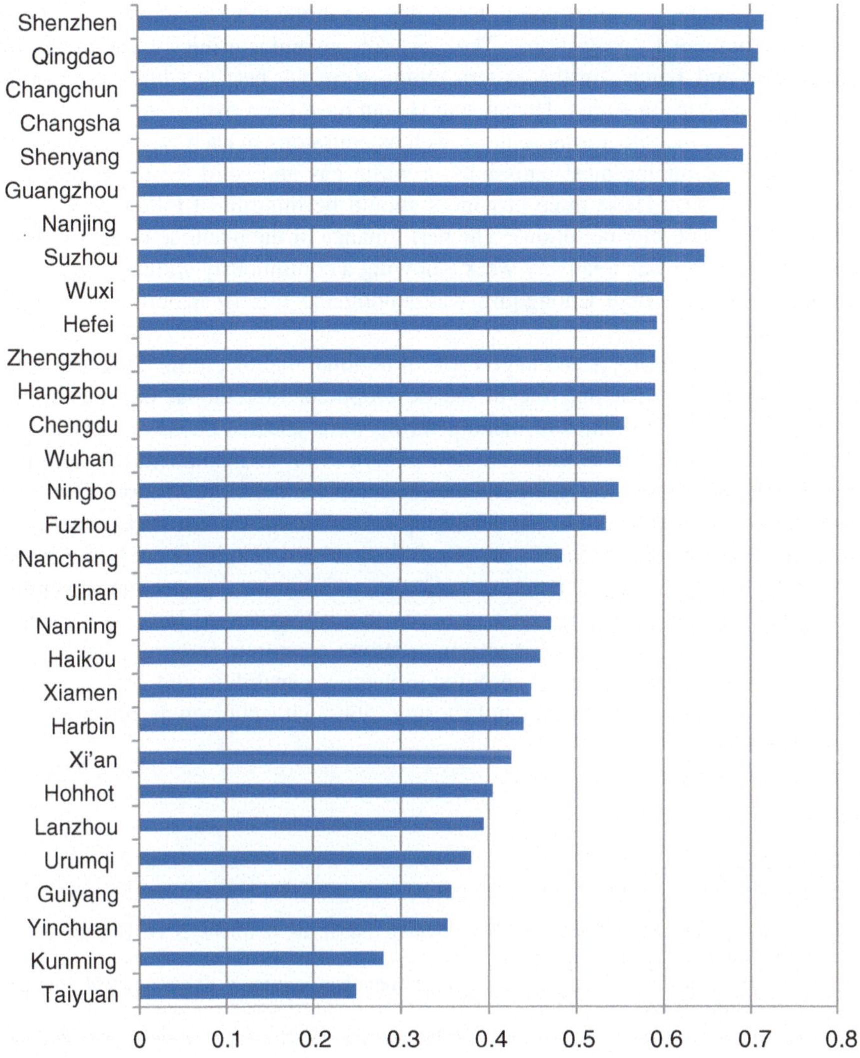

Fig. 4.10 Overall capability evaluation of urban manufacturing industries

4.4 Chapter Summary

Analyzing manufacturing development from a regional perspective is conducive for summarizing regional development experiences, and exploring and identifying development patterns in regional manufacturing development for leading regions to define reasonable positioning and set clear directions for sustainable development. This chapter employed objective data for regional manufacturing industries as the foundation for evaluating a new pattern of regional manufacturing from four areas namely, economic resourcefulness, technological innovation, resource utilization, and environmental protection. The overall manufacturing capabilities of Guangdong and Jiangsu in the eastern region were the best in China; the manufacturing industries in Anhui, Henan, and Hunan have been rising rapidly, showing rapid growth in manufacturing scale as well as emissions in each type of pollutant, and accounting for the most emissions in waste gas and solid waste incineration among all regions. These three provinces should be mindful of following the old path of "treatment after pollution." The performance of the manufacturing sector in the western region was relatively weak, showing a continuously widening gap with developed regions; only Chongqing was among the top 10 in terms of overall performance.

Urban manufacturing is an integral part of regional manufacturing industries and has a radiating and leading role in the development of regional manufacturing. A sample of 30 central cities of manufacturing were selected based on availability of manufacturing data to evaluate the new pattern of urban manufacturing from the perspectives of economic resourcefulness and environmental protection. The economic resourcefulness of the manufacturing sector in the eastern region was outstanding; in particular, in Jiangsu province, Suzhou and Nanjing were both ranked among the top three cities. On the contrary, the environmental protection capability of Jiangsu province is a cause for concern in that it was ranked last. On the other hand, Shenzhen, a manufacturing city with strong economic resourcefulness, also had outstanding performance in resource protection capability, and could provide lessons for Suzhou and other manufacturing cities with poor environmental protection capability.

References

Beijing Municipal Bureau of Statistics. 2014. *Beijing Statistical Yearbook 2014*. Beijing: China Statistics Press.

Changchun Municipal Bureau of Statistics. 2014. *Changchun Statistical Yearbook 2014*. Beijing: China Statistics Press.

Changsha Municipal Bureau of Statistics. 2014. *Changsha Statistical Yearbook 2014*. Beijing: China Statistics Press.

Chengdu Municipal Bureau of Statistics. 2014. *Chengdu Statistical Yearbook 2014*. Beijing: China Statistics Press.

Chongqing Municipal Bureau of Statistics. 2014. *Chongqing Statistical Yearbook 2014*. Beijing: China Statistics Press.

Dalian Municipal Bureau of Statistics. 2014. *Dalian Statistical Yearbook 2014*. Beijing: China Statistics Press.

Fujian Provincial Bureau of Statistics. 2014. *Fujian Statistical Yearbook 2014*. Beijing: China Statistics Press.

Fuzhou Municipal Bureau of Statistics. 2014a. *Fuzhou Statistical Yearbook 2014*. Beijing: China Statistics Press.

Fuzhou Municipal Bureau of Statistics. 2014b. *Fuzhou Statistical Yearbook 2014*. Beijing: China Statistics Press.

Gansu Provincial Bureau of Statistics. 2014. *Gansu Statistical Yearbook 2014*. Beijing: China Statistics Press.

Guangxi Bureau of Statistics. 2014. *Guangxi Statistical Yearbook 2014*. Beijing: China Statistics Press.

Guangzhou Municipal Bureau of Statistics. 2014. *Guangzhou Statistical Yearbook 2014*. Beijing: China Statistics Press.

Guiyang Municipal Bureau of Statistics. 2014. *Guiyang Statistical Yearbook 2014*. Beijing: China Statistics Press.

Guizhou Bureau of Statistics. 2014. *Guizhou Statistical Yearbook 2014*. Beijing: China Statistics Press.

Hainan Provincial Bureau of Statistics. 2014. *Hainan Statistical Yearbook 2014*. Beijing: China Statistics Press.

Hangzhou Municipal Bureau of Statistics. 2014. *Hangzhou Statistical Yearbook 2014*. Beijing: China Statistics Press.

Harbin Municipal Bureau of Statistics. 2014. *Harbin Statistical Yearbook 2014*. Beijing: China Statistics Press.

Hebei Provincial Bureau of Statistics. 2014. *Hebei Statistical Yearbook 2014*. Beijing: China Statistics Press.

Heilongjiang Provincial Bureau of Statistics. 2014. *Heilongjiang Statistical Yearbook 2014*. Beijing: China Statistics Press.

Henan Bureau of Statistics. 2014. *Henan Statistical Yearbook 2014*. Beijing: China Statistics Press.

Hohhot Municipal Bureau of Statistics. 2014. *Hohhot Statistical Yearbook 2014*. Beijing: China Statistics Press.

Hubei Bureau of Statistics Information Center. 2014. *Hubei Statistical Yearbook 2014*. Beijing: China Statistics Press.

Hunan Provincial Bureau of Statistics. 2014. *Hunan Statistical Yearbook 2014*. Beijing: China Statistics Press.

Inner Mongolia Provincial Bureau of Statistics. 2014. *Inner Mongolia Statistical Yearbook 2014*. Beijing: China Statistics Press.

Jiangsu Provincial Bureau of Statistics. 2014. *Jiangsu Statistical Yearbook 2014*. Beijing: China Statistics Press.

Jiangxi Bureau of Statistics. 2014. *Jiangxi Statistical Yearbook 2014*. Beijing: China Statistics Press.

Jilin Bureau of Statistics. 2014. *Hunan Statistical Yearbook 2014*. Beijing: China Statistics Press.

Jinan Municipal Bureau of Statistics. 2014. *Jinan Statistical Yearbook 2014*. Beijing: China Statistics Press.

Kunming Bureau of Statistics. 2014. *Kunming Statistical Yearbook 2014*. Beijing: China Statistics Press.

Lanzhou Bureau of Statistics. 2014. *Lanzhou Statistical Yearbook 2014*. Beijing: China Statistics Press.

Li, Lianshui, and Du Zhanyuan. 2004. *A research report on the development of China's Manufacturing Sector 2004*. Beijing: Science Press.

Li, Lianshui, and Du Zhanyuan. 2005. *A research report on the development of China's Manufacturing Sector 2005*. Beijing: Science Press.

Li, Lianshui, and Du Zhanyuan. 2006. *A research report on the development of China's Manufacturing Sector 2006*. Beijing: Science Press.

Li, Lianshui, and Du Zhanyuan. 2007. *A research report on the development of China's Manufacturing Sector 2007*. Beijing: Science Press.

Li, Lianshui, and Du Zhanyuan. 2008. *A research report on the development of China's Manufacturing Sector 2008*. Beijing: Science Press.

Li, Lianshui, and Du Zhanyuan. 2009. *A research report on the development of China's Manufacturing Sector 2009*. Beijing: Science Press.

Li, Lianshui, and Du Zhanyuan. 2010. *A research report on the development of China's Manufacturing Sector 2010*. Beijing: Science Press.
Li, Lianshui, and Du Zhanyuan. 2011. *A research report on the development of China's Manufacturing Sector 2011*. Beijing: Science Press.
Li, Lianshui, and Du Zhanyuan. 2012. *A research report on the development of China's Manufacturing Sector 2012*. Beijing: Science Press.
Li, Lianshui, and Du Zhanyuan. 2013. *A research report on the development of China's Manufacturing Sector 2013*. Beijing: Science Press.
Liaoning Provincial Bureau of Statistics. 2014. *Liaoning Statistical Yearbook 2014*. Beijing: China Statistics Press.
Nanchang Municipal Bureau of Statistics. 2014. *Nanchang Statistical Yearbook 2014*. Beijing: China Statistics Press.
Nanjing Municipal Bureau of Statistics. 2014. *Nanjing Statistical Yearbook 2014*. Beijing: China Statistics Press.
Nanning Municipal Bureau of Statistics. 2014. *Nanning Statistical Yearbook 2014*. Beijing: China Statistics Press.
National Bureau of Statistics. 2014a. *China Statistical Yearbook on Science and Technology 2014*. Beijing: China Statistics Press.
National Bureau of Statistics. 2014b. *China Statistical Yearbook 2014*. Beijing: China Statistics Press.
National Bureau of Statistics. 2014c. *China Statistical Yearbook 2014*. Beijing: China Statistics Press.
Ningbo Municipal Bureau of Statistics. 2014. *Ningbo Statistical Yearbook 2014*. Beijing: China Statistics Press.
Qingdao Municipal Bureau of Statistics. 2014. *Qingdao Statistical Yearbook 2014*. Beijing: China Statistics Press.
Qinghai Bureau of Statistics. 2014. *Qinghai Statistical Yearbook 2014*. Beijing: China Statistics Press.
Shaanxi Provincial Bureau of Statistics. 2014. *Shaanxi Statistical Yearbook 2014*. Beijing: China Statistics Press.
Shandong Provincial Bureau of Statistics. 2014. *Shandong Statistical Yearbook 2014*. Beijing: China Statistics Press.
Shanghai Bureau of Statistics. 2014. *Shanghai Statistical Yearbook 2014*. Beijing: China Statistics Press.
Shanxi Statistics Bureau. 2014. *Shanxi Statistical Yearbook 2014*. Beijing: China Statistics Press.
Shenyang Bureau of Statistics. 2014. *Shenyang Statistical Yearbook 2014*. Beijing: China Statistics Press.
Shenzhen Bureau of Statistics. 2014. *Shenzhen Statistical Yearbook 2014*. Beijing: China Statistics Press.
Shijiazhuang Bureau of Statistics. 2014. *Shijiazhuang Statistical Yearbook 2014*. Beijing: China Statistics Press.
Sichuan Statistics Bureau. 2014. *Sichuan Statistical Yearbook 2014*. Beijing: China Statistics Press.
Statistical Bureau of Anhui Province. 2014. *Anhui Statistical Yearbook 2014*. Beijing: China Statistics Press.
Statistical Bureau of Guangdong Province. 2014. *Guangdong Statistical Yearbook 2014*. Beijing: China Statistics Press.
Statistical Bureau of Ningxia Hui Autonomous Region. 2014. *Ningxia Statistical Yearbook 2014*. Beijing: China Statistics Press.
Statistical Bureau of Xinjiang Uygur Autonomous Region. 2014. *Xinjiang Statistical Yearbook 2014*. Beijing: China Statistics Press.
Suzhou Bureau of Statistics. 2014. *Suzhou Statistical Yearbook 2014*. Beijing: China Statistics Press.

Taiyuan Municipal Bureau of Statistics. 2014. *Taiyuan Statistical Yearbook 2014*. Beijing: China Statistics Press.
Tianjin Municipality Bureau of Statistics. 2014. *Tianjin Statistical Yearbook 2014*. Beijing: China Statistics Press.
Tibet Statistical Bureau. 2014. *Tibet Yearbook 2014*. Beijing: China Statistics Press.
Urumqi Municipal Bureau of Statistics. 2014. *Urumqi Statistical Yearbook 2014*. Beijing: China Statistics Press.
Wuhan Municipal Bureau of Statistics. 2014. *Wuhan Statistical Yearbook 2014*. Beijing: China Statistics Press.
Wuxi Municipal Bureau of Statistics. 2014. *Wuxi Statistical Yearbook 2014*. Beijing: China Statistics Press.
Xi'an Municipal Bureau of Statistics. 2014. *Xi'an Statistical Yearbook 2014*. Beijing: China Statistics Press.
Xiamen Municipal Bureau of Statistics. 2014. *Xiamen Statistical Yearbook 2014*. Beijing: China Statistics Press.
Yinchuan Municipal Bureau of Statistics. 2014. *Yinchuan Statistical Yearbook 2014*. Beijing: China Statistics Press.
Zhejiang Provincial Bureau of Statistics. 2014. *Zhejiang Statistical Yearbook 2014*. Beijing: China Statistics Press.
Zhengzhou Municipal Bureau of Statistics. 2014. *Zhengzhou Statistical Yearbook 2014*. Beijing: China Statistics Press.

Chapter 5
Development of China's Manufacturing Sector: Industry Research

Jun Liu, Zhonghua Cheng and Nian Zhong

This chapter is divided into six sections: The first analyzes the development status of China's manufacturing sector and evaluates new models. The second section covers the impact of innovations in science and technology on China's manufacturing sector. The third section discusses energy efficiency in China's manufacturing sector and its impact factors. The fourth section analyzes the impact of informationization on the green growth of China's manufacturing sector. The fifth section contains an impact study of industrial clustering on technological innovation in China's manufacturing sector. The sixth section is a chapter summary.

5.1 Status of Development of China's Manufacturing Sector and Evaluation of New Industry Model

The overall development status of China's manufacturing sector is analyzed in this section according to the requirements of new manufacturing development models and based on the latest industry classification of manufacturing in conjunction (see Chap. 5 for details). In addition, this section evaluates the new industry model that employs the four dimensions of economic resourcefulness, technological innovation capability, energy efficiency, and environmental pollution.`

5.1.1 Manufacturing Economic Resourcefulness Analysis

The most important index to measure the manufacturing sector is economic resourcefulness. Economic effectiveness is the issue with which enterprises and

J. Liu (✉) · Z. Cheng · N. Zhong
Nanjing University of Information Science and Technology, Nanjing, China
e-mail: llsh@nuist.edu.cn

L. Li and Z. Du (eds.), *A Research Report on the Development of China's Manufacturing Sector (2016)*, Current Chinese Economic Report Series,
DOI 10.1007/978-981-10-4445-8_5

Table 5.1 Indexes related to industrial scale in manufacturing

Name of index	Total value of 2012 (billion yuan)	Total value of 2013 (billion yuan)	Growth rate (%)
Total manufacturing output	80692.32	87046.40	7.87
Revenues from primary manufacturing business	80566.23	90194.15	11.95
Total manufacturing profits	4857.05	5070.57	4.40

Data source China Statistical Yearbook of 2013 and 2014, China Statistics Press

industrial development are most concerned. In order to determine the development of the economic resourcefulness of the manufacturing sector, it is analyzed in this section from various perspectives of economic development, such as industry scale, total profits, and total output growth.

1. Scale analysis of China's manufacturing industries

   (1) Analysis of the overall scale

To understand the development of manufacturing industries, growth in industrial scale has to be considered as the two are interlinked. In this section, the development scale of China's manufacturing industries is analyzed based on three indexes, namely, total industrial output, revenues from primary business, and total profits, as shown in Table 5.1.

As Table 5.1 shows, China's total manufacturing output was 87.046404 trillion yuan in 2013, an increase of 7.87% from 2012. Revenues from primary manufacturing business were 90.194151 trillion yuan, an increase of 11.95% from 2012. Total manufacturing profits were 50.70569 trillion yuan, an increase of 4.4% from 2012. Figure 5.1 shows China's total manufacturing output and annual growth between 2003 and 2013, indicating a general upward trend in China's total manufacturing output during the period. However, due to the global financial crisis in 2008, the annual growth rate in 2009 fell significantly from that in 2008 but rebounded vigorously thereafter. However, the growth rate of China's total manufacturing output significantly dropped from 2010 (see Fig. 5.1).

(2) Industry-by-industry analysis

First, the classification of manufacturing industries was described. According to the requirements of the "National Bureau of Statistics notice on the implementation of the new National Industry classification Standards," the new "National Industry classification Standard" (GB/T4755-2011) was implemented uniformly. This standard[1] adopted a linear classification method and a hierarchical coding method to classify national industries into types of industries, using the three sizes of big, medium, and small, represented by codes made up of one Latin letter and four Arabic numerals. The industry type was represented by one Latin letter, that is, A,

[1] National Bureau of Statistics, "National Industry classification Standards" (GB/T4745-2011).

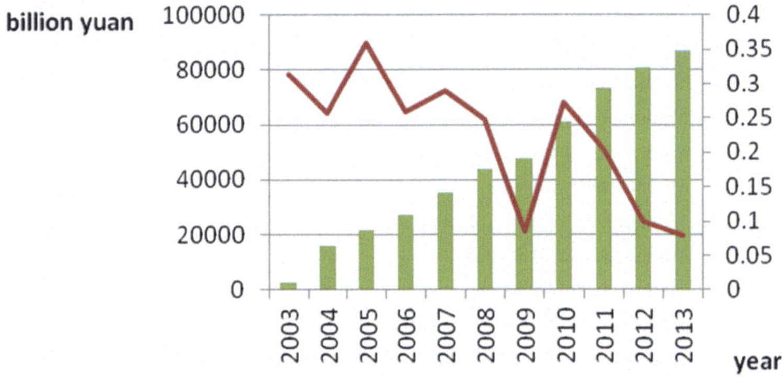

Fig. 5.1 China's total manufacturing output of "above-scale" industries and their respective growth rates in 2003–2013 by industry classification. *Data source* Sorted and calculated from related data in the China Statistical Yearbooks of 2004–2014

B, C, and so on. Then, big industry was represented by two Arabic numerals starting with 01; medium industry was represented by three Arabic numerals, with the first two digits signifying big and the third digit signifying medium; and small industry was represented by four Arabic numerals, with the first three digits signifying medium and the last digit signifying small. According to this standard, China's manufacturing sector is of type C, which is comprised of 31 different industries that are big by definition. This standard was adopted for the classification of manufacturing industries in this chapter, as shown in Table 5.2.

Since total industrial output by industry was not released in the China Statistical Yearbook 2014, the index of revenues from primary business was employed in this chapter using data from the China Statistical Yearbook 2014. Table 5.3 lists the 2013 ranking of revenues from primary businesses in above-scale industries by industry classification in China (Fig. 5.2).

We can observe that the top 10 above-scale manufacturing industries with the highest revenues from primary business were manufacture of computers, communication, and other electronic equipment; manufacture of raw chemical materials and chemical products; smelting and pressing of ferrous metals; manufacture of electrical machinery and equipment; manufacture of automobiles; processing of food from agricultural products; manufacture of non-metallic mineral products; smelting and pressing of non-ferrous metals; manufacture of general purpose machinery; and processing of petroleum, coking, processing of nuclear fuel. The revenues from primary businesses in these 10 industries accounted for 65.55% of the revenues from primary business in the entire manufacturing sector. The lowest ranked 10 industries on the list from the bottom up were metal products, machinery, and equipment repair; other manufactures; utilization of waste resources; printing, reproduction of recording media; manufacture of furniture; manufacture of chemical

Table 5.2 Classification of manufacturing industries in China

Industry code	Industry name
C	Manufacturing
C13	Processing of food from agricultural products
C14	Manufacture of foods
C15	Manufacture of wine, drinks, and refined tea
C16	Manufacture of tobacco
C17	Manufacture of textiles
C18	Manufacture of textile wearing and apparel
C19	Manufacture of leather, fur, feather, and related products and footware
C20	Processing of timber, manufacture of wood, bamboo, rattan, palm, and straw products
C21	Manufacture of furniture
C22	Manufacture of paper and paper products
C23	Printing, reproduction of recording media
C24	Manufacture of articles for culture, education, and sport activity
C25	Processing of petroleum, coking, processing of nuclear fuel
C26	Manufacture of raw chemical materials and chemical products
C27	Manufacture of medicines
C28	Manufacture of chemical fibers
C29	Manufacture of rubber and plastic
C30	Manufacture of non-metallic mineral products
C31	Smelting and pressing of ferrous metals
C32	Smelting and pressing of non-ferrous metals
C33	Manufacture of metal products
C34	Manufacture of general purpose machinery
C35	Manufacture of special purpose machinery
C36	Manufacture of automobiles
C37	Manufacture of railway, ship, aerospace, and other transportation equipment
C38	Manufacture of electrical machinery and equipment
C39	Manufacture of computers, communication, and other electronic equipment
C40	Manufacture of measuring instruments
C41	Other manufactures
C42	Utilization of waste resources
C43	Metal products, machinery, and equipment repair

Data source National Bureau of Statistics, National Industry classifications (GB/T 4755-2011)

fibers; manufacture of measuring instruments; manufacture of tobacco; processing of timber, manufacture of wood, bamboo, rattan, palm, and straw products; and manufacture of articles for culture, education, and sport activity. The revenues from primary businesses in these 10 industries accounted for only 7.28% of the revenues

Table 5.3 Ranking of revenues from primary business in above-scale industries by industry classification in China in 2013

Rank	Industry	Revenues from primary business (billion yuan)	Ratio of revenues from primary business to total manufacturing revenues from primary businesses (%)
1	Manufacture of computers, communication, and other electronic equipment	7722.63	8.56
2	Manufacture of raw chemical materials and chemical products	7632.98	8.46
3	Smelting and pressing of ferrous metals	7631.69	8.46
4	Manufacture of electrical machinery and equipment	6101.81	6.77
5	Manufacture of automobiles	6054.00	6.71
6	Processing of food from agricultural products	5949.71	6.60
7	Manufacture of non-metallic mineral products	5128.43	5.69
8	Smelting and pressing of non-ferrous metals	4653.63	5.16
9	Manufacture of general purpose machinery	4278.90	4.74
10	Processing of petroleum, coking, processing of nuclear fuel	4067.98	4.51
11	Manufacture of textiles	3616.06	4.01
12	Manufacture of metal products	3284.29	3.64
13	Manufacture of special purpose machinery	3205.75	3.55
14	Manufacture of rubber and plastic	2731.06	3.03
15	Manufacture of medicines	2059.29	2.28
16	Manufacture of textile wearing and apparel	1925.09	2.13
17	Manufacture of foods	1816.50	2.01
18	Manufacture of railway, ship, aerospace, and other transportation equipment	1654.51	1.83
19	Manufacture of wine, drinks, and refined tea	1518.52	1.68
20	Manufacture of paper and paper products	1347.16	1.49
21	Manufacture of leather, fur, feather, and related products and footware	1249.31	1.39

(continued)

Table 5.3 (continued)

Rank	Industry	Revenues from primary business (billion yuan)	Ratio of revenues from primary business to total manufacturing revenues from primary businesses (%)
22	Manufacture of articles for culture, education, and sport activity	1203.78	1.33
23	Processing of timber, manufacture of wood, bamboo, rattan, palm, and straw products	1202.19	1.33
24	Manufacture of tobacco	829.27	0.92
25	Manufacture of measuring instruments	768.19	0.85
26	Manufacture of chemical fibers	728.18	0.81
27	Manufacture of furniture	646.28	0.72
28	Printing, reproduction of recording media	529.13	0.59
29	Utilization of waste resources	334.00	0.37
30	Other manufactures	230.78	0.26
31	Metal products, machinery, and equipment repair	93.04	0.10

from primary business of the entire manufacturing sector, which indicates that large gaps exist in revenues from primary businesses among manufacturing industries.

2. China's total manufacturing profit analysis

Figure 5.3 shows China's total manufacturing profits and growth rates for the latest 6 years. The trend of the total output of the manufacturing sector for the latest 6 years was similar to the trend of total profits: total manufacturing profits grew rapidly in 2008–2010, then dropped rapidly in 2010–2012, and rebounded quickly in 2013.

3. China's manufacturing output growth analysis

During the process of industrialization, manufacturing typically goes through stages of growth led by non-durable consumer goods, investment in industrial and capital goods, and durable consumer goods. Therefore, we need to identify industries with the highest growth rates and their characteristics that contribute to high growth. Table 5.4 and Fig. 5.4 show the ranking of growth rates of revenues from primary businesses in above-scale manufacturing industries in 2013 to 2012 by industry classification.

According to the ranking in Table 5.4, the top 10 industries with the fastest growth were manufacture of medicines; manufacture of automobiles; manufacture of articles for culture, education, and sport activity; processing of timber, manufacture of wood, bamboo, rattan, palm, and straw products; printing, reproduction

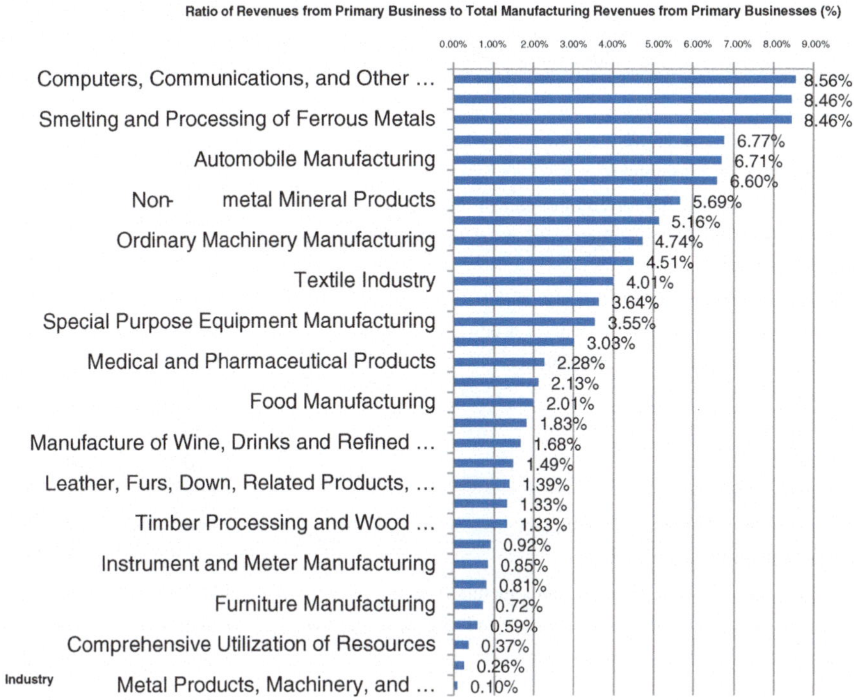

Fig. 5.2 Ranking of revenues from primary business in above-scale industries by industry classification ranking in 2013

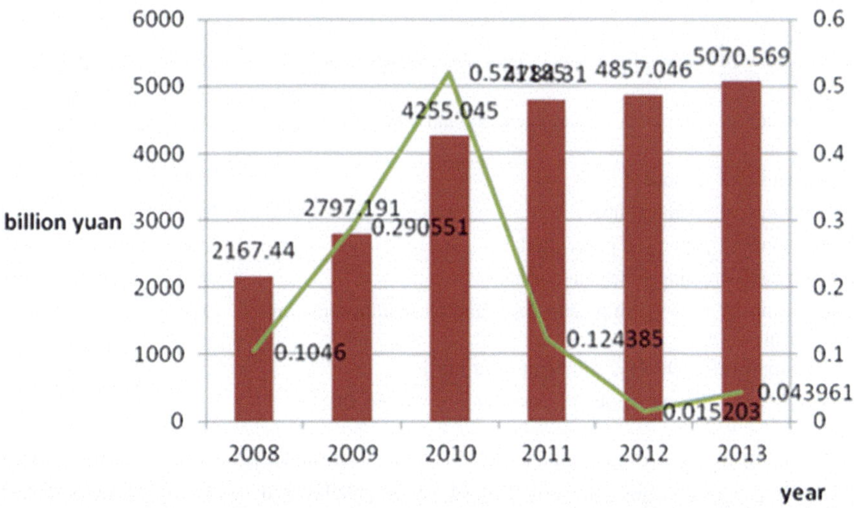

Fig. 5.3 Total profits and growth rates of China's manufacturing sector in 2008–2013. *Data source* Sorted and calculated from related data in the China Statistical Yearbooks of 2008–2014

Table 5.4 Ranking of growth rates of revenues from primary businesses in above-scale manufacturing industries in 2013–2012 by industry classification

Industry	Growth rate (%)	Ranking
Manufacture of medicines	18.78	1
Manufacture of automobiles	18.16	2
Manufacture of articles for culture, education, and sport activity	17.13	3
Processing of timber, manufacture of wood, bamboo, rattan, palm, and straw products	17.00	4
Printing, reproduction of recording media	16.67	5
Manufacture of non-metallic mineral products	16.58	6
Manufacture of measuring instruments	15.40	7
Manufacture of foods	14.72	8
Utilization of waste resources	14.36	9
Manufacture of foods	14.10	10
Manufacture of furniture	13.98	11
Manufacture of rubber and plastic	13.06	12
Manufacture of metal products	12.98	13
Smelting and pressing of non-ferrous metals	12.77	14
Manufacture of raw chemical materials and chemical products	12.65	15
Manufacture of general purpose machinery	12.47	16
Manufacture of textiles	12.16	17
Manufacture of wine, drinks, and refined tea	12.08	18
Manufacture of electrical machinery and equipment	11.91	19
Manufacture of special purpose machinery	11.65	20
Manufacture of textile wearing and apparel	11.37	21
Others	11.30	22
Manufacture of leather, fur, feather, and related products and footware	10.87	23
Manufacture of computers, communication, and other electronic equipment	9.65	24
Manufacture of tobacco	9.52	25
Manufacture of chemical fibers	7.97	26
Manufacture of paper and paper products	7.76	27
Smelting and pressing of ferrous metals	6.65	28
Manufacture of railway, ship, aerospace, and other transportation equipment	5.06	29
Metal products, machinery, and equipment repair	5.04	30
Processing of petroleum, coking, processing of nuclear fuel	3.25	31

of recording media; manufacture of non-metallic mineral products; manufacture of measuring instruments; manufacture of foods; utilization of waste resources; and processing of food from agricultural products. The average growth rate of these 10 industries was 16.29%. The bottom 10 industries with the lowest growth rate were, in descending order, other manufactures; manufacture of leather, fur, feather, and

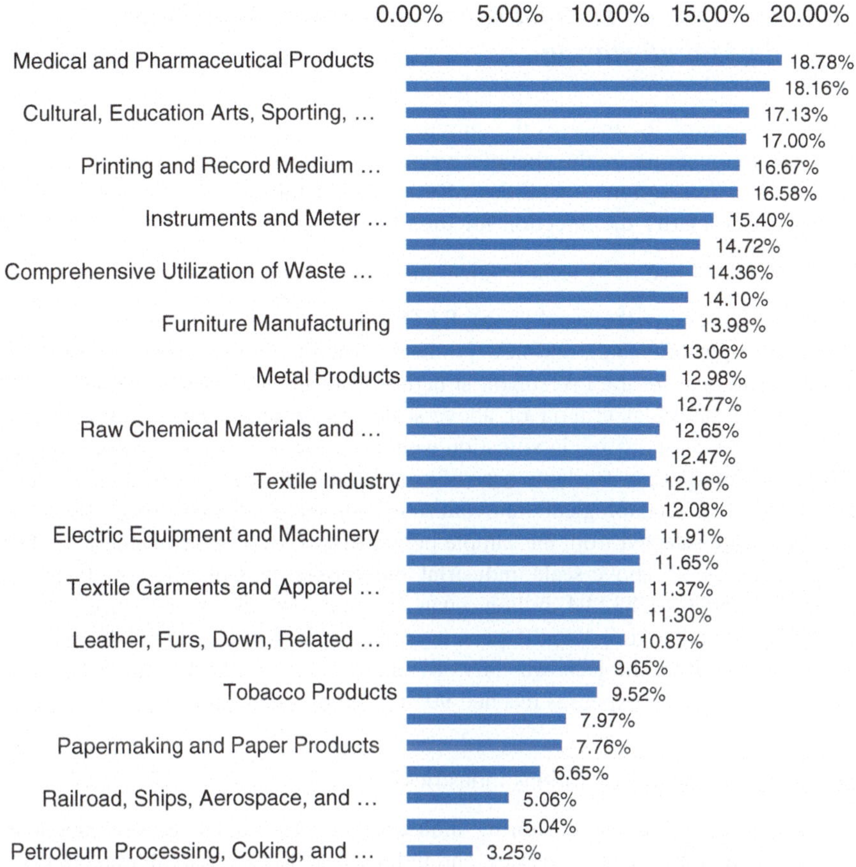

Fig. 5.4 Growth rates of revenues from primary business in above-scale manufacturing industries in 2013 to 2012. Data source sorted and calculated from related data in the China Statistical Yearbook of 2013–2014

related products and footware; manufacture of computers, communication, and other electronic equipment; manufacture of tobacco; manufacture of chemical fibers; manufacture of paper and paper products; smelting and pressing of ferrous metals; manufacture of railway, ship, aerospace, and other transportation equipment; metal products, machinery, and equipment repair; and processing of petroleum, coking, processing of nuclear fuel. The average growth rate of these 10 industries was 7.71%.

5.1.2 *Analysis of Technological Innovation Capabilities in Manufacturing*

This section primarily analyzes the current status, investment, and output of technological innovations in the manufacturing sector to enhance understanding of technological innovation capabilities of the manufacturing industries in China at present and to clarify the direction for future development.

1. Investment structure of technological innovations in manufacturing

R&D expenditure (thousand yuan), R&D full-time equivalent employees (FTE, measured in person-years), and new product development expenses were selected as indexes to analyze the investment structure of technological innovations.

Since the statistical criteria of above-scale industrial enterprise research and R&D activities by industry classification were not provided by the National Bureau of Statistics in 2010, to facilitate ease of comparison and analysis of the statistical results, data on technological innovation investment of China's manufacturing industries were selected from the sample period of 2012–2013 (see Table 5.5). The R&D expenses of above-scale industrial enterprises in China's manufacturing sector amounted to 832.94 billion yuan, an increase of 15.46% from 2012. R&D FTE amounted to 2,368,205 person-years, a growth rate of 11.1% from 2012. The number of R&D projects was 309,081, an increase of 12.6% from 2012. New product development expenses reached 905.62 billion yuan, an increase of 15.45% from 2012.

(1) R&D expenditures of various industries

R&D expenditures of manufacturing industries directly reflect the investment in technology manufacturing industries; and indirectly reflect the growth potential and scientific strength of each industry. The 2013 R&D expenses of manufacturing by industry are shown in Table 5.6. The R&D expenditures of the seven industries—namely, the computer, communications, and other electronic equipment manufacturing; manufacture of electrical machinery and equipment; manufacture of automobiles; manufacture of raw chemical materials and chemical products; smelting

Table 5.5 2012–2013 technological innovation investment of above-scale industrial enterprise in China's manufacturing sector

Index	2012	2013	Growth rate (%)
R&D expenses (thousand yuan)	721,397,850	832,937,750	15.46
Number of R&D projects (project)	274,504	309,081	12.60
R&D FTE (person-years)	2,131,537	2,368,205	11.10
New product development expenses (thousand yuan)	784,440,220	905,615,220	15.45

Data source Sorted and calculated from related data in the China Statistical Yearbooks of 2013 and 2014; and the China Statistical Yearbooks on Science and Technology of 2013 and 2014

Table 5.6 Ranking of R&D expenditures of above-scale industrial enterprises in manufacturing by industry classification in 2013

Rank	Industry	R&D funding (thousand yuan)	Ratio of R&D expenditures to total manufacturing R&D funding (%)
1	Manufacture of computers, communication, and other electronic equipment	130,933,542	15.72
2	Manufacture of electrical machinery and equipment	84,181,525	10.11
3	Manufacture of automobiles	75,332,276	9.04
4	Manufacture of raw chemical materials and chemical products	68,078,535	8.17
5	Smelting and pressing of ferrous metals	64,913,729	7.79
6	Manufacture of general purpose machinery	56,717,532	6.81
7	Manufacture of special purpose machinery	52,321,369	6.28
8	Manufacture of railway, ship, aerospace, and other transportation equipment	41,480,745	4.98
9	Manufacture of medicines	38,847,404	4.66
10	Smelting and pressing of non-ferrous metals	31,455,909	3.78
11	Manufacture of metal products	23,460,522	2.82
12	Manufacture of non-metallic mineral products	21,914,605	2.63
13	Manufacture of rubber and plastic	20,484,697	2.46
14	Processing of food from agricultural products	18,060,160	2.17
15	Manufacture of textiles	16,101,784	1.93
16	Manufacture of measuring instruments	15,560,534	1.87
17	Manufacture of foods	10,335,735	1.24
18	Processing of petroleum, coking, processing of nuclear fuel	9,541,430	1.15
19	Manufacture of paper and paper products	8,913,211	1.07
20	Manufacture of wine, drinks, and refined tea	8,557,737	1.03
21	Manufacture of textile wearing and apparel	7,135,060	0.86
22	Manufacture of chemical fibers	6,791,533	0.82
23	Manufacture of articles for culture, education, and sport activity	5,043,439	0.61
24	Manufacture of leather, fur, feather, and related products and footware	3,445,463	0.41

(continued)

Table 5.6 (continued)

Rank	Industry	R&D funding (thousand yuan)	Ratio of R&D expenditures to total manufacturing R&D funding (%)
25	Printing, reproduction of recording media	3,106,863	0.37
26	Manufacture of tobacco	2,819,665	0.34
27	Processing of timber, manufacture of wood, bamboo, rattan, palm, and straw products	2,789,476	0.33
28	Manufacture of furniture	2,312,965	0.28
29	Other manufactures	1,490,078	0.18
30	Metal products, machinery, and equipment repair	810,222	0.10

Data source Sorted and calculated from related data in the China Statistical Yearbook 2014

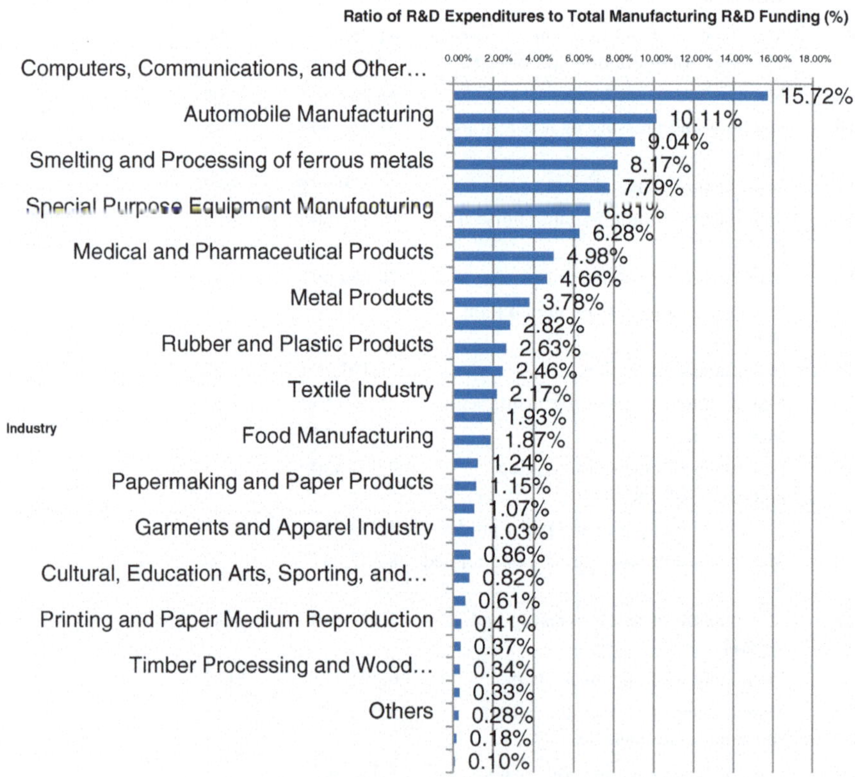

Fig. 5.5 Distribution of 2013 R&D expenditures of above-scale enterprises by industry classification. *Data source* Sorted and calculated from related data in the China Statistical Yearbook 2014

and pressing of ferrous metals; manufacture of general purpose machinery; and manufacture of special purpose machinery—accounted for more than half of the total R&D expenditure of the entire manufacturing sector, reaching 64.93%.

Figure 5.5 visually presents the ranking of R&D expenditures of the industrial enterprises in the manufacturing sector. As revealed from the input structure of R&D funding in 2013, investment in technology innovations was uneven in China's manufacturing sector. In 2013, the investment in technology innovations was mainly concentrated in the manufacture of computers, communication, and other electronic equipment (15.72%); manufacture of electrical machinery and equipment (10.11%); manufacture of automobiles (9.04%); manufacture of raw chemical materials and chemical products (8.17%), smelting and pressing of ferrous metals (7.79%); manufacture of general purpose machinery (6.81%), and manufacture of special purpose machinery (6.28%). The internal R&D expenditures of the seven industries accounted for 63.92% of the total R&D expenditures of the entire manufacturing sector. In addition, as observed from the table, the R&D expenditures of many industries were inadequate, such as processing of timber, manufacture of wood, bamboo, rattan, palm, and straw products (0.33%), manufacture of furniture (0.28%); metal products, machinery, and equipment repair (0.10%); and other traditional industries.

(2) Number of full-time R&D employees by industry

The number of FTEs in above-scale manufacturing enterprises reflects the input of technological employees by manufacturing industries. Table 5.7 shows the ranking of 2013 R&D FTEs by industry. The number of R&D FTEs of the seven industries —namely, manufacture of computers, communication, and other electronic equipment; manufacture of electrical machinery and equipment; manufacture of automobiles; manufacture of general purpose machinery; manufacture of special purpose machinery; manufacture of raw chemical materials and chemical products; and manufacture of medicines—accounted for more than half of the number of R&D FTEs in the manufacturing sector, which increased to 63.6%.

Figure 5.6 visually presents the ranking of R&D FTEs in above-scale industrial enterprises by manufacturing industry classification. Like for R&D expenditures of the above-scale industrial enterprises by manufacturing industry classification, there was an uneven number of R&D FTEs of above-scale enterprises among different manufacturing industries. Among them, the following five industries accounted for the least R&D FTEs, that is, none comprised even 0.5% of R&D FTEs of manufacturing: manufacture of furniture (0.4%); processing of timber, manufacture of wood, bamboo, rattan, palm, and straw products (0.35%); other manufactures (0.3%); metal products, machinery, and equipment repair (0.22%); and manufacture of tobacco (0.18%). As Fig. 5.6 shows, there was more R&D FTEs in the following six industries, which together accounted for 58.4% of R&D FTEs in manufacturing:

Table 5.7 2013 Ranking of number of R&D FTEs of above-scale industrial enterprises in manufacturing by industry classification

Rank	Industry	R&D FTEs (person-years)	Ratio of R&D FTEs to total manufacturing R&D FTEs (%)
1	Manufacture of computers, communication, and other electronic equipment	390,977	16.51
2	Manufacture of electrical machinery and equipment	255,835	10.80
3	Manufacture of automobiles	195,682	8.26
4	Manufacture of general purpose machinery	191,916	8.10
5	Manufacture of special purpose machinery	178,461	7.54
6	Manufacture of raw chemical materials and chemical products	170,087	7.18
7	Manufacture of medicines	123,200	5.20
8	Smelting and pressing of ferrous metals	107,190	4.53
9	Manufacture of railway, ship, aerospace, and other transportation equipment	105,869	4.47
10	Metal products, machinery, and equipment repair	79,315	3.35
11	Manufacture of non-metallic mineral products	73,646	3.11
12	Manufacture of measuring instruments	69,174	2.92
13	Manufacture of rubber and plastic	64,068	2.71
14	Smelting and pressing of non-ferrous metals	57,560	2.43
15	Manufacture of textiles	53,289	2.25
16	Processing of food from agricultural products	38,162	1.61
17	Manufacture of textile wearing and apparel	34,322	1.45
18	Manufacture of foods	27,389	1.16
19	Manufacture of wine, drinks and refined tea	21,113	0.89
20	Manufacture of articles for culture, education, and sport activity	20,909	0.88
21	Manufacture of paper and paper products	20,557	0.87
22	Manufacture of chemical fibers	16,563	0.70
23	Processing of petroleum, coking, processing of nuclear fuel	13,993	0.59
24	Manufacture of leather, fur, feather, and related products and footware	13,532	0.57

(continued)

Table 5.7 (continued)

Rank	Industry	R&D FTEs (person-years)	Ratio of R&D FTEs to total manufacturing R&D FTEs (%)
25	Printing, reproduction of recording media	11,363	0.48
26	Manufacture of furniture	9383	0.40
27	Processing of timber, manufacture of wood, bamboo, rattan, palm, and straw products	8208	0.35
28	Other manufactures	7082	0.30
29	Metal products, machinery, and equipment repair	5115	0.22
30	Manufacture of tobacco	4246	0.18

Data source Sorted and calculated from related data in the China Statistical Yearbook 2014

computer, communication, and other electronic equipment manufacturing (16.51%); manufacture of electrical machinery and equipment (10.8%); manufacture of automobiles (8.26%); manufacture of general purpose machinery (8.1%); manufacture of special purpose machinery (7.54%); and manufacture of raw chemical materials and chemical products (7.18%).

(3) R&D expenses on new products by industry

New product development refers to the series of decision-making processes that begin with selection of a marketable product followed by product design, manufacturing art design, and production input. In a broad sense, R&D of a new product includes both the development of a new product as well as an improved and upgraded version of the old product. R&D of new products is both an integral component of an enterprise's R&D, and one of its core strategies for survival and development. In addition, an enterprise's R&D expenses on new product development reflect its technology investment in manufacturing, which indirectly reflects the enterprise's growth potential and future development capabilities. Figure 5.8 shows the 2013 R&D ranking of expenses on new products in above-scale industrial enterprises in the manufacturing sector by industry classification.

Figure 5.7 visually presents the ranking of R&D expenses on new products of above-scale industrial enterprises in the manufacturing sector by industry classification.

As Table 5.8 and Fig. 5.8 reveal, the total R&D expenses on new products of the following seven industries amounted to 64.91% of total manufacturing R&D: manufacture of computers, communication, and other electronic equipment (17.18%); manufacture of electrical machinery and equipment (11.54%); manufacture of automobiles (8.81%); manufacture of general purpose machinery (7.24%); manufacture of raw chemical materials and chemical products (6.95%);

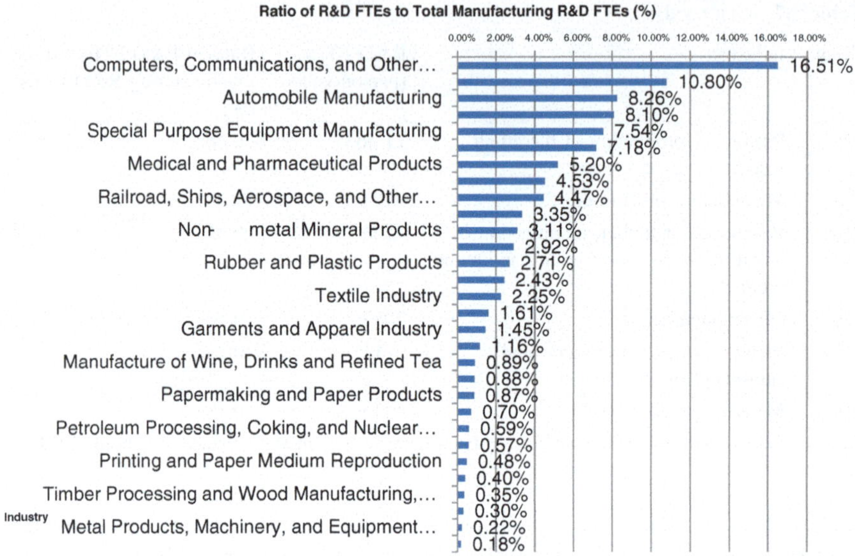

Fig. 5.6 Weight distribution of 2013 R&D FTEs in above-scale manufacturing enterprises by industry classification. *Data source* Sorted and calculated from related data in the China Statistical Yearbook 2014

smelting and pressing of ferrous metals (6.69%); and manufacture of special purpose machinery (6.5%). On the other hand, the total R&D expenses on new products of the following eight industries accounted for only 2.74% of total manufacturing R&D: manufacture of articles for culture, education, and sport activity (0.74%); manufacture of leather, fur, feather, and related products and footware (0.46%); printing, reproduction of recording media (0.36%); processing of timber, manufacture of wood, bamboo, rattan, palm, and straw products (0.33%); manufacture of furniture (0.31%); manufacture of tobacco (0.22%); other manufactures (0.2%); and metal products, machinery, and equipment repair (0.12%). As shown, the R&D expenses on new products among manufacturing industries varied significantly.

3. Output structure of technological innovations in manufacturing

Indexes, such as number of newly developed items, sales revenues of new products, and number of patents, were selected to analyze the output structure of technological innovations. Since the statistical criteria of number of patents owned by above-scale industrial enterprises by industry classification were not provided by the National Bureau of Statistics in 2010, to facilitate ease of analysis, data were also selected from 2012 and 2013, as shown in Table 5.9.

The 2013 output of technological innovations in China's manufacturing sector maintained rapid growth, which was consistent with the trend of the sector's technological input. The number of newly developed items reached 351,682, which

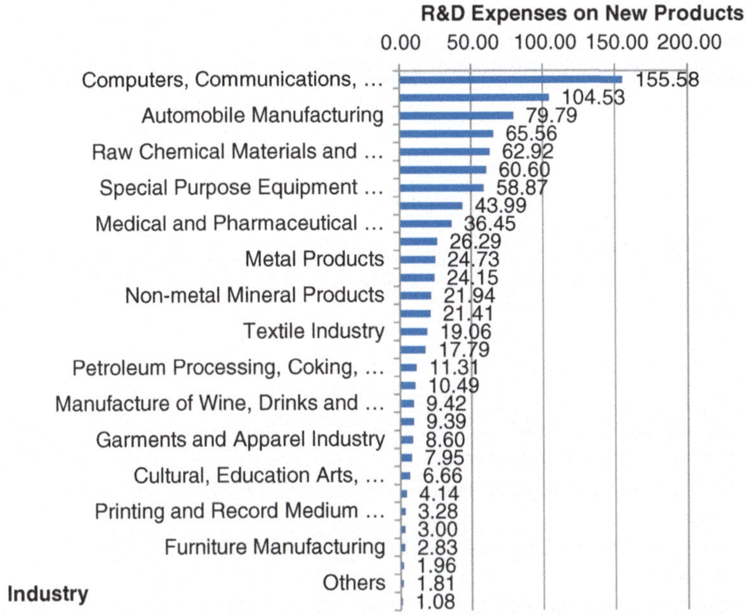

Fig. 5.7 Distribution of 2013 R&D expenses on new products of above-scale industrial enterprises in the manufacturing sector by industry classification

represented growth of 10.83% from 2012. There were 195,598 patents in 2013, which represented an increase of 15.46% from 2012. Sales revenue generated by new products was 12.65454673 trillion yuan, representing growth of 16.48% from 2012.

(1) Number of newly developed products by industry

The number of newly developed products in the manufacturing sector reflects its development status and the technological strength of the manufacturing sector. Table 5.10 shows the 2013 ranking of the number of newly developed products in above-scale industrial enterprises in the manufacturing sector.

Figure 5.8 visually presents a ranking of the number of newly developed products in above-scale industrial enterprises in the manufacturing sector by industry classification.

As observed from Table 5.10 and Fig. 5.8, each of the following seven industries had more than 20,000 newly developed items, which was significantly higher than other industries, accounting for 65.55% of the new products of the entire manufacturing sector: manufacture of computers, communication, and other electronic equipment (12.91%); manufacture of electrical machinery and equipment (12.51%); manufacture of general purpose machinery (9.76%); manufacture of special purpose machinery (8.9%); manufacture of medicines (7.54%); manufacture of automobiles (7.22%); and manufacture of raw chemical materials and chemical products (6.71%). On the other hand, the number of newly developed items of each

Table 5.8 2013 ranking of R&D expenses on new products of above-scale industrial enterprises in the manufacturing sector by industry classification

Rank	Industry	R&D expenses on new products (billion)	R&D expenses on new products to total manufacturing ratio R&D expenses in new (%)
1	Manufacture of computers, communication, and other electronic equipment	155.58	17.18
2	Manufacture of electrical machinery and equipment	104.53	11.54
3	Manufacture of automobiles	79.79	8.81
4	Manufacture of general purpose machinery	65.56	7.24
5	Manufacture of raw chemical materials and chemical products	62.92	6.95
6	Smelting and pressing of ferrous metals	60.60	6.69
7	Manufacture of special purpose machinery	58.87	6.50
8	Manufacture of railway, ship, aerospace, and other transportation equipment	43.99	4.86
9	Manufacture of medicines	36.45	4.02
10	Smelting and pressing of non-ferrous metals	26.29	2.90
11	Manufacture of metal products	24.73	2.73
12	Manufacture of rubber and plastic	24.15	2.67
13	Manufacture of non-metallic mineral products	21.94	2.42
14	Processing of food from agricultural products	21.41	2.36
15	Manufacture of textiles	19.06	2.10
16	Manufacture of measuring instruments	17.79	1.96
17	Processing of petroleum, coking, processing of nuclear fuel	11.31	1.25
18	Manufacture of foods	10.49	1.16
19	Manufacture of wine, drinks, and refined tea	9.42	1.04
20	Manufacture of chemical fibers	9.39	1.04
21	Manufacture of textile wearing and apparel	8.60	0.95
22	Manufacture of paper and paper products	7.95	0.88

(continued)

Table 5.8 (continued)

Rank	Industry	R&D expenses on new products (billion)	R&D expenses on new products to total manufacturing ratio R&D expenses in new (%)
23	Manufacture of articles for culture, education, and sport activity	6.66	0.74
24	Manufacture of leather, fur, feather, and related products and footware	4.14	0.46
25	Printing, reproduction of recording media	3.28	0.36
26	Processing of timber, manufacture of wood, bamboo, rattan, palm, and straw products	3.00	0.33
27	Manufacture of furniture	2.83	0.31
28	Manufacture of tobacco	1.96	0.22
29	Other manufactures	1.81	0.20
30	Metal products, machinery, and equipment repair	1.08	0.12

Data source Sorted and calculated from related data in the China Statistical Yearbook 2014

of the bottom 10 industries on the list accounted for not even 1% of the total of the entire manufacturing sector. The percentage of the total number of newly developed products of these bottom 10 industries accounted for only 4.43% of the entire manufacturing sector. In addition, the number of newly developed items dropped drastically from 2012, starting from the 8th-ranked industry on the list.

(2) Sales revenues of new products by industry

The output of new manufacturing products reflects the development status of new products and the effectiveness of technology transfer. Owing to the lack of such an index in the China Statistical Yearbook of 2013, sales revenues of new products were employed instead. Table 5.11 shows the ranking of sales revenues of new products in above-scale industrial enterprises in the manufacturing sector by industry classification.

Figure 5.9 is a visual presentation of the ranking in sales revenues of new products in above-scale industrial enterprises in the manufacturing sector by industry classification.

As Table 5.11 and Fig. 5.9 show, the sales revenues of new products in the following industries were ranked in the top five: manufacture of computers, communication, and other electronic equipment manufacturing (19.09%); manufacture of automobiles (11.92%); manufacture of electrical machinery and equipment (10.95%); manufacture of raw chemical materials and chemical products (7.22%); and smelting and pressing of ferrous metals (6.3%). The total sales revenues of new products in these top five industries accounted for more than half of that in the

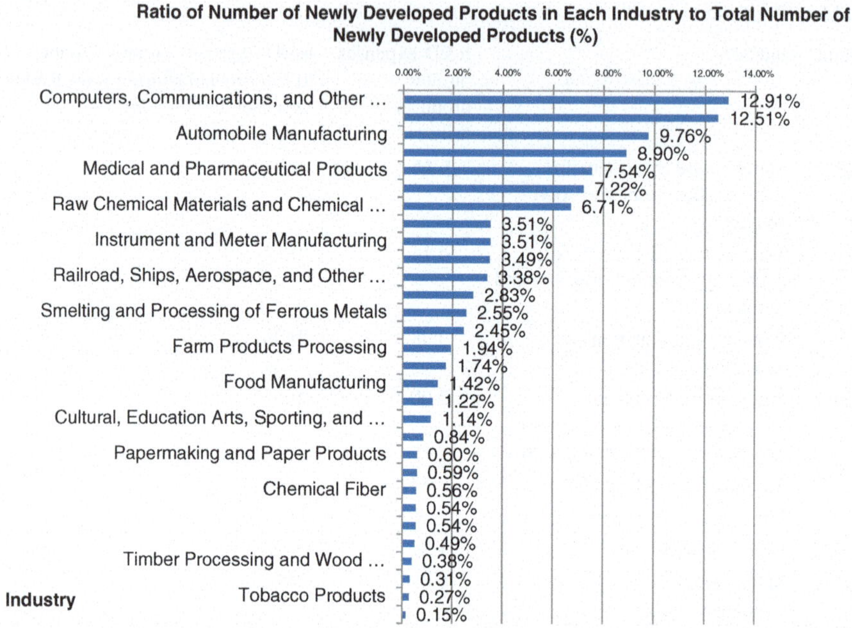

Fig. 5.8 Distribution of number of newly developed products in above-scale enterprises in the manufacturing sector in 2013 by industry classification. *Data source* Sorted and calculated from related data in the China Statistical Yearbook 2014

Table 5.9 2012–2013 output of technological innovations in above-scale industrial enterprises in the manufacturing sector

Output index of technological innovations	2012	2013	Growth rate (%)
Number of newly developed items (item)	317,317	351,682	10.83
Sales revenues of new product (ten thousand yuan)	1,086,427,500	1,265,454,673	16.48
Number of patents (item)	169,410	195,598	15.46

Data source Sorted and calculated from related data in the China Statistical Yearbook 2014

entire manufacturing industry, at 55.49%, whereas the total sales revenues of new products in the bottom 10 industries in the sample accounted for only 5.26% of the total in the entire manufacturing industry.

(3) Number of patents by industry

The number of valid patents in the manufacturing sector reflects the effectiveness of technological innovations in manufacturing. Table 5.12 shows the ranking of the number of valid patents in above-scale industrial enterprises in the manufacturing sector by industry classification.

Table 5.10 2013 ranking of the number of newly developed products in above-scale industrial enterprises in the manufacturing sector by industry classification

Rank	Industry	Number of newly developed products	Ratio of number of newly developed products in each industry to total number of newly developed products (%)
1	Manufacture of computers, communication, and other electronic equipment	45,390	12.91
2	Manufacture of electrical machinery and equipment	43,991	12.51
3	Manufacture of automobiles	34,325	9.76
4	Manufacture of special purpose machinery	31,313	8.90
5	Manufacture of medicines	26,523	7.54
6	Manufacture of automobiles	25,374	7.22
7	Manufacture of raw chemical materials and chemical products	23,590	6.71
8	Manufacture of metal products	12,353	3.51
9	Manufacture of measuring instruments	12,330	3.51
10	Manufacture of rubber and plastic	12,281	3.49
11	Manufacture of railway, ship, aerospace, and other transportation equipment	11,895	3.38
12	Manufacture of non-metallic mineral products	9936	2.83
13	Smelting and pressing of ferrous metals	8971	2.55
14	Manufacture of textiles	8614	2.45
15	Processing of food from agricultural products	6816	1.94
16	Smelting and pressing of non-ferrous metals	6131	1.74
17	Manufacture of foods	4982	1.42
18	Manufacture of textile wearing and apparel	4308	1.22
19	Manufacture of articles for culture, education, and sport activity	4022	1.14
20	Manufacture of wine, drinks and refined tea	2969	0.84
21	Manufacture of paper and paper products	2112	0.60
22	Manufacture of furniture	2059	0.59

(continued)

Table 5.10 (continued)

Rank	Industry	Number of newly developed products	Ratio of number of newly developed products in each industry to total number of newly developed products (%)
23	Manufacture of chemical fibers	1980	0.56
24	Printing, reproduction of recording media	1913	0.54
25	Manufacture of leather, fur, feather, and related products and footware	1886	0.54
26	Processing of petroleum, coking, processing of nuclear fuel	1707	0.49
27	Processing of timber, manufacture of wood, bamboo, rattan, palm, and straw products	1335	0.38
28	Other manufactures	1104	0.31
29	Manufacture of tobacco	948	0.27
30	Metal products, machinery, and equipment repair	524	0.15

Data source Sorted and calculated from related data in the China Statistical Yearbook 2014

Figure 5.10 visually presents the distribution of the number of valid patents in above-scale industrial enterprises in the manufacturing sector by industry classification.

As shown in Table 5.12 and Fig. 5.10, the number of valid patents in the manufacture of computers, communication, and other electronic equipment industry accounted for 29.88% of the total in the entire manufacturing industry. The top five manufacturing industries with the most valid patents were as follows: manufacture of computers, communication, and other electronic equipment manufacturing (29.88%); manufacture of electrical machinery and equipment (11.77%); manufacture of special purpose machinery (8.58%); manufacture of general purpose machinery (7.32%); and manufacture of raw chemical materials and chemical products (6.71%). The total number of valid patents of these five industries accounted for more than half of the total in the entire manufacturing industry at 64.25%. On the other hand, 15 industries accounted for less than 1% of the entire manufacturing industry in terms of number of valid patents. For example, the metal products, machinery, and equipment repair industry accounted for only 0.08%; manufacture of leather, fur, feather, and related products and footware accounted for 0.22%; and manufacture of furniture accounted for 0.27%. As observed in Fig. 5.10, the number of valid patents was significantly uneven among the manufacturing industries, with the number of valid patents of the first-ranked industry 371 times that of the last-ranked industry.

(4) Analysis of science, technology, and innovation indicator of each industry in the manufacturing sector

Table 5.11 2013 ranking of sales revenues of new products in above-scale industrial enterprises in the manufacturing sector by industry classification

Rank	Industry	Sales revenues of new products (ten thousand yuan)	Ratio of sales revenues of new products in each industry to total sales revenues of new products in the manufacturing industry (%)
1	Manufacture of computers, communication, and other electronic equipment	2,416,351,860	19.09
2	Manufacture of automobiles	1,508,409,600	11.92
3	Manufacture of electrical machinery and equipment	1,386,050,580	10.95
4	Manufacture of raw chemical materials and chemical products	913,762,980	7.22
5	Smelting and pressing of ferrous metals	797,191,660	6.30
6	Manufacture of general purpose machinery	726,936,130	5.74
7	Manufacture of special purpose machinery	589,470,650	4.66
8	Smelting and pressing of non-ferrous metals	519,156,430	4.10
9	Manufacture of railway, ship, aerospace, and other transportation equipment	475,615,950	3.76
10	Manufacture of textiles	405,125,710	3.20
11	Manufacture of medicines	360,616,740	2.85
12	Manufacture of rubber and plastic	293,166,160	2.32
13	Manufacture of metal products	272,196,660	2.15
14	Processing of petroleum, coking, processing of nuclear fuel	264,694,160	2.09
15	Manufacture of non-metallic mineral products	241,081,280	1.91
16	Processing of food from agricultural products	212,164,530	1.68
17	Manufacture of tobacco	159,131,300	1.26
18	Manufacture of chemical fibers	150,933,880	1.19
19	Manufacture of measuring instruments	148,987,170	1.18
20	Manufacture of textile wearing and apparel	147,661,360	1.17

(continued)

Table 5.11 (continued)

Rank	Industry	Sales revenues of new products (ten thousand yuan)	Ratio of sales revenues of new products in each industry to total sales revenues of new products in the manufacturing industry (%)
21	Manufacture of paper and paper products	138,225,200	1.09
22	Manufacture of wine, drinks and refined tea	113,376,920	0.90
23	Manufacture of foods	109,684,040	0.87
24	Manufacture of articles for culture, education, and sport activity	87,217,070	0.69
25	Leather, furs, down, related products, and footware	73,894,070	0.58
26	Printing, reproduction of recording media	43,600,060	0.34
27	Manufacture of furniture	39,106,850	0.31
28	Processing of timber, manufacture of wood, bamboo, rattan, palm, and straw products	33,556,740	0.27
29	Other manufactures	17,789,230	0.14
30	Metal products, machinery, and equipment repair	9,391,780	0.07

Data source Sorted and calculated from related data in the China Statistical Yearbook 2014

The science, technology, and innovation indicator of each industry in the manufacturing sector is shown in Table 5.13. According to K-means clustering, these 30 industries were classified into four types according to K-means clustering. There were 20 industries in type I: processing of food from agricultural products; manufacture of foods; manufacture of wine, drinks, and refined tea; manufacture of tobacco; manufacture of textiles; manufacture of textile wearing and apparel; manufacture of leather, fur, feather, and related products and footware; processing of timber, manufacture of wood, bamboo, rattan, palm, and straw products; manufacture of furniture; manufacture of paper and paper products; printing, reproduction of recording media; manufacture of articles for culture, education, and sport activity; processing of petroleum, coking, processing of nuclear fuel; manufacture of chemical fibers; manufacture of rubber and plastic; manufacture of non-metallic mineral products; smelting and pressing of non-ferrous metals; manufacture of metal products; manufacture of measuring instruments; and other manufactures. These 20 industries were the bottom 20 in terms of both R&D expenses and the number of R&D FTEs, that is, they were trailing industries in technology investment. However, the manufacture of metal products industry was ranked 8th in both

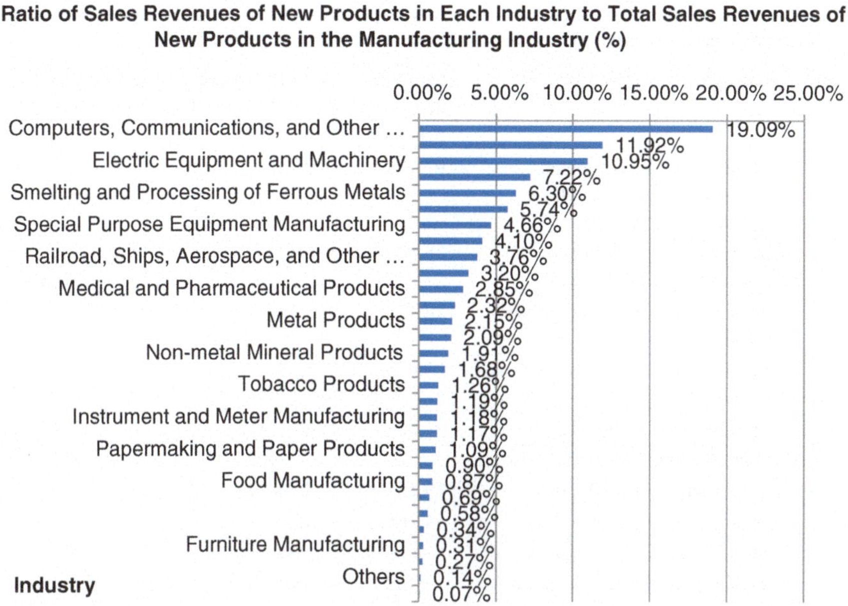

Fig. 5.9 Distribution of sales revenues of new products in above-scale enterprises in the manufacturing sector in 2013 by industry classification. *Data source* Sorted and calculated from related data in the China Statistical Yearbook 2014

the number of newly developed items and the number of valid patents, whereas the manufacture of measuring instruments industry was ranked 9th in the number of newly developed items and 10th in the number of valid patents. The performance of these two industries in this first category was noteworthy. There was only one industry classified as type II, namely, manufacture of computers, communication, and other electronic equipment manufacturing, which was ranked first in five different indexes with scores far higher than those of other industries. There were three industries in type III—manufacture of general purpose machinery; manufacture of automobiles; and manufacture of electrical machinery and equipment—which were ranked fourth, third, and second, respectively, for R&D FTEs; third, sixth, and second, respectively, for number of newly developed items; fourth, seventh, and second, respectively, for number of valid patents; and sixth, second, and third, respectively, for new product output. In general, the investment and output of technology innovations were higher in type III industries. There were five industries in type IV: manufacture of raw chemical materials and chemical products; manufacture of medicines; smelting and pressing of ferrous metals; manufacture of special purpose machinery; and manufacture of railway, ship, aerospace, and other transportation equipment. Among them, the investment of technology innovations in smelting and pressing of ferrous metals was higher than its output whereas the opposite was observed in manufacture of special purpose machinery.

Table 5.12 2013 ranking of number of valid patents in above-scale industrial enterprises in the manufacturing sector by industry classification

Rank	Industry	Number of valid patents	Ratio of number of valid patents to total number of valid patents in the manufacturing sector (%)
1	Manufacture of computers, communication, and other electronic equipment	97,994	29.88
2	Manufacture of electrical machinery and equipment	38,601	11.77
3	Manufacture of special purpose machinery	28,145	8.58
4	Manufacture of general purpose machinery	23,994	7.32
5	Manufacture of raw chemical materials and chemical products	22,005	6.71
6	Manufacture of medicines	19,558	5.96
7	Manufacture of automobiles	14,106	4.30
8	Manufacture of metal products	9656	2.94
9	Manufacture of railway, ship, aerospace, and other transportation equipment	9461	2.88
10	Manufacture of measuring instruments	9236	2.82
11	Manufacture of non-metallic mineral products	8941	2.73
12	Smelting and pressing of ferrous metals	7018	2.14
13	Smelting and pressing of non-ferrous metals	6753	2.06
14	Manufacture of rubber and plastic	6086	1.86
15	Manufacture of articles for culture, education, and sport activity	3355	1.02
16	Processing of food from agricultural products	3221	0.98
17	Manufacture of foods	3105	0.95
18	Manufacture of textiles	2587	0.79
19	Manufacture of textile wearing and apparel	1977	0.60
20	Processing of petroleum, coking, processing of nuclear fuel	1710	0.52
21	Manufacture of wine, drinks and refined tea	1538	0.47

(continued)

Table 5.12 (continued)

Rank	Industry	Number of valid patents	Ratio of number of valid patents to total number of valid patents in the manufacturing sector (%)
22	Printing, reproduction of recording media	1404	0.43
23	Manufacture of chemical fibers	1288	0.39
24	Manufacture of paper and paper products	1282	0.39
25	Manufacture of tobacco	1168	0.36
26	Processing of timber, manufacture of wood, bamboo, rattan, palm, and straw products	1011	0.31
27	Other manufactures	933	0.28
28	Manufacture of furniture	880	0.27
29	Leather, furs, down, related products, and footware	712	0.22
30	Metal products, machinery, and equipment repair	264	0.08

Data source Sorted and calculated from related data in the China Statistical Yearbook 2014

As indicated in the above analysis, manufacture of computers, communication, and other electronic equipment manufacturing was the industry with the highest investment and output of technology innovations. Industries with relatively high investment and output of technology innovations were manufacture of electrical machinery and equipment; manufacture of automobiles; manufacture of general purpose machinery; smelting and pressing of ferrous metals; manufacture of raw chemical materials and chemical products; manufacture of special purpose machinery; manufacture of railway, ship, aerospace, and other transportation equipment; and manufacture of medicines. On the other hand, the investment and output of technology innovations were generally low in all other industries. However, the manufacture of measuring instruments industry showed an advantage in both the number of newly developed products and the number of valid patents while the new product output was still relatively low.

5.1.3 Analysis of Energy Consumption in Manufacturing

China is currently undergoing rapid industrial development with a drastic increase in mineral resources and energy consumption. This section primarily discusses the following three areas: total energy consumption in manufacturing, electricity consumption in manufacturing, and energy utilization effectiveness. The overall index employed to measure economic effectiveness in energy consumption was tons of

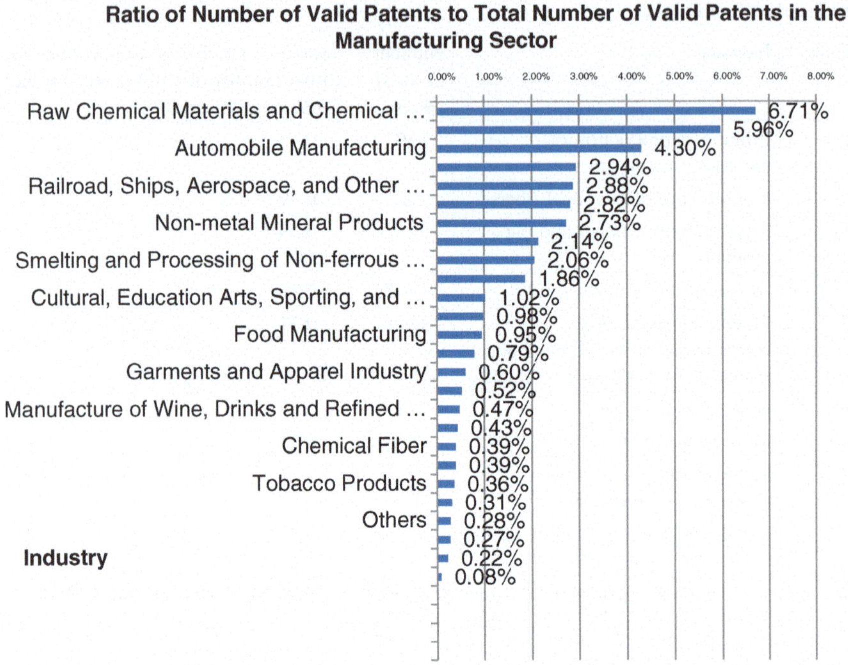

Fig. 5.10 Distribution of number of valid patents of above-scale enterprises in the manufacturing sector in 2013 by industry classification. *Data source*: Sorted and calculated from related data in the China Statistical Yearbook 2014

coal equivalent consumed per ten thousand yuan of output in manufacturing; lower values are better. As Table 5.14 indicates, the value was 0.28 tons of coal equivalent per 10,000 yuan of output in 2013, an increase of 7.99% from 2012. Compared to 2012, electricity consumption increased 8.07% in 2013.

1. Structure of energy consumption in manufacturing

According to the data, the top five manufacturing industries with the highest energy consumption in 2013 were the same five as those in the previous 7 years. They were smelting and pressing of ferrous metals; manufacture of raw chemical materials and chemical products; manufacture of non-metallic mineral products; processing of petroleum, coking, processing of nuclear fuel; and smelting and pressing of non-ferrous metals. The total energy consumption in these five industries accounted for 77.36% of the total energy consumption of the entire manufacturing industry. In addition, the bottom six industries with the least energy consumption were the same as those in 2012. They were metal products, machinery, and equipment repair; utilization of waste resources; manufacture of furniture; manufacture of tobacco; manufacture of articles for culture, education, and sport activity; and manufacture of measuring instruments. They accounted for

Table 5.13 2013 ranking of science, technology, and innovation indexes of manufacturing industries

Industry	R&D expenses (ten thousand yuan)	R&D FTE (person-years)	No. of new products	No. of valid patents	Sales revenues of new products
Manufacture of computers, communication, and other electronic equipment	1	1	1	1	1
Manufacture of electrical machinery and equipment	2	2	2	2	3
Manufacture of automobiles	3	3	6	7	2
Manufacture of raw chemical materials and chemical products	4	6	7	5	4
Smelting and pressing of ferrous metals	5	8	13	12	5
Manufacture of general purpose machinery	6	4	3	4	6
Manufacture of special purpose machinery	7	5	4	3	7
Manufacture of railway, ship, aerospace, and other transportation equipment	8	9	11	9	9
Manufacture of medicines	9	7	5	6	11
Smelting and pressing of non-ferrous metals	10	14	16	13	8
Manufacture of metal products	11	10	8	8	13
Manufacture of non-metallic mineral products	12	11	12	11	15
Manufacture of rubber and plastic	13	13	10	14	12
Processing of food from agricultural products	14	16	15	16	16
Manufacture of textiles	15	15	14	18	10
Manufacture of measuring instruments	16	12	9	10	19
Manufacture of foods	17	18	17	17	23
Processing of petroleum, coking, processing of nuclear fuel	18	23	26	20	14
Manufacture of paper and paper products	19	21	21	24	21
Manufacture of wine, drinks, and refined tea	20	19	20	21	22

(continued)

Table 5.13 (continued)

Industry	R&D expenses (ten thousand yuan)	R&D FTE (person-years)	No. of new products	No. of valid patents	Sales revenues of new products
Manufacture of textile wearing and apparel	21	17	18	19	20
Manufacture of chemical fibers	22	22	23	23	20
Manufacture of articles for culture, education, and sport activity	23	20	19	15	24
Leather, furs, down, related products, and footware	24	24	25	29	25
Printing, reproduction of recording media	25	25	24	22	26
Manufacture of tobacco	26	30	29	25	17
Processing of timber, manufacture of wood, bamboo, rattan, palm, and straw products	27	27	27	26	28
Manufacture of furniture	28	26	22	28	27
Other manufactures	29	28	28	27	29
Metal products, machinery, and equipment repair	30	29	30	30	30

Data source Sorted and calculated from related data in the China Statistical Yearbook 2014

Table 5.14 Energy consumption in China's manufacturing industry

Energy and environmental indexes	2012	2011	Growth rate (%)
Total energy consumption (10,000 tons of coal equivalent)	205667.69	200403.37	2.63
Electricity consumption (hundred million kWh)	26822.46	25526.84	5.08
Consumption per 10,000 yuan of output (tons of coal equivalent)	0.254879	0.273035	−6.65

Data source Sorted and calculated from related data in the China Statistical Yearbook 2014 and the China Energy Statistical Yearbook 2014

0.6% of the total energy consumption of the entire manufacturing sector (see Table 5.15 and Fig. 5.11). It is apparent that energy consumption was very uneven among the manufacturing industries.

Figure 5.11 visually presents the distribution of energy consumption in each manufacturing industry.

Table 5.15 Ranking of total energy consumption of manufacturing industries

Rank	Industry	Total electricity consumption (billion kWh)	Relative to total electricity (%)
1	Smelting and pressing of ferrous metals	6883.9	28.73
2	Manufacture of raw chemical materials and chemical products	4408.1	18.40
3	Manufacture of non-metallic mineral products	3656.1	15.26
4	Processing of petroleum, coking, processing of nuclear fuel	1925.5	8.04
5	Smelting and pressing of non-ferrous metals	1661.7	6.94
6	Manufacture of textiles	736.6	3.07
7	Manufacture of rubber and plastic	470.4	1.96
8	Manufacture of metal products	435	1.82
9	Manufacture of paper and paper products	415.3	1.73
10	Manufacture of general purpose machinery	390.5	1.63
11	Manufacture of automobiles	360.9	1.51
12	Processing of food from agricultural products	357.1	1.49
13	Manufacture of computers, communication, and other electronic equipment	280.2	1.17
14	Manufacture of electrical machinery and equipment	260.6	1.09
15	Manufacture of special purpose machinery	217.9	0.91
16	Manufacture of foods	191.4	0.80
17	Other manufactures	190.9	0.80
18	Manufacture of medicines	189	0.79
19	Manufacture of chemical fibers	161	0.67
20	Manufacture of wine, drinks and refined tea	159.7	0.67
21	Processing of timber, manufacture of wood, bamboo, rattan, palm, and straw products	152.2	0.64
22	Manufacture of railway, ship, aerospace, and other transportation equipment	104.5	0.44
23	Manufacture of textile wearing and apparel	97.1	0.41
24	Leather, furs, down, related products, and footware	65.2	0.27
25	Printing, reproduction of recording media	44.8	0.19
26	Manufacture of measuring instruments	36.8	0.15
27	Manufacture of articles for culture, education, and sport activity	32.9	0.14
28	Manufacture of tobacco	25.6	0.11
29	Manufacture of furniture	24.7	0.10
30	Utilization of waste resources	16.9	0.07
31	Metal products, machinery, and equipment repair	6.6	0.03

Data source Sorted and calculated from related data in the China Statistical Yearbook 2014

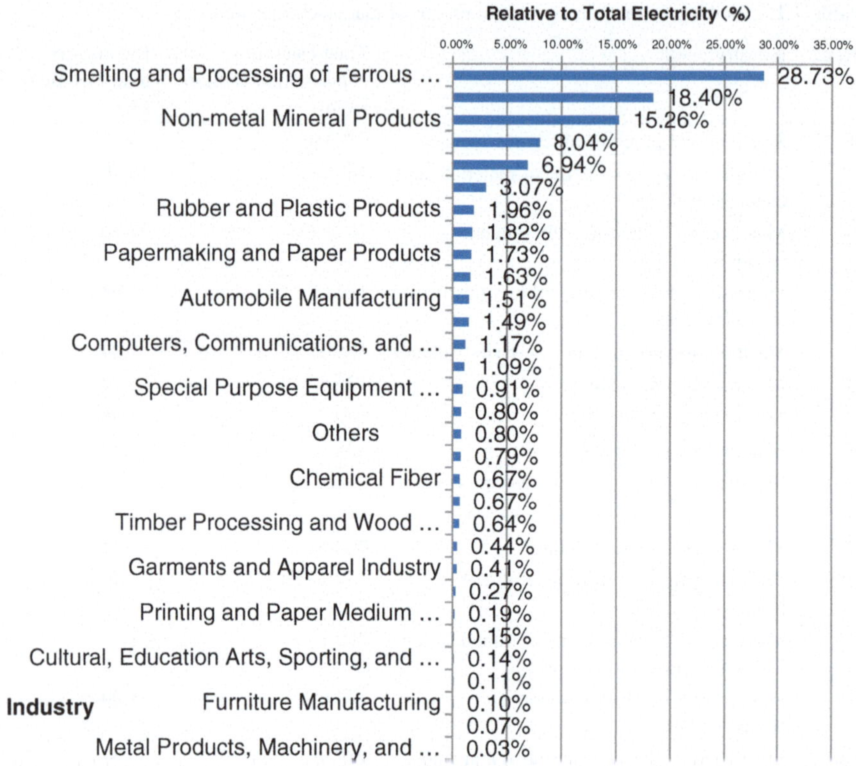

Fig. 5.11 Distribution of total energy consumption in the manufacturing sector by industry classification. *Data source* Sorted and calculated from related data in the China Energy Statistical Yearbook 2014

As observed in Fig. 5.11, total energy consumption was high mainly in heavy industries, such as smelting and pressing of ferrous metals, manufacture of non-metallic mineral products, and petroleum processing. Energy consumption in smelting and pressing of ferrous metals accounted for 28.73% of that in the entire manufacturing sector, giving this industry the highest ranking for total energy consumption. This is because China is the world's largest producer and consumer country of iron and steel, which forms part of the smelting and pressing of ferrous metals industry. However, other issues, such as the unreasonable industry structure, and low technology content, were also apparent. Energy consumption in the manufacture of raw chemical materials and chemical products accounted for 18.4% of the total, ranking this industry second in total energy consumption, which was also very high. Energy conservation and emissions reduction are problems that require urgent attention. Energy consumption in the manufacture of non-metallic mineral products accounted for 15.26% of the total, ranking this industry third in total energy consumption. Therefore, in the production of non-metal minerals, higher technology content and lower environmental load that are more adaptive to society's developmental needs should be pursued.

2. Structure of electricity consumption in the manufacturing sector

Economic development cannot occur without energy. As one of the most basic forms of energy, electricity is widely used in almost all areas of the national economy. Electricity usage is a barometer of economic growth and the two move in the same direction. When analyzing the relationship between the manufacturing sector and its electricity usage, measures related to optimizing adjustment to the structure of the manufacturing sector as well as energy conservation and emissions reduction are crucial to facilitate better and faster economic development in China. As Table 5.16 and Fig. 5.12 show, electricity consumption of four industries—smelting and pressing of ferrous metals; manufacture of raw chemical materials and chemical products; smelting and pressing of non-ferrous metals; and manufacture of non-metallic mineral products—accounted for 60% of the entire manufacturing sector. Apparently, volatility in electricity consumption was strongly related to the high-energy consuming nature of production in such industries as iron and steel, cement, and petrochemicals.

Table 5.16 Ranking of total electricity consumption of manufacturing sector

Industry	Total electricity consumption (billion kWh)	Relative to total electricity (%)	Rank
Smelting and pressing of ferrous metals	570.40	19.68	1
Manufacture of raw chemical materials and chemical products	434.10	14.98	2
Smelting and pressing of non-ferrous metals	411.40	14.19	3
Manufacture of non-metallic mineral products	314.80	10.86	4
Manufacture of textiles	153.30	5.29	5
Manufacture of metal products	121.30	4.18	6
Manufacture of rubber and plastic	109.90	3.79	7
Manufacture of computers, communication, and other electronic equipment	80.90	2.79	8
Manufacture of general purpose machinery	74.60	2.57	9
Processing of petroleum, coking, processing of nuclear fuel	67.70	2.34	10
Manufacture of automobiles	67.40	2.33	11
Manufacture of electrical machinery and equipment	65.00	2.24	12
Manufacture of paper and paper products	59.90	2.07	13
Processing of food from agricultural products	57.40	1.98	14
Other manufactures	41.60	1.44	15
Manufacture of special purpose machinery	40.90	1.41	16

(continued)

Table 5.16 (continued)

Industry	Total electricity consumption (billion kWh)	Relative to total electricity (%)	Rank
Manufacture of chemical fibers	35.00	1.21	17
Manufacture of medicines	28.30	0.98	18
Processing of timber, manufacture of wood, bamboo, rattan, palm, and straw products	26.90	0.93	19
Manufacture of foods	23.00	0.79	20
Manufacture of textile wearing and apparel	21.40	0.74	21
Manufacture of railway, ship, aerospace, and other transportation equipment	21.10	0.73	22
Manufacture of wine, drinks and refined tea	16.80	0.58	23
Leather, furs, down, related products, and footware	15.20	0.52	24
Printing, reproduction of recording media	11.00	0.38	25
Manufacture of measuring instruments	8.30	0.29	26
Manufacture of articles for culture, education, and sport activity	6.90	0.24	27
Manufacture of tobacco	5.40	0.19	28
Manufacture of furniture	5.00	0.17	29
Utilization of waste resources	2.40	0.08	30
Metal products, machinery, and equipment repair	570.40	0.04	31

Data source Sorted and calculated from related data in the China Energy Statistical Yearbook 2014

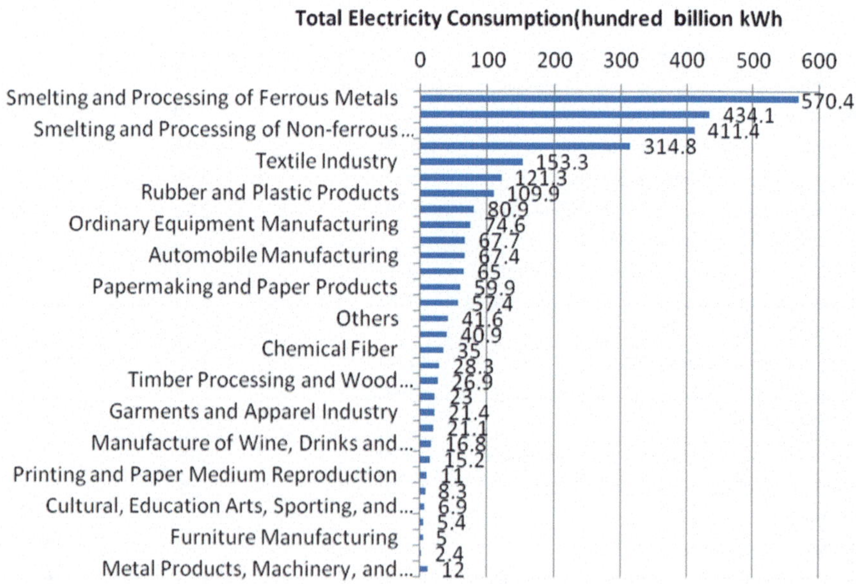

Fig. 5.12 Total electricity consumption

3. Energy utilization effectiveness analysis

Energy intensity is one of the indexes generally used to measure energy utilization effectiveness and is the amount of energy needed for each unit of GDP (or product) produced. Since the data of total output by industry were not published in the China Statistics Yearbook 2014, in this report, we adopted the following: energy intensity = total energy consumption (ten thousand tons of standard coal equivalent)/revenues from primary businesses (hundred million yuan). Generally, the higher the energy intensity is, the lower is the effectiveness of energy utilization. Table 5.17 shows the energy utilization effectiveness of the manufacturing sector by industry in 2013.

As Table 5.17 and Fig. 5.13 show, energy utilization effectiveness was lower in such industries as smelting and pressing of ferrous metals; manufacture of non-metallic mineral products; other manufactures; manufacture of raw chemical materials and chemical products; processing of petroleum, coking, processing of nuclear fuel; and smelting and pressing of non-ferrous metals. On the other hand, energy utilization effectiveness was higher in such industries as manufacture of computers, communication, and other electronic equipment manufacturing and manufacture of tobacco.

Table 5.17 Ranking of energy utilization effectiveness of manufacturing industries

Industry	Energy intensity	Ranking of energy utilization effectiveness
Smelting and pressing of ferrous metals	0.9020	1
Manufacture of non-metallic mineral products	0.7129	2
Other manufactures	0.6920	3
Manufacture of raw chemical materials and chemical products	0.5775	4
Processing of petroleum, coking, processing of nuclear fuel	0.4733	5
Smelting and pressing of non-ferrous metals	0.3571	6
Manufacture of paper and paper products	0.3083	7
Manufacture of chemical fibers	0.2622	8
Manufacture of textiles	0.2037	9
Manufacture of rubber and plastic	0.1593	10
Manufacture of metal products	0.1432	11
Processing of timber, manufacture of wood, bamboo, rattan, palm, and straw products	0.1266	12
Manufacture of wine, drinks and refined tea	0.1060	13
Manufacture of medicines	0.1058	14
Manufacture of foods	0.1040	15
Printing, reproduction of recording media	0.0847	16
Manufacture of general purpose machinery	0.0835	17
Metal products, machinery, and equipment repair	0.0709	18

(continued)

Table 5.17 (continued)

Industry	Energy intensity	Ranking of energy utilization effectiveness
Processing of food from agricultural products	0.0656	19
Manufacture of railway, ship, aerospace, and other transportation equipment	0.0632	20
Manufacture of special purpose machinery	0.0597	21
Manufacture of automobiles	0.0596	22
Leather, furs, down, related products, and footware	0.0522	23
Utilization of waste resources	0.0506	24
Manufacture of textile wearing and apparel	0.0504	25
Manufacture of measuring instruments	0.0428	26
Manufacture of electrical machinery and equipment	0.0427	27
Manufacture of furniture	0.0382	28
Manufacture of computers, communication, and other electronic equipment	0.0363	29
Manufacture of tobacco	0.0309	30
Manufacture of articles for culture, education, and sport activity	0.0306	31

Data source Sorted and calculated from related data in the China Energy Statistical Yearbook 2014

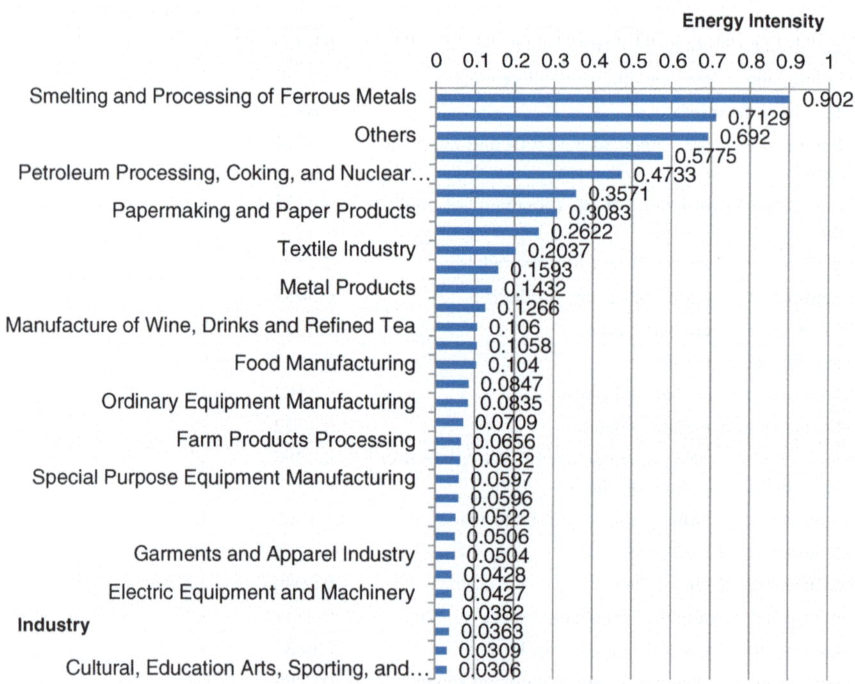

Fig. 5.13 Energy utilization effectiveness of the manufacturing sector by industry classification. *Data source* Sorted and calculated from related data in the China Energy Statistical Yearbook 2014

5.1.4 *Analysis of Environmental Pollution in the Manufacturing Industry*

The three types of waste emissions are important indexes that measure environmental pollution. In this section, the three types of waste emissions were adopted for the primary analysis of environmental pollution in China's manufacturing sector. Table 5.18 shows the three types of waste emissions in China's manufacturing sector in 2011–2013. As observed from the table, industrial wastewater emissions increased drastically from 17.03 billion tons in 2012 to 40.07 billion tons in 2013, an increase of 135.27%. Industrial sulfur dioxide emissions dropped slightly to 9.4 million tons in 2013 from 9.53 million tons in 2012. General solid waste incineration dropped 5.87% to 933,940 tons in 2013 from 2012.

Figures 5.14, 5.15, and 5.16 visually present the structure of the three types of waste emissions in manufacturing.

Table 5.18 Changes in emissions of three types of waste in manufacturing

Environmental index	2011	2012	2013	Growth rate (%)
Total industrial wastewater emissions (thousand tons)	17317350.0	17030150.0	40067290.0	135.27
Industrial sulfur dioxide emissions (thousand tons)	9680.3	9532.1	9443.6	-0.93
General solid waste incineration (thousand tons)	968818.0	992196.0	103692.0	11.03

Data source Sorted and calculated from related data in the China Energy Statistical Yearbook 2014

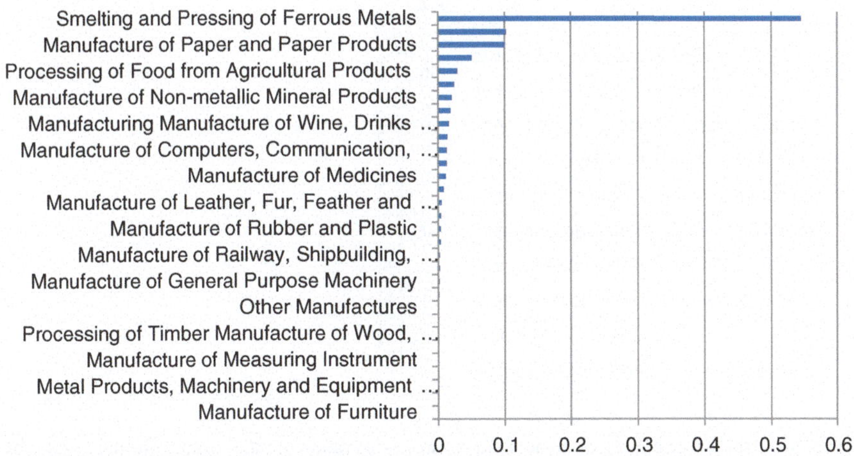

Fig. 5.14 2013 Distribution of total industrial wastewater emissions in the manufacturing sector by industry classification. *Data source* Sorted and calculated from related data in the China Statistical Yearbook on Environment 2014

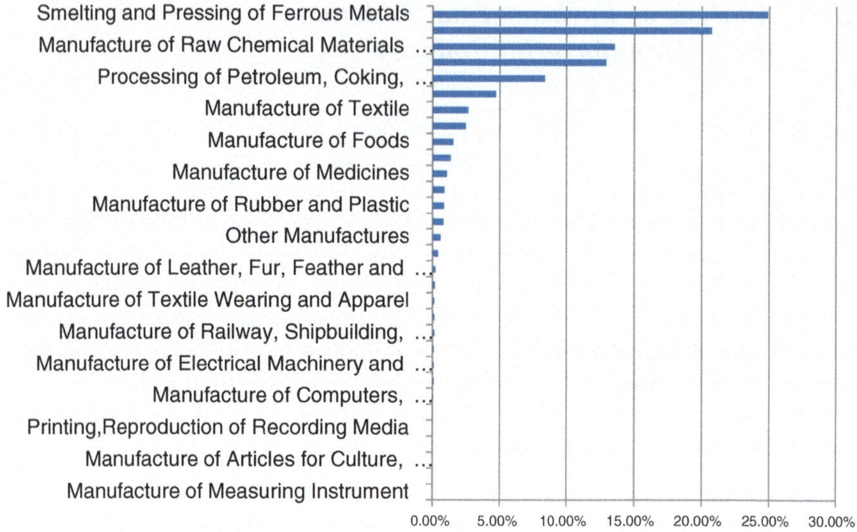

Fig. 5.15 Distribution of total waste gas (SO_2) emissions in the manufacturing sector by industry classification in 2013. *Data source* Sorted and calculated from related data in the China Statistical Yearbook on Environment 2014

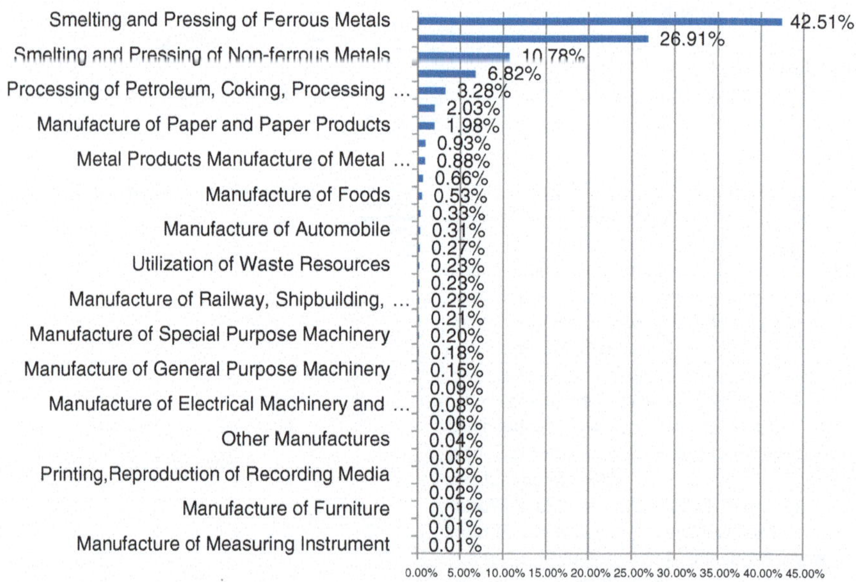

Fig. 5.16 2013 Distribution of total solid waste incineration in the manufacturing sector by industry classification. *Data source* Sorted and calculated from related data in the China Statistical Yearbook on Environment 2014

As observed from the 2013 statistical data on emissions and handling of wastewater emissions in the manufacturing sector by industry classification (see Fig. 5.14), the following five industries incurred the most wastewater emissions: smelting and pressing of ferrous metals (54.57%); manufacture of raw chemical materials and chemical products (10.22%); the manufacture of paper and paper products (10%); manufacture of textiles (5.06%); and processing of food from agricultural products (2.91). Their total wastewater emissions accounted for 82.77% of the manufacturing sector's total. On the other hand, the following five industries had the least wastewater emissions: manufacture of furniture (0.01%); printing, reproduction of recording media (0.03%); metal products, machinery, and equipment repair (0.03%); manufacture of articles for culture, education, and sport activity (0.04%); and manufacture of measuring instruments (0.04%). Their total wastewater emissions accounted for only 0.15% of the manufacturing sector's total.

As shown in Fig. 5.15, sulfur dioxide emissions in 2013 were concentrated in the following six industries: smelting and pressing of ferrous metals (24.9%); manufacture of non-metallic mineral products (20.76%); manufacture of raw chemical materials and chemical products (13.58%); smelting and pressing of non-ferrous metals (12.95%); processing of petroleum, coking, processing of nuclear fuel (8.39%); and the manufacture of paper and paper products (4.75%). Their total emissions accounted for 85.335% of the manufacturing sector's total. On the other hand, the six industries with the least sulfur dioxide emissions were utilization of waste resources (0.06%); printing, reproduction of recording media (0.04%); manufacture of furniture (0.03%); manufacture of articles for culture, education, and sport activity (0.02%); metal products, machinery, and equipment repair (0.01%); and manufacture of measuring instruments (0.17%).

As observed from Fig. 5.16, solid waste incineration in 2013 was highly concentrated in the following five industries: smelting and pressing of ferrous metals (42.51%); manufacture of raw chemical materials and chemical products (26.91%); manufacture of non-metallic mineral products (10.78%); smelting and pressing of non-ferrous metals (6.82%); and processing of petroleum, coking, processing of nuclear fuel (3.28%). Their total general industrial solid waste incineration accounted for 90.30% of the manufacturing sector's total. On the other hand, the following five industries had the lowest levels of total general industrial solid waste incineration: Manufacture of measuring instruments (0.01%); manufacture of articles for culture, education, and sport activity (0.01%); manufacture of furniture (0.01%); metal products, machinery, and equipment repair (0.02%); and printing, reproduction of recording media (0.02%). Their total general industrial solid waste incineration accounted was a negligible proportion of the manufacturing sector's total.

5.1.5 Comprehensive Evaluation of China's New Manufacturing Sector Patter

By referring to the method used in Chap. 5 for the overall evaluation of China's manufacturing sector, this chapter employed the four dimensions of economic resourcefulness, innovation capabilities, energy-saving potential, and environmental protection to construct a comprehensive index for evaluating the new industry patter. Specific indexes are shown in Table 5.19.

The evaluation of the existing and potential development in each of China's manufacturing industries was made possible by assessing the new industry patter in China's manufacturing sector. The maximizing deviation method was employed (see related content in Chap. 5 for specifics) to evaluate the comprehensive indexes for the new industry patter in China's manufacturing sector (see Table 5.20).

As observed from the ranking, the top five industries in the overall evaluation of the new industry pattern were manufacture of computers, communication, and other electronic equipment; manufacture of electrical machinery and equipment; manufacture of automobiles; manufacture of general purpose machinery; and manufacture of special purpose machinery. Relatively speaking, the economic resourcefulness and technology innovation capabilities of these industries were higher whereas their energy consumption and environmental pollution were lower. Therefore, the Chinese government should devote more resources to develop these types of industries. The five trailing industries in the ranking were smelting and pressing of ferrous metals; manufacture of paper and paper products; other manufactures; utilization of waste resources; and metal products, machinery, and equipment repair. The industrial pollution and energy consumption of these five industries were relatively higher whereas their economic resourcefulness and technology innovation capabilities were lower. Limiting development of these industry types is worth consideration.

5.2 Impact of Technology Innovations on China's Manufacturing Sector

Innovation drives the knowledge economy, The 18th *Communist* Party Congress Work Report proposed policy guidance for deepening technology reform; promoting close relations between technology and economy; establishing an institutional mechanism for improving and motivating original innovations; integrating innovations; continuing to innovate from the transfer of knowledge absorbed; establishing innovation collaboration mechanisms between industries and academia; and clearly defining technology innovations as a core part of China's overall development. The manufacturing industry has been the engine of development for a long time and this development has been impacted by many factors. China is currently at a critical stage of economic transformation and reform; the

Table 5.19 Evaluation index system for China's new industry patter in the manufacturing sector

Overall index	Serial No.	Main index	Serial No.	Sub-index	
Evaluation index system for China's new industry model in manufacturing	A	Economic index	A1	Output	Total manufacturing output (hundred million yuan)
			A2		Ratio of total manufacturing output to total industrial output (%)
			A3	Profit	Total profits of the manufacturing enterprise (hundred million yuan)
			A4		Profit margin per manufacturing worker (yuan/person)
			A5	Efficiency	Labor productivity rate per manufacturing worker (ten thousand yuan/person)
			A6	Employment	Number of manufacturing workers (ten thousand people)
			A7		Ratio of number of manufacturing workers to total number of industrial workers (%)
			A8	Taxes	Total taxes and profits of manufacturing enterprise (hundred million yuan)
			A9		Average income tax rate of manufacturing worker (ten thousand/person)
	B	Technological index	B1	R&D	Manufacturing R&D expenditures (ten thousand yuan)
			B2		Manufacturing R&D FTEs (person-years)
			B3		Manufacturing R&D input-intensity (%)
			B4		Ratio of manufacturing R&D workers to number of employed (%)
			B5	Product development	Number of newly developed manufacturing products (item)
			B6		R&D expenses on newly developed manufacturing products (ten thousand yuan)
			B7	Patent	Number of industrial patents applied (item)
			B8		Number of industrial patents owned (item)

(continued)

Table 5.19 (continued)

Overall index	Serial No.	Main index	Serial No.	Sub-index	
			B9	Technology transfer	Output of new manufacturing products (ten thousand yuan)
			B10		Output rate of new manufacturing products (%)
			B11		Industrial technology innovation input–output coefficient
	C	Energy index	C1	Total consumption	Industrial energy consumption (ten thousand tons of coal equivalent)
			C2		Industrial energy consumption per output unit (ten thousand tons of coal equivalent/hundred million yuan)
			C3	Electricity consumption	Ratio of industrial electricity consumption (%)
			C4	Consumption coal	Industrial energy consumption (ten thousand tons of coal equivalent)
	D	Environmental protection index	D1	Wastewater	Industrial pollutant emissions (wastewater) (ten thousand tons)
			D2		Industrial pollutant emissions (wastewater) per output unit (ten thousand tons/hundred million yuan)
			D3	Waste gas	Industrial pollutant emissions (waste gas) (hundred million cubic meters)
			D4		Industrial pollutant emissions (waste gas) per output unit (hundred million cubic meters/hundred million yuan)
			D5	Solid waste	Industrial emissions (solid wastes) (tons)
			D6		Industrial emissions (solid wastes) per output unit (tons/hundred million yuan)
			D7		Industrial solid waste disposal rate (%)

Table 5.20 Ranking of China's new manufacturing patter by industry classification in 2013

Industry	New evaluation score	Rank
Manufacture of computers, communication, and other electronic equipment	1.283671	1
Manufacture of electrical machinery and equipment	1.092249	2
Manufacture of automobiles	1.039817	3
Manufacture of general purpose machinery	0.940573	4
Manufacture of special purpose machinery	0.910022	5
Manufacture of tobacco	0.858218	6
Manufacture of railway, ship, aerospace, and other transportation equipment	0.811531	7
Manufacture of medicines	0.802509	8
Manufacture of raw chemical materials and chemical products	0.794434	9
Manufacture of measuring instruments	0.755043	10
Manufacture of metal products	0.752765	11
Processing of food from agricultural products	0.736394	12
Manufacture of rubber and plastic	0.734979	13
Manufacture of textiles	0.723666	14
Smelting and pressing of non-ferrous metals	0.690856	15
Manufacture of textile wearing and apparel	0.677882	16
Manufacture of articles for culture, education, and sport activity	0.629578	17
Leather, furs, down, related products, and footware	0.627215	18
Manufacture of chemical fibers	0.617823	19
Manufacture of foods	0.617216	20
Manufacture of wine, drinks and refined tea	0.609681	21
Printing, reproduction of recording media	0.606659	22
Manufacture of non-metallic mineral products	0.60557	23
Manufacture of furniture	0.603539	24
Processing of petroleum, coking, processing of nuclear fuel	0.59243	25
Processing of timber, manufacture of wood, bamboo, rattan, palm, and straw products	0.582348	26
Metal products, machinery, and equipment repair	0.576422	27
Utilization of waste resources	0.540358	28
Other manufactures	0.52753	29
Manufacture of paper and paper products	0.508718	30
Smelting and pressing of ferrous metals	0.459529	31

Data source Sorted and calculated from related data in the China Statistical Yearbook 2014, China Statistical Yearbook on Science and Technology 2014, China Statistical Yearbook on Environment 2014, and China Energy Statistical Yearbook 2014 demographic dividend is gradually decreasing while pressure from the resource environment continues to increase. Manufacturing will be able to stay healthy and thrive only if technology innovation capabilities can be improved and a technology innovation path well suited to China's social system and economic development can be identified.

5.2.1 Literature Review

Research on technology innovations and economic growth in the manufacturing sector can be divided into three categories. The first proposes that technology innovation is an important factor in the promotion of economic growth. Arundel (2001) employed patent data to assess the relationship between technology innovations and economic growth and found a significantly positive relationship between the two. Huang and Yu (2007) employed a number of patent licenses as a proxy variable for technology innovation activities to perform empirical analysis on the relationship between China's technology innovation activities and economic growth. The results showed that patents of new practical technologies have the biggest impact on economic growth, followed by patents on outlook design; meanwhile, invention patents have the least impact on economic growth, indicating a low utilization rate of technology innovation results in China. Further analysis showed that the effect of technology innovation on economic growth weakens from east to west. In a comprehensive survey on technology advances, Kennedy and Thirlwall (1972) found that increased R&D expenditures significantly impact overall national economic growth. Lu and Jin (2011) conducted an empirical analysis on the long-term equilibrium relationship between R&D investment and economic growth using a unit root test and a co-integration test. The results showed that R&D investment has a significant promoting effect on economic development and that the output elasticity of R&D worker input is greater than the output elasticity of R&D investment input. In addition, both input factors of R&D investment and R&D workers have certain lagging effects during the process of directly promoting technology advances and enhancing the production rate. Liang et al. (2006) employed co-integration theory and an error correction model to study the correlation between economic growth and innovation activities. The results showed that innovation activities significantly impacted short-term economic growth in China but the extent was far lower compared with the long-term impact. From the long-term point of view, basic research on various kinds of innovation activities helps achieve and promote higher economic growth. Guo (2006) analyzed the contribution of technological advances to the industrial structure in China. Guo (2006) considered that technological advances promote adjustment, optimization, and upgrade of industrial structure, and have played a positive role in promoting the rise of the tertiary sector in China. Xu conducted an empirical study on the regression relationship between R&D investment and technology innovations on the one hand, and GDP on the other, and found that R&D investment and technology innovations can promote GDP growth. In addition, Xu proposed that R&D investment promotes economic growth because as R&D investment promotes technology innovations, the input and diffusion thereof further improves economic growth. Zhu and Zhang employed a panel data research method to study the impact of technology innovation on the regional economic growth of the eight biggest economic regions in China. The conclusions indicated there is higher explanatory power in the contribution of technology innovation capabilities to economic growth; that China lacks the technology innovation level found in developed regions; and the contribution of technology innovation to economic growth in China is lower than that in developed regions, resulting in a continuously growing gap in economic development between China and developed regions.

According to the second research view, technology innovation has no impact on economic growth. Jiang and Shao (2010) employed a t-test to calculate and analyze the correlation between technology input and economic growth, and found that technology input has very little impact on economic growth while the input of technological workers has an insignificant impact on economic growth. Wang and Yang (2007) primarily adopted an econometrics analysis method to estimate the degree of contribution made by technological advances to economic growth in China since the reform and opening period. The results showed that China's economic growth is primarily dependent on large capital investment. Although the overall technology level is improving, the degree of contribution remains low. Li and Huang (2009) constructed a model combining a GES production function and C–D production function to estimate the contribution made by capital investment, labor input, and technological advances to economic growth. The results showed that the annual average growth rate of technological advances is relatively low and the contribution rate of technological advances to economic growth is also low. China's economy primarily depends on production inputs, such as capital and labor, which exemplify the traditional mode of extensive economic growth. Niu and Sun (2013) employed a generalized C–D production function to analyze the contribution of capital, labor, and technological advances to the economic growth of Gansu province. The study found that Gansu remains at the development stage, requiring high capital and labor inputs to promote its economic growth; in other words, Gansu is still in the extensive growth mode in which the technology transfer effect is not significant. Since the economic development of less developed regions is affected by factors like the level of marketization, efficiency in resource allocation, and foundation of industrial development, the production level from the transformation of technological advance is not high and neither is the effect of economic growth. Lawrence (1999) compared economic growth among East Asian countries and pointed out that economic growth in a majority of developed countries indeed came from innovations. However, in the newly industrialized countries or regions, such as the so-called Four Little Dragons in Asia, no correlation was observed between growth and innovations. Nonetheless, from a strategic point of view, Lawrence (1999) recommended that developing countries, such as China, consider focusing more on education and innovations when economic growth begins to take off.

According to the last research view, technology innovation inhibits economic growth. Based on endogenous economic growth theory, Dou and Zhao (2005) employed an empirical analysis to study the impact of technology input on economic growth. The results showed that the impact of technology input on economic growth is insignificant in large and medium-sized industrial enterprises; technology inputs in R&D institutes have significantly negative impact on economic growth; and technology inputs in higher education institutes promote economic growth. Under the assumption of variable returns to scale, Li and Xie (2007) analyzed the growth factors of technological advances in China. The results indicated that China is currently at a stage of increasing returns to scale, and is primarily dependent on capital and labor inputs to achieve economic growth, while the impact on economic output due to technological advances is very low or even negative, showing an average rate of technological advance of −0.39%. Qin et al. (2014) employed economic growth as the threshold variable to construct a panel threshold model and performed an empirical analysis on the impact of technology advances on economic growth. The results showed that the impact of technological advances have a "double threshold effect" on economic growth; technological advances will have a positive impact on economic growth only beyond a given threshold of economic growth. Due to uneven development among regions in China, the positive impact of technological advances on economic growth weakens gradually from east to west in China.

To a certain extent, Chinese and foreign scholars have made some achievements in studying the impact of technology innovations on economic growth, although there have been fewer analyses related to the function of independent R&D and technology introduction in the economic growth of the manufacturing sector using the technology innovation model. Technology innovations are mainly divided into two major types: independent innovations and others. Generally, independent innovation is further divided into three types: original innovations, integrated innovations; and introduction, absorption, and re-innovation. Original innovations refer to independent R&D and are the primary path undertaken by an enterprise to acquire its core technology. Technology introduction, digestion, absorption, and re-innovation serve as crucial supplements. Using the manufacturing point of view, we explored the impact of technology innovations on manufacturing, and in conjunction with the perspectives of independent R&D and technology introduction, we empirically analyzed their impacts on economic growth.

5.2.2 Empirical Analysis

1. Model specification

There are many factors that impact economic growth in the manufacturing sector. They are mainly physical capital investment, labor input, technology innovations, investment, output, industry openness, and policies. In order to accurately explore the impact factors on growth in manufacturing while simultaneously considering the characteristics of industry panel data, we attributed the impact factors to physical capital investment, labor input, foreign direct investment, technology innovations, and industry openness where technology innovations were sub-divided into independent R&D and technology introduction. The quantitative model is defined as follows:

$$
\begin{aligned}
1n\,Y_{it} = \alpha + \beta_1 1n\,R\&D_{it} + \beta_2 1n\,Tec_{it} + \beta_3 1n\,K_{it} + \beta_4 1n\,L_{it} \\
+ \beta_5 1n\,FDI_{it} + \beta_6 Trade_{it} + \varepsilon_{it}
\end{aligned}
$$

where Y_{it}, $R\&D_{it}$, Tec_{it}, K_{it}, L_{it}, FDI_{it}, and $\beta_6 Trade_{it}$ represent economic growth in manufacturing, R&D funding, technology introduction, physical capital investment, labor input, foreign direct investment, and industry openness, respectively, in the t-th year; α represents a constant term; β represents the variable coefficients; and ε represents the residual.

2. Description of variables

(1) Economic output of manufacturing sector (Y)

The economic output of the manufacturing sector reflects the production capabilities of manufacturing, which are generally represented by total industrial output. However, owing to the lack of such data in the China Statistical Yearbooks 2013 and 2014, and in order to preserve data continuity, the index of revenues from primary businesses was adopted as a replacement. Revenues from primary businesses reflect the industrial product sales of an enterprise in its daily operations, and are most meaningful and representative of an enterprise when total output information is missing. The revenue data from the primary businesses of manufacturing industries were taken from the China Statistical Yearbooks (2006 to 2014).

(2) Independent R&D (R&D)

Technology innovation is a crucial factor that impacts the economic resourcefulness of the manufacturing sector. Technology is the first production force in which improvement of production efficiency in the manufacturing sector relies on advances in science and technology, of which independent R&D plays a particularly integral role in technology innovations. Independent R&D is the primary means undertaken by an enterprise in acquiring its core technology. Only after acquiring core technology can an enterprise secure its leading position in a business. Therefore, we drew on existing research and selected internal R&D expenses to measure independent R&D. We expected that independent R&D would significantly impact economic growth in the manufacturing sector.

(3) Technology introduction (Tec)

In addition to R&D activities, other technology activities also impact technology innovation activities in manufacturing. Advances in science and technology in manufacturing enterprises originate from independent R&D and technology introduction. Although technology introduction can improve the scientific and technological level in manufacturing enterprises in the short run, long-term technology introduction will lead to dependency; core technology will be controlled by people; international competitiveness will not be strong; and the promoting effect of technology introduction on economic growth in manufacturing will be insignificant. We drew on existing research (Deutsche Bank 2013) to reflect technology introduction by employing the sum of technology introduction expenses, and digestion and absorption expenses. We expected that technology introduction would positively impact economic growth in manufacturing.

(4) Physical capital investment (K)

Physical capital investment directly impacts growth in the manufacturing sector. Physical capital investment has always been an integral factor in China's economic growth. Economic growth in China primarily relies on capital investment and is sustained by resource consumption, which is typical of the extensive growth mode (Wu 2006, 2007). Under certain conditions, the greater the physical capital investment and production scale of an enterprise are, the stronger is productivity and the higher is total output. Considering data availability, net investment in fixed assets was adopted to measure physical capital investment. We expected that physical capital investment would positively impact economic growth in the manufacturing sector.

(5) Labor input (L)

Labor input is another factor that impacts growth in the manufacturing sector. Manufacturing development is inseparable from people's participation. According to classic economics and new theories of economic growth, human labor plays a crucial role in economic growth. The number of personnel directly impacts production scale in manufacturing. In addition, personnel quality directly impacts production efficiency in manufacturing. Human labor has a decisive role in the growth of the manufacturing sector. We drew on existing research (Zhu 2005; Hu and Shao 2010) and selected the total annual average number of manufacturing personnel as the labor input. We expected that as labor input increases, production innovation capabilities of manufacturing would strengthen.

(6) Foreign direct investment (FDI)

Investment is one of three drivers of economic growth. Investment can promote development of output scale in the manufacturing sector and plays a positive role in enhancing the economic resourcefulness in the manufacturing sector. Yao et al. (2006) considered that foreign direct investment could minimize inefficiency in China's domestic production and expedite China's technological advances. Therefore, foreign direct investment can drive economic growth. Wang et al. (2010) employed panel data from seven internal departments of the equipment manufacturing industry in 31 provinces from 2001 to 2007 and used an ordinary least squares model and a fixed-effects model in the study to find that foreign direct investment significantly promotes growth in the manufacturing sector. We selected the amount of foreign direct investment to measure FDI. We expected that foreign direct investment would positively impact the economic resourcefulness of the manufacturing sector.

(7) Industry openness (Trade)

It has been 30 years since China started opening up, and the level of its openness has been increasing gradually over the years and the manufacturing sector has been developing rapidly. The level of openness is an integral factor that impacts economic growth in the manufacturing sector. The more open an industry is, the higher its exports are, and hence, the more exports drive the economy. Wu (2012) argued that the level of openness is primarily improved by technological efficiency, which promotes total factor productivity (TFP). Li (2011) considered that an increase in the level of openness significantly promoted economic growth. To reflect the factors of economic growth in the manufacturing sector fully, we incorporated exports into the model and employed export dependency in the manufacturing sector to reflect the level of openness, represented by the value of export delivery divided by manufacturing output. Since the index of total industrial output was not published in the China Statistical Yearbooks of 2013 and 2014, we employed the value of export delivery divided by revenues from primary business in manufacturing to show the level of external dependency in the manufacturing sector. We expected that the level of openness would positively impact growth in manufacturing.

3. Data processing Classification of data acquired by industry classification

(1) Consolidation of industries

In the China Statistical Yearbooks on Science and Technology for 2006–2012, the manufacturing sector is divided into 30 industries. Among them, rubber and plastics were classified as two different industries; and transportation equipment manufacturing was classified as an individual industry. On the other hand, in the China Statistical Yearbooks of 2013 and 2014, the manufacturing industry was divided into 31 industries. Among them, manufacture of automobiles was classified as a separate industry from manufacture of railway, ship, aerospace, and other transportation equipment, whereas the manufacture of rubber and plastic was consolidated as one industry. Therefore, in order to ensure data consistency, the individual industries of the manufacture of rubber and plastic industries listed in 2006–2012 were consolidated into one industry. In addition, the manufacture of automobiles industry and the manufacture of railway, ship, aerospace, and other transportation equipment industry listed in 2013 and 2014 were consolidated into one industry, renamed transportation equipment manufacturing.

(2) Elimination of industries

Since there was variance in 2013 and 2014 in the classification of "others" and "arts, crafts, and others," these two industries were eliminated. In addition, there were no statistical data available on R&D funding in waste resources and waste materials recycling and processing as well as manufacture of tobacco. Furthermore, there are no data on the R&D funding. Therefore, these two industries were eliminated.

(3) Revision of industry names

Since some of the industry names were revised in the China Statistical Yearbook of 2013 and 2014, to ensure accurate reflection of the current industry classification in the manufacturing industry, we revised the following industry names accordingly: beverage manufacturing was renamed manufacture of wine, drinks, and refined tea; garment, footware, and headgear manufacturing was renamed manufacture of textile wearing and apparel; leather, furs, down, and related products was renamed manufacture of leather, fur, feather, and related products and footware; cultural, education, and sporting products was renamed manufacture of articles for culture, education, and sport activity; instrument, meter, cultural, and clerical machinery manufacturing was renamed manufacture of measuring instruments.

(4) Handling of missing data

Since the average annual number of manufacturing workers was not published in the China Statistical Yearbook 2013, an average over the years of 2006–2008 was adopted instead. The final sample comprised 26 industries over 9 years.

The 26 industries were as follows: processing of food from agricultural products; manufacture of foods; manufacture of wine, drinks, and refined tea; manufacture of textiles; manufacture of textile wearing and apparel; manufacture of leather, fur, feather, and related products and footware; processing of timber, manufacture of wood, bamboo, rattan, palm, and straw products; manufacture of furniture; manufacture of paper and paper products; printing, reproduction of recording media; manufacture of articles for culture, education, and sport activity; processing of petroleum, coking, processing of nuclear fuel; manufacture of raw chemical materials and chemical products; manufacture of medicines; manufacture of chemical fibers; manufacture of rubber and plastic; manufacture of non-metallic mineral products; smelting and pressing of ferrous metals; smelting and pressing of non-ferrous metals; manufacture of metal products; manufacture of general purpose machinery; manufacture of special purpose machinery; transportation equipment manufacturing; manufacture of electrical machinery and equipment; manufacture of computers, communication, and other electronic equipment; and manufacture of measuring instruments.

4. Statistical description of main variable indexes

Table 5.21 provides a brief statistical description of the characteristics of the major variables.

5. Quantitative results and analysis

(1) Selection of quantitative methods

In order to obtain reliable conclusions, for our estimation, we used methods that are widely adopted by Chinese and foreign scholars, that is, fixed-effects (FE),

Table 5.21 A brief statistical description of the characteristics of the major variables

Main variable	Sample size	Average	Standard deviation	Minimum	Maximum
lnY	234	9.4677	0.9917	7.2346	11.2545
lnR&D	234	4.3493	1.3788	1.1225	7.1329
lnTEC	234	1.8840	1.6373	−2.5121	5.2185
lnK	234	7.3905	1.0579	4.4841	9.5293
lnL	234	5.3904	0.7320	3.7245	6.7805
lnFDI	234	4.6969	1.0534	2.5349	7.4028
Trade	234	0.1740	0.1625	0.0096	0.6758

Table 5.22 Quantitative regression analysis

Explanatory variable	Fixed effects (FE)	Random effects (RE)	Feasible generalized least squares (FGLS)
	Equation (1)	Equation (2)	Equation (3)
lnR&D	0.0520* (1.90)	0.0582** (2.27)	0.1623*** (6.55)
lnTEC	0.0259 (1.61)	0.0347** (2.21)	0.1027*** (5.76)
lnK	0.6364*** (20.90)	0.6387*** (23.20)	0.6497*** (27.91)
lnL	0.4040*** (5.79)	0.3335*** (5.72)	0.2085*** (6.84)
lnFDI	0.0269 (0.85)	0.0222 (0.72)	−0.1384*** (−5.04)
Trade	0.4721** (2.39)	0.4079** (2.20)	0.6803*** (7.05)
Constant term	2.1030*** (6.66)	2.4554*** (9.36)	3.1447*** (25.65)
R^2	0.8786	0.8903	
Sample size	234	234	234

Note The subscript values for the coefficients are t-values; *** and **, and * indicate the relevant variables have passed significance tests of 1, 5, and 10%, respectively

random-effects, and feasible generalized least square (FGLS) models. Stata 12.0 software was adopted for the analysis.

(2) Analysis of quantitative results

The estimation results of the explanatory variable coefficients are shown in Table 5.22 where the first column shows the estimation results using the FE model, the second column the results using the RE model, and the third column the results using the FGLS model.

According to the Hausman test, p was 0.041; whenpis less than 0.05, the null hypothesis is rejected. Therefore, we selected the FE model, not the RE model as according to the Hausman test, the results from the FE model were better than those from the RE model. Comparing Eqs. (1) and (3), we observe that except for FDI, the coefficients of the explanatory variables generated from both the FE model and the FGLS model were fully consistent. Since the possible existence of heteroscedasticity and serial correlation were eliminated in the FGLS model, we considered that the results generated in the FGLS model were robust. By observing Eq. (3), we arrive at the following conclusions.

First, under the condition that physical capital investment, labor input, foreign direct investment, and level of openness were controlled, the coefficient of independent R&D was 0.1623 and passed the 1% significance test, indicating that independent R&D positively impacts economic growth in manufacturing, which is consistent with our expectation. In addition, for every 1% increase in independent R&D, there was approximately a 0.1623% increase in the economic growth of the manufacturing industry. The coefficient of technology introduction was 0.1027 and passed the 1% test, indicating that technology introduction has a significantly positive impact on the economic growth of the manufacturing sector, which is consistent with our expectation. In addition, for every 1% increase in technology introduction, there was approximately a 0.1027% increase in the economic growth of the manufacturing sector. Based on this analysis, the following two conclusions were drawn. 1. Technology innovations significantly promote economic growth in the manufacturing sector. This is because through independent R&D in the manufacturing sector, technological advances are strengthened and levels of science and technology in manufacturing enterprises can be significantly enhanced, improving the competitiveness of enterprises. 2. Independent R&D investment is stronger than investment of technology introduction in terms of enhancing economic growth in manufacturing enterprises. This is possibly because an enterprise can only acquire its core technology through independent R&D; although technology introduction can enhance the technology level of an enterprise, core technology can hardly be acquired through technology introduction and this impacts the competitiveness of an enterprise's products in the market. Therefore, technology introduction has a weaker power to enhance economic growth in manufacturing, which is why China has been encouraging independent R&D. Through independent R&D, manufacturing enterprises can master core scientific technology and improve competitiveness. In the current knowledge era, enterprises can survive, develop, and expand only after mastering core technology. In addition, other technology activities impact differently on the level of technology enhancement in manufacturing enterprises. The sources of enterprise technology are divided into two parts: first, independent R&D, and second, technology introduction and cooperative development. Technology introduction is an effective way to improve the technology level of manufacturing enterprises. During technology introduction, an enterprise first digests and absorb the technology introduced, followed by purposeful changes, enhancement of technology level in the enterprise, and final transformation into productivity. Technology transformation is the final stage in the process. Only if an enterprise can fully digest and absorb the technology introduced can it become the foundation for both future technology transformation and independent innovations that will finally lead to the acquisition of the enterprise's own core technology and a competitive advantage in the market.

Second, the coefficient of physical capital was 0.6497 and passed the 1% significance test, indicating that at present, physical capital investment has a significantly positive impact on economic growth in manufacturing. This is consistent with our expectation. In addition, for every 1% increase in physical capital investment, there was an approximately 0.6497% increase in economic growth in manufacturing. Physical capital investment is one of the crucial factors that impact economic growth in manufacturing. The coefficient of labor input was 0.2085 and passed the 1% significance test, indicating that at present, labor input has a significantly positive impact on economic growth in manufacturing. In addition, for every 1% increase in labor input, there was an approximately 0.2085% increase in economic growth in manufacturing. This shows that labor input is one of the crucial factors that impact growth in manufacturing. The number of manufacturing workers in China has been increasing every year and the education level of manufacturing workers has been improving, indicating that the skills quality of workers is improving, there by promoting economic growth in manufacturing.

In addition, the coefficient of foreign direct investment was −0.1384 and passed the 1% significance test, indicating that foreign direct investment negatively impacts growth in manufacturing, that is, foreign direct investment inhibits economic growth in manufacturing. This is contrary to our expectation. Chinese scholars are divided about whether foreign direct investment promotes economic growth, leading to two different views on the impact of foreign direct investment on the economy. The first view is that foreign direct investment promotes capital accumulation and technological advances, thereby enhancing economic growth. The other view is that foreign direct investment generates a crowding-out effect on domestic investment, which is not conducive to economic structural adjustment, thereby inhibiting economic growth. Foreign direct investment per se brings technologies that are missing in China's manufacturing industries and can, in the short run, enhance the technology level of a manufacturing enterprise, improve production efficiency, and strengthen productivity. However, these kinds of technologies are controlled by people and core technologies are difficult to transform. Hence, long-term technological advances in manufacturing enterprises are difficult to maintain. If enterprises rely solely on an inflow of foreign technologies but do not undertake independent innovations, enterprises will gradually lose their independent innovation capabilities, and hence, will lose their competitiveness as well.

5.2.3 Conclusions and Policy Recommendations

Panel data of the manufacturing sector from 2005 to 2013 were employed to study independent R&D and the impact of technology introduction on total manufacturing output. The conclusions are as follows. When physical capital investment, labor input, foreign direct investment, and level of openness were controlled, both independent R&D and technology introduction showed positive impacts on economic growth in manufacturing, that is, independent R&D and technology introduction could promote economic growth in manufacturing. In addition, physical capital investment and labor input can promote economic growth in the manufacturing sector. In terms of the extent of the impact, the impact of independent R&D and technology introduction on the manufacturing economy is smaller than the impact from physical capital investment and labor input. Foreign direct investment has an inhibitory effect on economic growth in the manufacturing sector. The level of openness can significantly improve economic growth in the manufacturing sector, which can be interpreted indirectly as exports significantly promoting economic growth in manufacturing.

Based on these conclusions, the following recommendations are proposed.

(1) China should increase science research and development funding and promote independent R&D. According to our analysis, when physical capital investment, labor input, foreign direct investment, and level of openness were controlled, both independent R&D and technology introduction were conducive to economic growth in China's manufacturing sector. As observed, technology innovations have provided tremendous support to the development of China's manufacturing. Therefore, on the one hand, science research and development funding should be increased, especially independent R&D funding. The marginal benefits generated by independent R&D investment are greater than those generated by technology introduction, which is supplemental to independent R&D. In addition, we should recognize that it is difficult to acquire real core technology through technology introduction, and that only by increasing the intensity of independent R&D, can an enterprise improve its technology level, become more efficient in its production, and hence, be able to produce products that are more competitive. Consequently, we recommend selective introduction of technologies in order to form a complementary relationship between independent R&D and technology introduction, such that China's manufacturing industry slogan can be transformed from "made in China" to "wisely made in China." At present, China has already implemented tax preferential policies for high-tech enterprises. In addition, qualified local governments should provide appropriate subsidies to enterprises for technology innovations.

(2) China should increase physical capital investment and promote the skills quality of workers. As observed from the results, physical capital investment and labor input have strong promoting effects on the manufacturing economy, which indicates that China's manufacturing sector is still at the factor-driven stage. For physical capital investment, China's manufacturing is under a macro environment of energy savings and emissions reduction. Only by eliminating outdated machinery and equipment, can China strengthen its manufacturing infrastructure and enhance and modernize its physical capital, such that productivity efficiency can be effectively improved and self-competitiveness can be strengthened. At this stage, China is gradually losing its human capital advantage and it is only with improvements in the

skills and qualities of workers that China will be able to promote production efficiency, thereby facilitating rapid development in the manufacturing sector.

(3) China should ensure there is wise use of foreign direct investment. As observed from the results, foreign direct investment has a negative impact on China's manufacturing economy, indicating that foreign direct investment inhibits the economic growth of China's manufacturing sector. Foreign direct investment per se brings certain advanced technology and management experience that can generate technology spillover effects for China's manufacturing sector, enabling development in manufacturing to a certain extent. However, we should recognize that advanced technology cannot be bought, and core technology can be acquired only through independent R&D. Therefore, China should gradually decrease foreign direct investment, particularly foreign investment in industries with high energy consumption and high pollution. In addition, China should increase independent R&D funding, attract idle funds to manufacturing, and lower dependence on foreign investment to better promote the healthy development of China's manufacturing sector.

(4) China should improve its level of exports and raise its level of openness. As observed from the results, the marginal effect from the level of openness to the economic growth of the manufacturing sector is the highest among all factors considered. This indicates that foreign trade strongly promotes economic growth of China's manufacturing sector and reflects that China's manufacturing products are highly competitive internationally. Although manufacturing exports are conducive to the rapid development of the manufacturing sector, exports rely on a stable international environment. Therefore, China should not rely overly on exports. While strengthening exports, domestic demand should also be boosted to strike a balance between the two. Thus, when crisis arrives, manufacturing enterprises will be able to remain strong and resilient.

5.3 Energy Efficiency in China's Manufacturing and Its Impact Factors

Manufacturing development is the key to a country activating industrial development, smoothly integrating into the globalized economy, and achieving rapid national economic development. Since China's reform and opening up, its economy has been developed rapidly, with manufacturing comprising almost one-third of its GDP. The fact that manufacturing is integral to China's economy is self-evident as is its contribution to economic growth. However, while manufacturing is developing rapidly, its energy consumption ratio also remains top-ranked among all other sectors, accounting for approximately 58% of the total industrial energy consumption in 2004–1012. Stan (2014) pointed out that while China's GDP has been growing at an average rate of 10% in the past 20 years, its energy consumption has also shown a huge and fast-growing trend. In addition, according to the experiences of developed countries, the total energy consumption in China will continue to rise and the peak of per capita energy consumption is yet to arrive.

"Green development" was proposed in China's National "12th Five-Year Plan." The model is an innovation of the traditional development foundation, which emphasizes green development, resource conservation, and construction of an environmentally friendly society. China is a large manufacturing country that is highly energy-consuming and seriously polluting. Therefore, energy conservation and emissions reduction are of crucial significance to China's manufacturing sector. To achieve sustainable development in manufacturing, besides factor inputs, China also needs to improve the carrying capacity and efficiency of its resource environment. Therefore, studies related to energy efficiency in manufacturing under the constraints of energy conservation and emissions reduction are very relevant and significant. This study analyzes the factors that impact energy efficiency in China's manufacturing industries under the environmental constraint.

5.3.1 Literature Review

Lovell (1993) argued that energy efficiency is a technique for efficiency in which fewer inputs, including energy, are used to generate the highest possible output. Patterson (1996) defined energy efficiency as the ratio between expected output and energy inputs in production. In recent years, there have been two types of research on energy efficiency in China. The first involves single-factor energy efficiency, which does not take into account other impact factors, and the second involves energy efficiency based on total factors.

Unit GDP energy consumption, an internationally representative single-factor energy efficiency index, is the ratio between energy consumption intensity and the energy production rate, which reflects the relationship between energy consumption in economic activities and effective output. The concept of total factor energy efficiency was first proposed by Hu and Wang (2006) in reference to the production relationship between inputs, such as resources, labor force, and energy, and GDP output. The evaluation index of total factor energy efficiency is the ratio of target energy input to actual energy input.

Although the single-factor energy efficiency index is intuitive and can be calculated easily, other input factors are not taken into account. The single-factor energy efficiency index aims only to observe the relationship between energy as a single input factor and GDP output. While GDP output is the result of energy and the combination of many substitutable input factors, such as labor force and capital, single-factor energy efficiency reflects only the ratio between energy and economic output, thereby exaggerating energy efficiency and failing to reflect the substitution effect on energy of other production factors, such as capital and labor force. On the other hand, the total-factor energy efficiency index, under the condition of keeping constant all input factors (e.g., capital and labor force), except for energy, follows the best production practice, which considers the ratio between target energy input and the actual input required for a certain output volume, thereby effectively compensating for the weakness in the traditional single-factor energy efficiency research method (Wang 2011). It was also revealed that for regions with abundant resources in particular, when evaluating the impact of energy efficiency, the total-factor energy efficiency index has an irreplaceable advantage over the single-factor method (Wang 2011). Therefore, the total-factor energy efficiency index is most widely adopted by scholars at present for research on energy efficiency.

Many Chinese and foreign scholars have conducted relevant research on the impact factors of energy efficiency, summarized as follows.

1. Industry rivalry

Industry rivalry can impact energy efficiency to a certain extent. As competition becomes more intense, enterprises are more motivated to engage in R&D that can stimulate improvement of production technology, bring about innovations, and enhance energy efficiency. Some scholars believe that competition among industries promotes energy efficiency (Arrow 1962; Nickell 1996; Ahn 2002) but others argue that competition among industries negatively impacts energy efficiency (Schumpeter 1942; Scharfstein 1988; Raith 2003).

2. Technology

Technological advance is a crucial factor that impacts energy efficiency. However, due to the rebound effect in energy, the impact of technological advances on energy efficiency has become uncertain. Some scholars have found that technological advances are conducive to improvements in energy efficiency (Li and Zhou 2006; Long and Li 2009; Jiang et al. 2011; Tang et al. 2014). On the other hand, some scholars have found that technological advances may cause a rebound effect in energy, thereby making the enhancement effect in energy efficiency insignificant (Khazzom 1980; Birol and Keppler 2000; Wang et al. 2008; Ouyan 2010).

3. Energy

Many scholars argue that structural change in energy consumption, such as lower proportion of coal consumption, has a positive impact on energy consumption. For example, the ratio of China's coal consumption to total consumption in 2013 was 66%, while oil and natural gas accounted for 24.2% and wind, water, and other energy sources accounted for 9.8%. Coal has been the major component of consumption for a long time. In addition, a lot of energy loss is generated during transportation and use (Sinton and Fridley 2000; Hu 2009). However, other scholars have arrived at different conclusions, that is, the higher the coal proportion is, the lower is the energy efficiency (Dong 2008; and Tan and Zhang 2010).

4. Level of openness

Many scholars have analyzed energy efficiency from the perspective of openness but their conclusions are vastly different. Some scholars believe that openness promotes energy efficiency (Sinton and Fridley 2000; Stan 2002; Zhang et al. 2007, 2009) while others argue that the promoting effect of openness to energy efficiency is insignificant or even negative effect (He 2000; Dong 2008; Liu 2008; Xiong 2009; Yang 2009).

5. Some scholars propose that environmental regulations can enhance energy efficiency (Bai and Song 2009; Wan and Tong 2010; and Zhang 2013) while others postulate that environmental regulations cannot promote energy efficiency, and in fact lower it (Li et al. 2011; Chen and Zhang 2012).

By reviewing the abovementioned studies, we observe that fruitful results have been achieved in research on the impact factors on energy efficiency but the research remains insufficient in the following areas. First, many studies have adopted data envelopment analysis (DEA) to explore total factor energy efficiency. DEA is an evaluation method that assesses the relative efficiency between similar types of input and output of certain decision-making units (DMUs). The efficiency measures in Farrell (1957) were confirmed by calculating the distance between the actual production points and the production frontier surface in each DMU. However, DEA might cause input slack in each DMU (Guo 2015). Second, many studies have ignored the undesirable output caused by energy input and have incorporated energy into the DEA model by treating energy only as one of the input factors. Hence, the value of efficiency measured reflects only the maximum output capacity, which will lead to errors in estimation and analysis (Zhang 2014). For example, when undesirable output was accounted for in calculating the total factor energy efficiency of China's regional ecosystem, the value of energy efficiency was significantly lower (Li et al. 2012). At present, very few Chinese studies have incorporated environmental constraints into the research framework of total factor energy efficiency. Research of total factor energy efficiency in China that has considered undesirable output is still in the early stage and these studies are less involved with total factor energy efficiency in manufacturing and more focused on the overall national level (Wang 2011). Third, many studies have focused on provincial regions and ignored the fact that industries have more practical means to save energy and are the basic starting point of energy policies in a country. Research that analyzes the industrial energy consumption rate and its impact factors provides better support for achieving energy conservation and emissions reduction (Sun et al. 2012a, b). In addition, using manufacturing industries as study subjects provides more enriched data that can identify the impact factors of energy efficiency and environmental factors better than using provinces as the study subjects can (Tang et al. 2009).

Based on the existing research foundation, this chapter attempts to study the following two areas in depth. First, based on the slacks-based measure (SBM) model proposed by Tone (2001), which applies to multi-input and multi-output complex production systems, the undesirable output accompanying production was accounted for, that is, the objective function of the input–output slack model was directly utilized to measure the energy efficiency in the manufacturing industries under the environmental constraints. Second, the majority of the literature has adopted panel data at a provincial level for analysis and discussion, which has often overlooked heterogeneity in energy efficiency among industries. Therefore, if energy efficiency of manufacturing industries were analyzed based on different levels of technology, consumption, and environmental pollution, and the respective impacts on energy efficiency were explored, the industrial heterogeneity of impact factors on energy efficiency in manufacturing would be better reflected as well. The present study classified the sample's 29 manufacturing industries in China based on "energy efficiency–environmental pollution" and was aimed at proposing relevant policy recommendations based on the specific industrial characteristics of China's manufacturing sector.

5.3.2 Measurement of Total-Factor Energy Efficiency Under Environmental Constraints and Analysis of Industries

1. Undesirable SBM model

In the existing research, the majority of scholars has adopted DEA analysis to measure total-factor energy efficiency but often have overlooked undesirable outputs, such as environmental pollution. Tone (2001) proposed the SBM model—which, even though it resolves the slackness between input factors and output in the DEA model and applies to complex production, which involves multi-inputs and multi-outputs—ignores the undesirable output that accompanies the production process. Fukuyama (2009) proposed that the point at which both the undesirable and desirable outputs were at zero, output would be accompanied simultaneously by the undesirable output. Yang et al. (2009) indicated that errors would be created in total-factor energy efficiency without the consideration of undesirable output leading to inflated efficiency levels. Therefore, the total-factor energy efficiency in the present study considered undesirable output based on industrial heterogeneity.

The undesirable output SBM model is

$$\min_{\lambda, s^+, s^-} \rho = \frac{1 - \left(\frac{1}{m}\right)\sum_{i=1}^{m} \frac{s_i^-}{x_{i0}}}{1 + \left(\frac{1}{n_1 + n_2}\right)\left[\sum_{j=1}^{n_1}\left(\frac{s_j^g}{y_{j0}^g}\right) + \sum_{j=1}^{n_2}\left(\frac{s_j^b}{y_{j0}^b}\right)\right]}$$

$$\begin{cases} x_0 = X\lambda + S^-, \\ y_0^g = Y^g\lambda - S^g, \\ y_0^b = Y^b\lambda + S^b, \\ \lambda \geq 0, S^- \geq 0, S^g \geq 0, S^b \geq 0, \end{cases}$$

where m represents types of input factor; n_1 is types of output; n_2 is types of undesirable output; X_0, Y_0, S_i^-, S_j^g, and S_j^b represent input, output, input slack, output slack, and undesirable output slack, respectively; X, Y, S^-, S^g, and S^b are the input, output, input slack, output slack, and undesirable output slack matrix of the DMU; λ is a series of vectors representing the weight of each input factor; and ρ is the efficiency score of the DMU (Guo et al. 2013).

2. Selection of variables

Energy is an important production factor; the study of energy efficiency has become necessary as energy conservation and emissions reductions are receiving more attention. The present study selected capital investment, energy, and labor as input factors and aimed to generate the following mean values according to the industry classification in manufacturing: revenues from primary businesses, and the undesirable outputs of industrial waste gas emissions, industrial wastewater emissions, and industrial solid waste incineration. Panel data of China's 29 manufacturing industries in 2003–2013 were selected as study samples, totaling 319 observations, from the *China Statistical Yearbooks (2004–2014)*, *China Energy Statistical Yearbook 2014*, and *China Statistical Yearbooks on Environment (2004–2014)*. A statistical description of the study samples is provided in Table 5.23.

3. Measurement results and analysis

MaxDEA 5.2 software was employed in this study. The model was defined as an undesirable SBM model that was input–output bidirectional and scale-invariant to measure the overall total-factor energy efficiency of China's 29 manufacturing industries as well as the individual total-factor energy efficiency of each industry. Table 5.24 lists the total-factor energy efficiency of the 29 manufacturing industries in 2003–2013 by industry classification and the overall energy efficiency of the entire sector.

Based on the observed results, the analysis was first performed from an overall manufacturing sector perspective. Figure 5.17 shows the energy efficiency trend of the overall manufacturing sector in 2003–2013.

As shown in Fig. 5.17, the overall energy efficiency of the manufacturing sector showed a steady upward trend in 2003–2013. It increased steadily from 0.625 in 2003 to 0.7516 in 2013 at a relatively slow pace. The overall energy efficiency in 2003–2013 was 0.6250, 0.6376, 0.6525, 0.6642, 0.6752, 0.6874, 0.7005, 0.7119, 0.7245, 0.7389, and 0.7516, respectively. The minimum was 0.6250, the maximum was 0.7516, the average energy efficiency was 0.6881, and the standard deviation was 0.0415. The annual average overall energy efficiency of China's manufacturing industry was 0.69, that is, under an unchanged output condition, the annual energy savings rate was 24.8–37.5%, which is similar to the results of Guo et al. (2015).

The annual average energy efficiency of the manufacturing sector by industry classification is shown in Fig. 5.18. There were significant differences in the total-factor energy efficiency among China's manufacturing industries. Among them, the highest total-factor energy efficiency (more than 0.825) was found in the following

Table 5.23 China's manufacturing sample data descriptive statistics

Category	Variable	Proxy variable	Unit	Maximum	Minimum	Standard deviation	Mean value
Input	Capital input	Investment in fixed assets	One hundred million yuan	13756.58	6.08	2471.89	2017.95
	Energy input	The total energy consumption	Ten thousand tons of standard coal	68839.00	33.00	12662.58	6579.13
	Labor input	Industry professionals	Ten thousand people	880.50	1.36	179.53	243.84
Output	Industry main business income	–	One hundred million yuan	77 226.31	50.89	17132.89	15906.16
	Industry industrial waste emissions	–	Ten thousand tons	2186335.00	235.00	151070.00	65589.55
	Industry industrial emissions	–	One hundred million cubic meters	173215.00	8.00	25674.45	9896.03
	Industry discharge of industrial solid waste	–	Ten thousand tons	44076.00	2.00	6716.53	2412.66

Table 5.24 Total factor energy efficiency measurement by industry classification in manufacturing

Industry	2003	2004	2005	2006	2007	2008	2009	2010	2011	2012	2013
Processing of food from agricultural products	0.6729	0.7071	0.7264	0.7559	0.7764	0.7930	0.8155	0.8283	0.8409	0.8594	0.8675
Manufacture of foods	0.5574	0.5710	0.5956	0.6257	0.6408	0.6583	0.6738	0.6949	0.7114	0.7285	0.7423
Manufacture of wine, drinks, and refined tea	0.7516	0.7701	0.7853	0.8056	0.8177	0.8269	0.8417	0.8584	0.8710	0.8846	0.8994
Manufacture of tobacco	0.7306	0.7496	0.7613	0.7892	0.8077	0.8216	0.8239	0.8376	0.8502	0.8565	0.8633
Manufacture of textiles	0.7987	0.8112	0.8243	0.8326	0.8461	0.8526	0.8648	0.8696	0.8720	0.8962	0.9057
Manufacture of textile wearing and apparel	0.5461	0.5527	0.5549	0.5628	0.5782	0.5898	0.6044	0.6176	0.6343	0.6572	0.6686
Manufacture of leather, fur, feather and related products and footware	0.4214	0.4358	0.4436	0.4629	0.4867	0.5063	0.5233	0.5471	0.5639	0.5964	0.6115
Processing of timber, manufacture of wood, bamboo, rattan, palm, and straw	0.4438	0.4591	0.4576	0.4769	0.4837	0.5042	0.5184	0.5320	0.5492	0.5541	0.5707
Manufacture of furniture	0.4178	0.4437	0.4724	0.4872	0.5093	0.5265	0.5455	0.5669	0.5616	0.5768	0.5913
Manufacture of paper and paper products	0.6382	0.6474	0.6543	0.6616	0.6685	0.6728	0.6736	0.6821	0.6878	0.6980	0.7076
Printing, reproduction of recording media	0.4573	0.4637	0.4755	0.4728	0.4832	0.4948	0.5082	0.5174	0.5206	0.5334	0.5399
Manufacture of artworks, and articles for culture, education, sports and recreation	0.5259	0.5343	0.5477	0.5585	0.5691	0.5760	0.5817	0.5963	0.6204	0.6422	0.6537
Processing of petroleum, coking and processing of nucleus fuel	0.7735	0.7844	0.7351	0.8026	0.8117	0.8194	0.8343	0.8476	0.8418	0.8636	0.8811
Manufacture of chemical raw material and chemical products	0.5773	0.5860	0.6251	0.6466	0.6552	0.6738	0.6728	0.6698	0.6876	0.7088	0.7196
Manufacture of medicines	0.5638	0.5705	0.5796	0.5738	0.5826	0.5933	0.6064	0.6179	0.6256	0.6392	0.6423
Manufacture of chemical fibers	0.7296	0.7378	0.7529	0.7600	0.7657	0.7661	0.7786	0.7894	0.8105	0.8243	0.8379
Manufacture of rubber and plastic	0.5427	0.5542	0.5528	0.5717	0.5749	0.5785	0.5806	0.5855	0.6086	0.6211	0.6373
Manufacture of non-metallic mineral products	0.7319	0.7422	0.7537	0.7536	0.7714	0.7886	0.8024	0.8153	0.8294	0.8436	0.8612
Smelting and pressing of ferrous metals	0.7462	0.7538	0.7517	0.7654	0.7786	0.7807	0.7985	0.8021	0.8206	0.8322	0.8481
Smelting and pressing of non-ferrous metals	0.7406	0.7553	0.7763	0.7815	0.7796	0.7816	0.7887	0.7906	0.7939	0.8047	0.8124

(continued)

Table 5.24 (continued)

Industry	2003	2004	2005	2006	2007	2008	2009	2010	2011	2012	2013
Manufacture of metal products	0.4384	0.4503	0.4625	0.4760	0.4813	0.5092	0.5174	0.5323	0.5564	0.5713	0.5886
Manufacture of general purpose machinery	0.5475	0.5513	0.5675	0.5732	0.5833	0.6062	0.6179	0.6232	0.6347	0.6394	0.6511
Manufacture of special purpose machinery	0.7498	0.7535	0.7782	0.7985	0.8081	0.8124	0.8357	0.8460	0.8573	0.8612	0.8694
Manufacture of transportation equipment	0.6483	0.6563	0.6785	0.6826	0.6915	0.7023	0.7174	0.7322	0.7419	0.7587	0.7792
Manufacture of electrical machinery and equipment	0.7468	0.7614	0.7735	0.7828	0.8013	0.8256	0.8348	0.8512	0.8680	0.8816	0.8994
Manufacture of computers, communication, and other electronic equipment	1.0000	1.0000	1.0000	1.0000	1.0000	1.0000	000	1.0000	1.0000	1.0000	1.0000
Manufacture of measuring instruments	0.5478	0.5771	0.6176	0.6385	0.6519	0.6726	0.7092	0.7243	0.7585	0.7775	0.8009
Manufacture of artwork and other manufacturing	0.4758	0.4916	0.5033	0.5264	0.5228	0.5369	0.5580	0.5627	0.5696	0.5784	0.5831
Utilization of waste resources	0.6023	0.6184	0.6275	0.6382	0.6526	0.6633	0.6872	0.7065	0.7224	0.7406	0.7643
Industry as a whole	0.6250	0.6376	0.6526	0.6642	0.6752	0.6874	0.7005	0.7119	0.7245	0.7389	0.7516

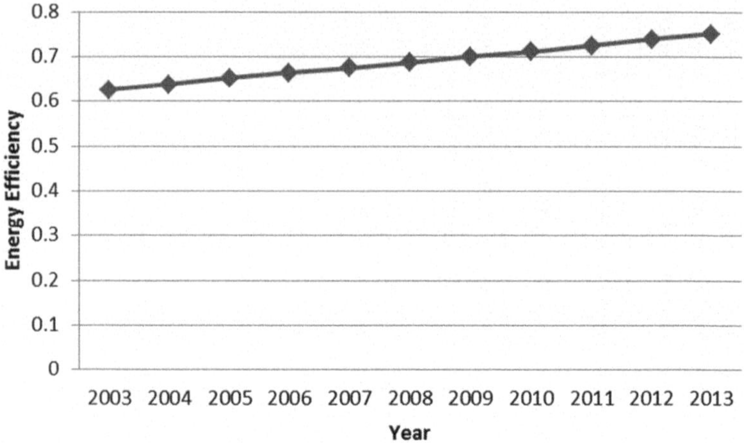

Fig. 5.17 Overall energy efficiency trend of the manufacturing sector in 2003–2013

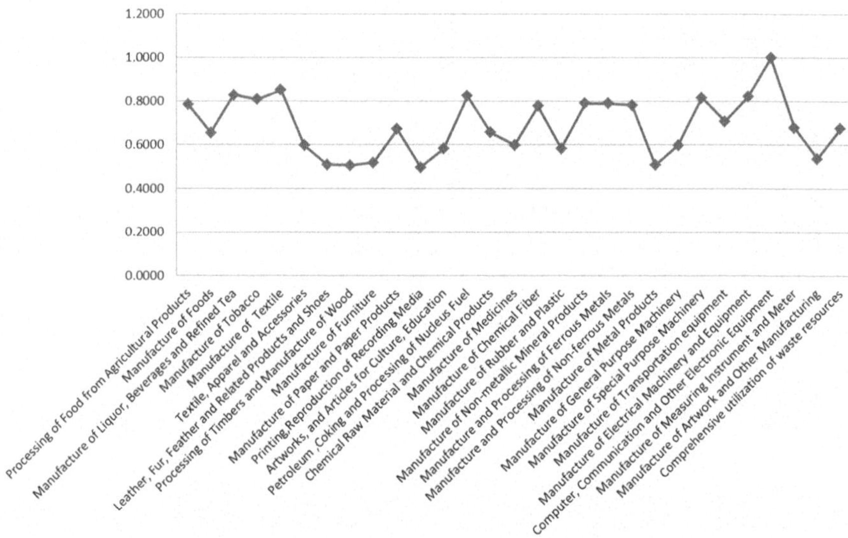

Fig. 5.18 Annual average energy efficiency distribution in manufacturing by industry classification from 2003 to 2013

three industries: manufacture of textiles; manufacture of wine, drinks, and refined tea; and manufacture of computers, communication, and other electronic equipment. Industries that had lower total-factor energy efficiency (below 0.5250) were the manufacture of leather, fur, feather, and related products and footware; processing of timber, manufacture of wood, bamboo, rattan, palm, and straw products; manufacture of furniture; printing, reproduction of recording media; and manufacture of metal products. The difference in energy efficiency between these two groups of industries was approximately 0.3, indicating that the latter group had higher potential for energy conservation. Compared to the results of the existing literature, the energy efficiency measurements obtained in this study by industry classification were generally lower. This is because energy efficiency was measured by considering undesirable output under environmental constraints in the present study. Therefore, the energy efficiency values were lower and more in line with the actual economic situation.

As observed from the comprehensive analysis above, the energy efficiency of the overall manufacturing sector and its individual industries increased steadily during 2003–2013. However, low energy efficiency exists in China's manufacturing sector. Therefore, improving the energy efficiency of the manufacturing sector would help achieve energy conservation and emissions reduction in China. Since undesirable output was accounted for under environmental constraints, the energy efficiency values generated in the present study were lower, which is more consistent with the expected results in reality.

2. Cluster analysis

Based on the abovementioned energy efficiency values obtained for the manufacturing industries, we classify the manufacturing industries based on the two perspectives of energy efficiency and environmental pollution. The calculated total-factor energy efficiency was adopted as the energy efficiency index. The environmental index was represented by the amount of individual industrial waste gas emissions divided by revenues from the primary business of each industry. The fast clustering method of squared Euclidean distance in SPSS 17.0 was adopted as the classification method. Figure 5.19 shows the classification based on energy efficiency and environmental pollution.

(1) High efficiency–high pollution industries were the manufacture of textiles; manufacture of wine, drinks, and refined tea; processing of petroleum, coking, processing of nuclear fuel; manufacture of non-metallic mineral products; smelting and pressing of ferrous metals; smelting and pressing of non-ferrous metals; and manufacture of chemical fibers.
(2) High efficiency–low pollution industries were the manufacture of computers, communication, and other electronic equipment; manufacture of electrical machinery and equipment; manufacture of special purpose machinery; manufacture of tobacco; transportation equipment manufacturing; manufacture of measuring instruments; and processing of food from agricultural products. China should focus on developing these industries.
(3) Low efficiency–high pollution industries were waste resources and waste materials recycling and processing; manufacture of raw chemical materials and chemical products; manufacture of foods; manufacture of medicines; processing of timber, manufacture of wood, bamboo, rattan, palm, and straw products; manufacture of metal products; and manufacture of paper and paper products. These types of industries have huge energy consumption and cause serious

Manufacture of Textile;Manufacture of Liquor, Beverages and Refined Tea; Processing of Petroleum,Coking and Processing of Nucleus Fuel; Manufacture of Chemical Fiber; Manufacture and Processing of Non-ferrous Metals; Manufacture and Processing of Ferrous Metals; Manufacture of Non-metallic Mineral Products	Medicines;Manufacture of Foods; Comprehensive utilization of waste resources; Processing of Timbers and Manufacture of Wood,Bamboo, Rattan, Palm and Straw; Manufacture of Metal Products; Manufacture of Chemical Raw Material and Chemical Products; Manufacture of Paper and Paper Products
Manufacture of Electrical Machinery and Equipment; Manufacture of Computer, Communication and Other Electronic Equipment; Manufacture of Tobacco; Manufacture of Special Purpose Machinery;Transportation equipment; Manufacture of Measuring Instrument and Meter; Processing of Food from Agricultural Products	Textile, Apparel and Accessories; Manufacture of Artworks, and Articles for Culture, Education, Sports and Recreation; Leather, Fur, Feather and Related Products and Shoes; Printing,Reproduction of Recording Media; General Purpose Machinery; Furniture; Rubber and Plastic;Artwork and Other Manufacturing

Environmental Pollution (high ↑ low)

Energy Efficiency (← high … low)

Fig. 5.19 Classification based on energy efficiency and environmental pollution

environmental pollution, and the government should restrict their development in favor of promoting technologies in energy savings and emissions reduction.

(4) Low efficiency–low pollution industries were the manufacture of textile wearing and apparel; manufacture of general purpose machinery; manufacture of articles for culture, education, and sport activity; manufacture of rubber and plastic; arts, crafts, and others; manufacture of furniture; manufacture of leather, fur, feather, and related products and footware; and printing, reproduction of recording media.

Types (1) and (2) are high-efficiency industries that generally have large-scale initial investment, are economically stronger, and possess more R&D funding in technology, more foreign technological introduction opportunities, higher technology levels, and higher value-added output. Types (3) and (4) are low-efficiency industries generally dominated by traditional labor-intensive processing characterized by low capital density level, slow technology upgrade, low value-added output, serious environmental pollution, and extensive economic growth.

5.3.3 Analysis of Impact Factors on Energy Efficiency in China's Manufacturing Sector

1. Quantitative model and data source

(1) Quantitative model

Based on the current economic development characteristics of China's manufacturing sector, the present study selected five basic impact factors in order to analyze the differences in total-factor energy efficiency in the manufacturing sector under environmental constraints. These five main impact factors are industry rivalry, technological advances, energy consumption structure, degree of openness, and environmental regulations. We defined the quantitative model as follows:

$$\begin{aligned}\ln Ene_{it} &= a + b_1 \ln Amou_{it} + b_2 \ln RD_{it} + b_3 \ln Struc_{it} \\ &\quad + b_4 \ln Expo_{it} + b_5 \ln Envir_{it} + x_{it}\end{aligned}$$

where InEne_{it} is energy efficiency of the individual manufacturing industry, and In$Amou_{it}$, InRD_{it}, In$Struc_{it}$, In$Expo_{it}$, and In$Envir_{it}$ represent industry rivalry, technological advances, energy consumption structure, degree of openness, and environmental regulations, respectively. Subscripts I and t represent industry and year, respectively; the other letters represent the coefficient and residual of the constant term and variable.

(2) Description of variables (see Table 5.25)

① Industrial energy efficiency (*Ene*)

Table 5.25 Description of variables

Variable	Variable declaration	Data sources	Expected direction
Amou	Quantity measure of industry competition in the industry	*China statistical yearbook* (2004–2014)	+
RD	R&D expenses within budget, measuring technology progress	*Science and technology of China statistical yearbook* (2004–2014)	+
Struc	Industry of coal consumption accounted, measure energy consumption structure	*China energy statistical yearbook* 2014	–
Expo	Industry export value of proportion, measure the opening to the outside world	*China statistical yearbook* (2004–2014)	+
Envir	Industry industrial solid waste comprehensive utilization, measure the environmental regulation	*China statistical yearbook on environment* (2004–2014)	+

The measurement results and analysis of the total-factor energy efficiency of the 29 manufacturing industries under environmental constraints are shown in Tables 5.22 and 5.23, and Figs. 5.17, 5.18, 5.19, 5.20, 5.21, 5.22. Using clustering analysis, we classified the manufacturing industries into four major types, namely, high energy efficiency–high pollution, high energy efficiency–low pollution, low energy efficiency–high pollution, and low energy efficiency–low pollution. We then studied the impact factors on energy efficiency in manufacturing from the following

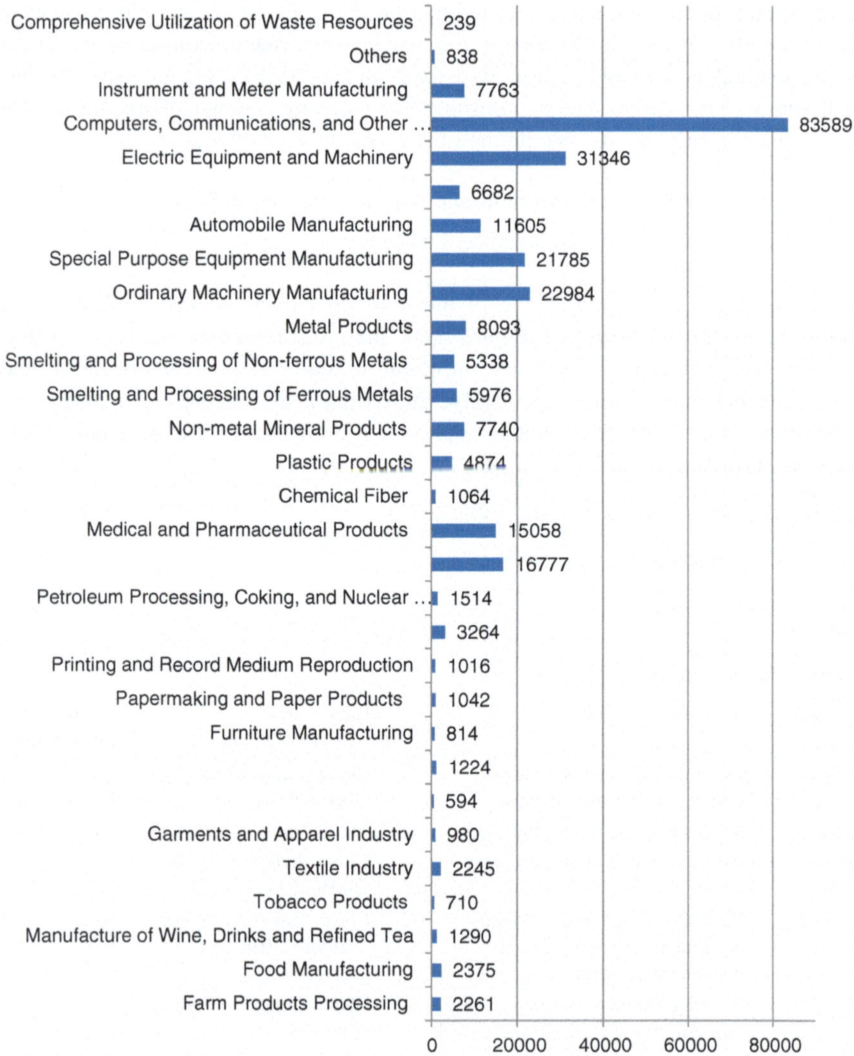

Fig. 5.20 Valid invention patent distribution among 20 industries of China's manufacturing sector in 2012

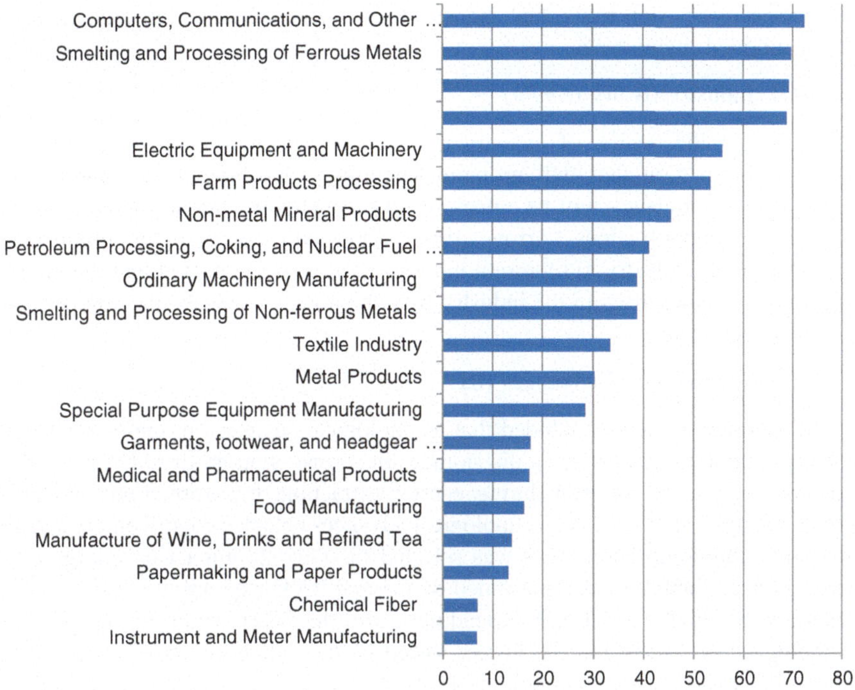

Fig. 5.21 Industrial output density distribution of 20 industries of China's manufacturing sector in 2012

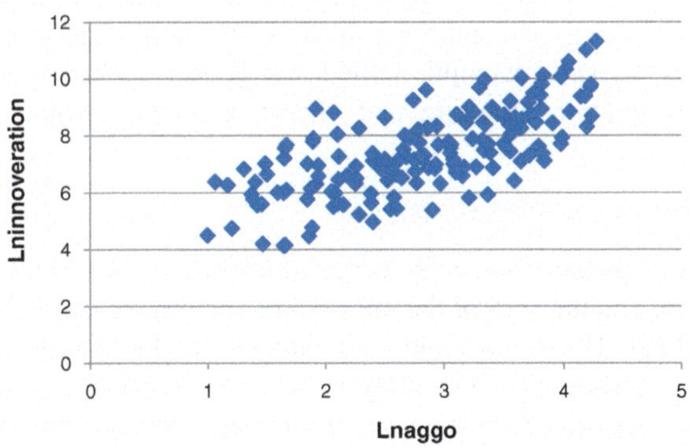

Fig. 5.22 Scatterplot diagram of industrial clustering levels and capacities for technological innovation of industries of China's manufacturing sector in 2005–2012

two perspectives: each individual industry-by-industry classification and the four aforementioned major types.

② Industry rivalry (*Amou*)

The higher the degree of industry rivalry is, the more incentive there is for enterprise innovation that will promote technology development in enterprises. Hence, energy efficiency will be enhanced, as competition tends to motivate R&D more than monopoly does (Arrow, 1962). This study adopted the number of enterprises in an industry to represent industry rivalry. We expected that the higher the number of enterprises in the industry was, the higher would be the competition and energy efficiency.

③ Technological advances (RD)

The literature review concluded that technological advances promote efficiency in energy extraction, transformation, storage, and transportation. In addition, energy consumption can be lowered during energy extraction, transformation, and processing (Tang et al. 2014). As technological advances enhance energy efficiency, an index of technological advances was selected to study its impact on energy efficiency in manufacturing, as represented by internal R&D expenditure in this study. Generally, the higher the R&D expenditures are, the faster would be the pace of technological advances and the higher would be the enhancement of energy efficiency. We expected that technological advances and energy efficiency would be directly correlated.

④ Energy consumption structure (Struc)

Research has shown that the higher coal consumption is, the lower is energy efficiency. This study adopted the ratio between total industrial coal consumption and total industrial energy consumption to show the energy consumption structure, that is, the higher is coal consumption, the lower is energy efficiency. We expected that energy consumption structure and energy efficiency would be negatively correlated.

⑤ Degree of openness (Expo)

The degree of openness impacts energy efficiency. Stan (2002) compared the period 1949–1960 and the start of the reform and opening up in 1978 to the closed-door period in 1961–1977 and found a significantly higher energy efficiency level in China during 1961–1977. This study selected the ratio of the value of export deliveries to revenues from the primary business of an industry to represent the degree of openness. The higher the ratio is, the higher is the degree of openness. We expected that the degree of openness and energy efficiency would be positively correlated.

⑥ Environmental regulations (Envir)

The famous "Porter Hypothesis" suggests that reasonable environmental regulations can encourage technological innovations, thereby achieving innovations that can compensate or even exceed the compliance costs related to environmental regulations. In other words, when the compensatory effect of innovations is greater than the total regulation costs, energy efficiency in enterprises will be enhanced. This study selected the comprehensive industrial solid waste utilization rate to measure environmental regulations. The higher the comprehensive industrial solid waste utilization rate is, the more stringent are the environmental regulations. We expected that environmental regulations would promote energy efficiency.

(1) Data source

We selected the 2003–2013 variable data of 28 manufacturing industries in China (waste resources and waste materials recycling and processing was eliminated due to missing data in many years) from the *China Statistics Yearbooks*, *China Energy Statistical Yearbooks*, and *China Statistical Yearbooks on Environment* as the sample data.

2. Analysis of impact factors of the manufacturing industries

(1) Statistical analysis

Brief descriptive statistics of the data characteristics of variables were computed and the results are shown in Table 5.26.

In order to generate more accurate conclusions, this study used the method widely adopted by Chinese and foreign scholars for the estimation, that is, FE, RE, and FGLS, and software Stata 2.0 for the analysis. The results of the coefficient estimates of the explanatory variables are shown in columns (A), (B), and (C) of Table 5.26.

The p-value in the Hausman test was 0.5274, and as this is greater than 0.05, the RE model should be employed. In other words, based on the Hausman test, the RE results were superior to the FE. As seen from Table 5.27, the signs of the coefficients in the RE and FGLS estimations were consistent. In addition, since the possible existence of heteroscedasticity and serial correlation in the FGLS

Table 5.26 Descriptive statistics of main variable index

Main variable	Sample size	Average	Standard deviation	Minimum	Maximum
ln *Ene*	308	−0.3797	0.2013	−0.8290	0
ln *Amou*	308	8.9145	1.0793	4.8675	10.5891
ln *RD*	308	1.1212	3.5322	−6.3200	7.1329
ln *Struc*	308	−0.8888	0.6888	−2.8567	0.9061
ln *Expo*	308	−2.1614	1.0853	−5.4402	−0.3504
ln *Envir*	308	−0.1772	0.1997	−1.1237	0.8912

(24) Quantitative results analysis

Table 5.27 Quantitative regression results

Explanatory variables	Fixed effects (FE)	Random effects (RE)	Feasible generalized least squares (FGLS)
	(1)	(2)	(3)
ln *Amou*	0.0624*** (8.21)	0.0615*** (8.38)	0.0452*** (9.34)
ln *RD*	0.0010 (1.10)	0.0010 (1.15)	0.0024* (1.78)
ln *Struc*	−0.0533*** (−4.43)	−0.0528*** (−4.49)	−0.0465*** (−4.50)
ln *Expo*	−0.1200*** (−12.48)	−0.1202*** (−13.06)	−0.1189*** (−21.05)
ln *Envir*	−0.0380** (−2.22)	−0.0384** (−2.25)	−0.0700*** (−3.80)
Constant term	−1.2506*** (−18.09)	−1.2422*** (−16.45)	−1.0952*** (−18.87)
R^2	0.239 6	0.240 6	–
OBS	308	308	308

Note The values in the subscripts of the coefficients are t-values; ***, **, and *indicate that the relevant variables have passed the significance tests of 1, 5, and 10%, respectively

estimation were eliminated to a certain extent, we consider the results generated in the FGLS estimation to be robust. Based on the observations in column (C), we arrived at the following conclusions.

First, the coefficient of industry rivalry was positive (0.0452) and passed the 1% significance test, indicating that the greater the number of enterprises is, the higher is the energy efficiency in the enterprises, that is, industry rivalry can significantly improve industrial energy efficiency, which is consistent with our expectation.

Second, the coefficient of technological advances was positive (0.0024) and passed the 1% significance test, indicating that technological advances and energy efficiency were positively correlated. In other words, when enterprises invest more resources in order to acquire technological advances, more energy-saving technologies can be developed, thereby promoting energy efficiency.

Furthermore, the coefficient of the impact of energy consumption structure to energy efficiency was negative (−0.0465) and passed the 1% significance test, indicating that the higher coal consumption is, the lower is energy efficiency. Coal has been the major component of energy consumption for a long time in China, reaching 66% in 2013. Research indicates that China is rich in coal resources and energy efficiency tends to be lower in inland areas, which primarily rely on coal; the coal consumption of some of those areas averaged more than 80% (Yang 2009). Energy consumption structure and energy efficiency were negatively correlated, which is consistent with our expectation.

In addition, the coefficient of the impact related to the degree of openness to energy efficiency was negative (−0.1189) and passed the 1% significance test, indicating that the higher is the value of export delivery of an industry, the lower is the energy efficiency of the industry. This is consistent with some previous results.

For example, Dong (2008) revealed that exports lower energy efficiency and imports do not significantly improve energy efficiency. Liu and Yin (2008) showed that for every 1% increase in exports of high energy-consuming products, China's average industrial energy efficiency was lowered by 4.4%, that is, a higher proportion of exports of high energy-consuming products would lead to lower industrial energy efficiency in China.

Finally, the coefficient of environmental regulations was negative (−0.0700) and passed the 1% significance test, indicating that the higher the comprehensive industrial solid waste incineration rate is, the more stringent are environmental regulations, which would not enhance energy efficiency but rather have a negative effect. This is because when innovations resulting from environmental regulations cannot compensate for the compliance costs incurred in environmental regulations, energy efficiency would be lowered rather than improved. This is consistent with the findings of some scholars, and although it is not consistent with our expectation, is in line with reality. Analysis of impact factors in manufacturing industries by industry classification

Based on clustering analysis of energy efficiency–environmental pollution, China's manufacturing industries were classified into four major types, namely, high efficiency–high pollution, high efficiency–low pollution, low efficiency–high pollution, and low efficiency–low pollution industries.

Since it was known from the above analysis that the FGLS model could eliminate the possible existence of heteroscedasticity and serial correlation to a certain extent, and that the results from the FGLS estimation are robust, the estimation results of the FGLS model by industry type are shown in Table 5.28.

Table 5.28 FGLS regression results by industry type

Industry / Explanatory variables	High efficiency-high pollution industry	High efficiency-low pollution industry	Low efficiency-high pollution industry	Low efficiency-low pollution industry
ln *Amou*	0.0203^{***} (4.38)	0.0514^{***} (3.86)	0.0422^{***} (4.21)	0.0814^{***} (9.93)
ln *RD*	0.0063^{***} (6.47)	0.0029 (0.90)	-0.0074^{***} (−3.80)	0.0161^{***} (8.44)
ln *Struc*	0.0641^{***} (6.77)	-0.1005^{***} (−4.13)	0.0396^{**} (2.27)	−0.015 6 (−0.82)
ln *Expo*	0.0136^{*} (1.73)	-0.0812^{***} (−4.32)	-0.2166^{***} (−8.72)	0.0681^{***} (6.04)
ln *Envir*	0.0033 (0.30)	0.0218 (0.42)	−0.0458 (−1.63)	0.0058 (0.13)
Constant term	-0.3238^{***} (−6.77)	-0.9699^{***} (−5.47)	-1.3587^{***} (−16.49)	-1.2419^{***} (−15.40)
OBS	77	77	66	88

Note The values in the subscripts of the coefficients are t-values; ***, **, and * indicate that the relevant variables have passed the significance tests of 1, 5, and 10%, respectively

Quantitative results by industry type

(2) Quantitative results analysis

We found from observations in Table 5.28 that heterogeneity exists in the impact factors on energy efficiency among manufacturing industries. The degree of impact and its correlation to energy efficiency vary across industry types. By analyzing the quantitative results of the different industry types, we arrived at the following conclusions.

First, for high energy efficiency–high pollution industries, such as the manufacture of textiles, the impact coefficients of industry rivalry and technological advances were both positive on energy efficiency (0.0203 and 0.0063, respectively) and passed the 1% significance test. This indicates that as the number of enterprises in the industry increases, the level of competition increases, technology investment intensity in the industry is greater, technological advances occur more rapidly, and thus, there is higher energy efficiency in this type of industry. The energy consumption structure coefficient was positive (0.0641) and passed the 1% significance test, which was similar to the results of some previous works in which a higher percentage of coal consumption could promote energy efficiency. The environmental regulations coefficient was positive (0.0033) but did not pass the significance test, indicating that for high pollution industries, the greater are environmental regulations, the higher is energy efficiency.

Second, for high efficiency–low pollution industries, such as the manufacture of computers, communications, and other electronic equipment, the impact coefficients of industry rivalry, technological advances, and environmental regulations to energy efficiency were all positive (0.0514, 0.0029, and 00218) and industry rivalry passed the 1% significance test. The energy consumption structure coefficient was negative (−0.1005) and passed the 1% significance test. This indicated that the more enterprises there are in an industry, the higher are the internal R&D expenditures, the more stringent are environmental regulations, while the higher coal consumption is, the higher is energy efficiency in that industry. This is because industries of the high efficiency–low pollution type are mostly high-tech industries with high technology barriers, quick technology upgrade, strong R&D capabilities, high value-added, low environmental pollution, and stringent environmental regulations, such that the higher is investment in technology, the greater is competition, and the higher is energy efficiency. The impact coefficient for the degree of openness was negative (–0.0812) and passed the 1% significance test because exports would lower energy efficiency whereas imports would not significantly improve energy efficiency (Dong, 2008). Increased export proportion of high energy-consuming products would lower China's industrial energy efficiency (Liu et al. 2008). In particular, processing of food from agricultural products, which belongs to this industry type, is among China's six highest energy-consuming industries. Therefore, our conclusion is consistent with that of some previous studies.

Third, for low efficiency–high pollution industries, such as the manufacture of raw chemical materials and chemical products, the impact coefficient of industry rivalry was positive (0.0422) and passed the 1% significance test. The impact coefficients of technological advances and industrial exports were negative (−0.0074 and −0.2166) and passed the 1% significance test, indicating that the higher are the internal R&D expenditures and export proportion, the lower is energy efficiency. This is because industries of this type mostly undertake traditionally labor-intensive processing, which has high energy consumption, causes severe environmental pollution, possesses low technology levels with low value-added output, and is in the extensive growth mode. Moreover, exports of high energy-consuming products will lower energy efficiency. In addition, the impact coefficient of environmental regulations to energy efficiency was negative (−0.0458) and did not pass the significance test because the innovation compensation resulting from environmental regulations cannot compensate for the compliance costs incurred in environmental regulations, thereby lowering energy efficiency.

In addition, for low efficiency–low pollution industries, such as the manufacture of textile wearing and apparel, industry rivalry, technological advances, degree of openness, and environmental regulations have positive impacts on energy efficiency. Except for the insignificant impact of environmental regulations, all other factors passed the 1% significance test. Energy consumption structure was negative (–0.0156) and did not pass the significance test. Since the output of this industry type is primarily low energy-consuming products, when the export ratio of high energy-consuming products is relatively lower, energy efficiency can be improved. Moreover, the more intense industry rivalry is, the greater the investment in technology is; while the lower the proportion of coal consumption is and the more stringent environmental regulations are, the higher energy efficiency is.

5.3.4 Conclusions and Implications

This study employed the 2003–2013 panel data of 29 manufacturing industries in China and adopted the SBM model to measure total-factor energy efficiency under environmental constraints, and classified these 29 industries into four major types based on energy efficiency and environmental pollution. Then, this study performed empirical analysis on the impacts of industry rivalry, technological advances, energy consumption structure, degree of openness, and their respective industrial heterogeneity. The study conclusions show the following. (1) The 2003–2013 energy efficiency of China's manufacturing industry improved steadily with energy conservation potential of 24.8–37.5%. Due to the consideration of undesirable output in this study, the measured energy efficiency was relatively lower, which is consistent with our expectation as well as the actual situation. (2) Industry rivalry and technological advances have positive impacts on energy efficiency in manufacturing, the latter being more significant. Energy consumption structure, degree of openness, and environmental regulations have significantly negative impacts on energy efficiency in manufacturing. (3) Four types of industries are defined according to cluster analysis, through which it is found that the extent of the impact and direction of the five explanatory variables vary across industry types, that is, impact heterogeneity exists among different industry types.

Based on these conclusions, the following implications are derived in this section.

High energy efficiency–high pollution industries should raise industry rivalry, increase technology investment, develop energy-saving technologies, develop clean energy, and lower pollution. The impact of environmental regulations on this type of industry is positive, which means that the higher the comprehensive industrial solid waste utilization is, the more extensive are the environmental regulations and the higher is energy efficiency. Generally, if industry rivalry, investment in technology R&D, and environmental regulation intensity were increased, energy efficiency in this industry could be improved.

For high efficiency–low pollution industries, industrial exports have a significantly negative impact on the energy efficiency of the industry because exports of farm products, which are by nature high energy consuming, significantly lowers the energy efficiency of China's manufacturing sector (Zhang et al. 2015). Therefore, on the one hand, trade structure should be optimized, which would lower the export proportion of high energy-consuming products and increase that of lower energy-consuming products. On the other hand, there should be development of energy-saving technologies, increased technology R&D, introduction of more advanced production technology equipment, and lower energy consumption during production of high energy-consuming products in order to improve energy efficiency from the source. Technological advances positively impact energy efficiency of this industry type because the majority of industries of this type are high-tech industries that should actively learn advanced technologies from foreign countries, increase R&D intensity in science and technology, and undergo timely technology upgrades while raising industry competition. China should focus on developing this industry type.

Low efficiency–high pollution industries mostly engage in traditionally labor-intensive processing, which is typical of intensive economic growth with lower technology level and slower upgrade, producing mostly high energy-consuming products and serious environmental pollution problems. Therefore, internal R&D expenditures should be restrained, technology investment should be reduced, industrial investment structure should be optimized, and the export proportion of high energy-consuming products should be limited to improve energy efficiency. China should restrict development of this industry type.

Low efficiency–low pollution industries should increase industry competition, increase the number of enterprises in each industry, raise R&D investment in science and technology, reasonably introduce more advanced production equipment, reasonably arrange energy consumption structure to reduce coal consumption proportion, actively develop clean energy, encourage exports, and increase environmental regulations intensity, all of which could improve energy efficiency in this type of manufacturing industry.

5.4 Informationization and Green Growth in China's Manufacturing Industry

Since the Chinese economic reform, China's manufacturing sector has continued to develop rapidly. This has been a relatively positive catalyst for China's industrialization and modernization. At present, however, China's manufacturing sector is facing many challenges and risks, the major bottlenecks of which come from weak independent innovation capabilities and environmental resource constraints. In order to overcome these bottlenecks, China should persevere with informationization (which is the extent to which its economy or society is becoming information based), expedite the informationization level, and deepen integration of industrialization using informationization to promote adjustments to and upgrades of the industrial structure. The Communist Party Central Committee and the State Council have attached great importance to informationization and the integration of informationization and industrialization. The 17th National Congress of the Communist Party of China clearly proposed the development of a modern industrial system, vigorous integration of informationization and industrialization, and transformation of scale-oriented industries into strength-oriented industries. The 18th National Congress of the Communist Party of China clearly stipulated that China should adhere to its unique manner of industrialization and informationization to deepen the integration of the two and to accelerate the transformation and upgrading of traditional industries. "Made in China by 2025" clearly proposed the acceleration of integrating a new generation of information technology in manufacturing to deepen adjustments to the manufacturing structure. It is apparent that, informationization has become China's national strategy for steering its innovation-driven development and making adjustments to its industrial structure. The question of how to take advantage of informationization to further promote green growth in manufacturing and to achieve sustainable, steady, and coordinated development of China's economy is a crucial issue that has yet to be resolved.

5.4.1 Literature Review

Informationization is a development process that primarily focuses on enhancing the innovation capabilities of enterprises and organizations, mostly by using digital products and services in conjunction with full utilization of information technologies and resources, such that collaboration and sharing of information, knowledge, and technologies is promoted for better quality and efficiency; in the process, economic and social transformation and upgrading would be enhanced (Guan et al. 2012).

Informationization is executed by means of high-tech information technologies that facilitate improvement, restructuring, or redefinition of trends in the industrial structure as well as the economic and social structures. By improving the information content and technological contribution level of economic growth, economic and social development can be propelled to higher levels with better organization and higher efficiency (National Bureau of Statistics of China, Science and Technology Statistical Evaluation Group 2011). The keys to achieve informationization include the following: 1. develop and make use of information resources and cultivate talent in the field of informationization to provide support for continual enhancement; and 2. encourage widespread usage of information technologies; and develop information technologies and industries to facilitate sustainable application capabilities of informationization enhancement. Therefore, informationization will have a profound impact on China's manufacturing sector and its economic growth.

Many foreign scholars have extensively discussed the correlation between informationization and economic growth. Some scholars have found that informationization only slightly promotes economic growth or is not conducive to economic growth at all. Roach (1987) revealed that although the United States invested several resources in the computer industry in the 20th century, productivity growth in the United States dropped significantly after 1973. Solow (1987) arrived at a similar conclusion—that huge investment in the computer industry did not lead to improvement in productivity growth, calling this phenomenon a "productivity paradox," which some scholars later verified. Inklaar et al. (2005) compared the contribution rate of informationization to economic growth of the United States and four major countries in Europe and found that informationization promoted the TFP rate in the United States but the impact was not significant in the four European countries. Stiroh and Botsch (2007) employed panel data of industries in the United States to verify the correlation between informationization and labor productivity after 2000, and found that informationization did not significantly promote the growth of labor productivity in the United States. Badescu and Garces-Ayerbe (2009) employed panel data of 342 enterprises in Spain to analyze the contribution of informationization to labor productivity and found that although labor productivity was enhanced to a certain extent during the observation period, the contribution of informationization was not significant. More recent research of Alma (2014) and Acemoglu (2014) also supported the "productivity paradox" to a certain extent.

On the other hand, more studies support that the view informationization positively contributes to economic growth. Stephen and Daniel (2000) analyzed the contribution of informationization on labor productivity in the 20th century in the United States. These studies found that informationization significantly promoted labor productivity in the United States at a contribution rate exceeding 0.4%. Schreyer (2001) analyzed the contribution of informationization on productivity growth in the major OECD member countries and found medium contribution levels in England and Japan at approximately 0.2% and lower contribution levels in Germany and France at less than 0.2%. Basu and Fernald (2007) employed industrial data in the United States to analyze the correlation between informationization and TFP growth and found a significantly positive correlation between the two. Jorgenson (2008) analyzed industries that caused increased productivity in the United States after 1985 and found that informationization promoted TFP growth in industries that employed information and communications technologies. Ahmed (2010) analyzed approaches that drove productivity growth in Asian countries and found that China's productivity growth was input driven while productivity growth in Japan and Korea were mostly driven by TFP. In addition, the study found that informationization was conducive to the growth of TFP in Asian countries but with varying contribution levels. Dahl et al. (2001) employed panel data of 31 industries in European countries to verify the impact of informationization on TFP and found that informationization was significantly conducive to growth in TFP. Lovric (2012) employed the panel data of 25 developed and developing countries in Europe to verify the impact of informationization on growth in labor productivity and found that in both developed and developing countries, informationization was conducive to labor productivity growth to a similar extent across these countries. More recent studies by Chou et al. (2014) and Wang (2015) support this positive conclusion.

Chinese scholars started to investigate the correlation between informationization and economic growth relatively later, and there are only a few such studies available. However, their research conclusions are essentially consistent with the international research—that informationization is conducive to economic growth or TFP. Wang and Yu (2004) analyzed the effect of informationization on economic structure based on the level of informationization development measured and found that informationization drove secondary industries more strongly but was weaker in primary and tertiary industries. Li and Hu (2008) adopted double extrapolation to estimate the contribution of information technologies on China's TFP growth and found that information technologies significantly promoted TFP growth. Li and Liu (2009) constructed an informationization evaluation index system to estimate the informationization level in Hunan province so as to verify the contribution of informationization to economic growth. The study found that informationization has become the third source after labor and capital to promote economic growth in Hunan. Sun et al. (2002) employed panel data of China's industries to analyze the contribution of informationization to China's economic growth and found that informationization was significantly conducive for economic growth and TFP growth in China. Based on this literature review, we observe that fruitful results have been achieved in research on the impact of informationization on economic growth but it remains insufficient in the following three areas. First, most of the literature has adopted the labor productivity rate or economic growth rate to characterize economic growth. However, under China's current industrial structure, which is primarily labor-intensive processing, this could underestimate efficiency (Fan et al. 2014). TFP can better reflect the contribution of informationization to economic growth mainly because this contribution can be reflected fully in TFP growth. Therefore, some of the literature has adopted TFP as a proxy variable for economic growth. However, when measuring TFP, these studies have considered only traditional input and output variables, such as capital, labor, and desirable output, without incorporating energy and environmental constraints into the TFP measurement, making it impossible to fully reflect and relate excessive versus normal energy consumption and good versus bad output. The evaluation of economic performance and social welfare may be biased (Chen 2001; Shadbegian and Gray 2005; Li et al. 2013), thereby leading to misinterpretation of results that call for excessive use of resources in certain areas of seriously low economic growth, resulting in excessive energy shortage, serious environmental pollution, and rapid deterioration of the ecosystem. These measures have deviated from the "sound and rapid" orientation of economic development as promoted by the national government and cannot fully reflect the impact of informationization on green growth in manufacturing. The second deficiency in Chinese research on the topic is that the majority of the existing literature measuring level of informationization has adopted a single index as the proxy variable, which cannot scientifically reflect the level of informationization in an industry. Some studies have employed websites, phones, number of base stations, per capita emails, and telecommunications sales value, among others, to represent the level of informationization. Other studies have adopted the ratio of workers employed in information industries to the total number of industrial workers or GDP ratios to measure the level of informationization. However, these indexes can only briefly describe the level of informationization in the industrial sector without scientific, reasonable, comprehensive, or systematic characterization of the actual level (Han et al. 2014). The third research deficiency is that most studies have employed regional data for empirical analysis and have overlooked heterogeneity in the impact of informationization on green growth among different manufacturing industries. Due to differences in development characteristics, technology levels, and the need for informationization among manufacturing industries, informationization has different impacts on the green growth of different industries. Macro data processing that aggregates all regional data is bound to overshadow industry heterogeneity and thus, cannot accurately reflect the impact of informationization on green growth in different industries (Sun et al. 2012a, b).

Based on the existing research foundation, more efforts should be made in the following areas. 1. Research should introduce the Malmquist–Luenberger (ML) productivity index to estimate TFP (i.e., green TFP) to better reflect the transformation and development of green manufacturing. 2. Research should construct an informationization evaluation index system and adopt a combination weighting approach to assign weights and perform a scientific, reasonable, comprehensive, and systematic evaluation of the informationization level to avoid possible measurement errors resulting from employing a single index. 3. Based on enterprises of different technology levels, industry size, energy consumption, environmental pollution, and profitability, research should construct a panel data model to analyze the varying impact of informationization on the green TFP of different industries. Thus, the impact of heterogeneity of informationization on the green growth of different manufacturing industries can be reflected better.

5.4.2 Theoretical Mechanism and Research Hypotheses

1. Transmission mechanism of informationization that promotes manufacturing green growth

The estimation of green TFP includes the input and output portions. Inputs include traditional factors of production, such as capital, labor, and energy. Outputs include desirable output GDP and undesirable environmental pollution emissions. As observed from the estimation method adopted for green TFP, when there are fewer inputs and the desirable output is higher than the undesirable output, then green TFP is higher. Therefore, the transmission mechanism discussed in this section primarily revolves around analysis of the following four areas.

(1) Substitution effect of informationization on the traditional factors of production

Informationization can lower R&D expenditures and transaction costs of enterprises through knowledge and technology spillover, help optimize combination of factors of production, restructure production and business processes, and continue improving their level of equipment and technology. At the same time, informationization can provide enterprises with highly efficient and intelligent production platforms and information platforms (Shahiduzzaman and Alam 2014), and improve efficiency in the management and decision making of enterprises, thereby enabling the fundamental transformation of traditional manufacturing production operations that require high inputs of capital and labor; these results result in the substitution effect of informationization on traditional production factors.

(2) Energy-saving effect of informationization on energy consumption

Informationization can significantly improve technology innovation levels and the innovation efficiency of enterprises through penetration and diffusion of knowledge and technology with backward- and forward-affiliated enterprises, as well as upstream and downstream industries (Han et al. 2014). In addition, the feedback and accumulative effect of the cycle between the manufacturing R&D department and the production department can be strengthened significantly. Therefore, growth of the cumulative effect of the cycle between informationization and technological advances is promoted. Since technological advances are a source of energy efficiency by nature (Li and Zhou 2006; Zhao 2013), the energy-saving effect of informationization on energy consumption is generated.

(3) Emission-reduction effect of informationization on environmental pollutants

On the one hand, informationization is conducive for industrial structural adjustment and promotes the upgrading of the internal industrial structure of manufacturing, thereby lowering environmental pollutant emissions from high energy-consuming industries and high pollution-emission industries in manufacturing. On the other hand, informationization is conducive for the transformation, upgrading, and updating of equipment and processes in manufacturing enterprises to more advanced and environmentally friendly technologies for green production and pollutant treatment, such that the energy utilization rate and environmental pollutant handling level is improved, resulting in the emissions-reduction effect of informationization on environmental pollutants (Zhang et al. 2014).

(4) Promoting effect of informationization on economic growth

The widespread application and development of information technologies has led to increasing numbers of enterprises making capital investments in the field of informationization, deepening the capital available in the field and directly promoting output growth. In addition, informationization can significantly enhance the technology level and labor productivity of enterprises, increase the information content of products and technology value-added, improve the contribution level of technology to overall social and economic growth, and result in a promoting effect of informationization on economic growth (Sun et al. 2012a, b).

However, informationization cannot always lead to an energy-saving effect of enterprises on energy consumption and an emissions effect on environmental pollutants. Some scholars have found that informationization can promote technological advances but may lead to an energy rebound effect in which the energy-saving effect and pollutant emission effect resulting from the technological improvement is offset by a new round of energy consumption and pollutant emissions from capital deepening and production growth (Shao et al. 2013; Hu 2014). Therefore, the promoting effect of informationization on green growth of manufacturing may not necessarily occur. At present, compared with developed countries, China's informationization level is significantly low but the energy-saving effect resulting from informationization is higher than its energy rebound effect (Zhang and Zhang 2014). Based on the above discussion, Hypothesis 1 is proposed as follows.

Hypothesis 1: Informationization is conducive to the green TFP growth of manufacturing, that is, informationization has significant knowledge and technology spillover effects.

2. Heterogeneous transmission mechanism of informationization that promotes manufacturing green growth

The performance of the heterogeneous transmission mechanism of informationization that promotes manufacturing green growth also depends on matching compensatory inputs, which include professional labor, R&D input, and industrial organization format. Different compensatory inputs may lead to bigger differences in the promoting effect of informationization on the green growth of different industries. In addition, differences in conditions, such as economic policies, industrial development environment, natural resource endowment, and the carrying capacity of the environment, impact the promoting effect among industries.

Therefore, there might be a particular heterogeneous transmission mechanism for the impact of informationization on manufacturing green growth. First, from the perspective of technology level, since there are vast differences among manufacturing industries in the complexity of technologies and products, innovation resource allocation, innovation risks and difficulties, and innovation willingness, willingness in the supply and demand of informationization in these industries is significantly different, resulting in different promoting effects of informationization on manufacturing green growth. Second, from the perspective of industry size, the transmission of the promoting effect of informationization is closely related to manufacturing green growth. Generally, emerging industries or high-tech industries are relatively smaller than other industries and the supply and demand for informationization in these industries are higher. The speed of penetration and diffusion of information technologies in these industries is much higher than in larger traditional manufacturing industries; the impact on green growth is also higher (Han et al. 2014). Therefore, significant size heterogeneity may exist in the transmission mechanism of the promoting effect of informationization on manufacturing green growth. Third, from the energy consumption perspective, generally speaking, energy consumption in heavy industries is relatively higher than in light industries. In particular, energy consumption, R&D of green technologies, and application capabilities vary greatly among different industries, resulting in significant differences in the supply and demand for informationization, such as energy-saving and emission-reduction technologies. Thus, the promoting effect of informationization on the energy consumption of green growth in manufacturing varies. Fourth, from the perspective of the degree of environmental pollution, since significant differences exist in the level of environmental pollution among different manufacturing industries, so there are differences in levels of energy saving and emissions reduction, environmental sensitivity, and environmental regulations (Chen 2014), generating differences in the selection, R&D, and application of technologies. Thus, the environmental effect of informationization on manufacturing green growth varies. Fifth, from the perspective of profitability, significantly uneven profitability exists among different manufacturing industries and the level of profitability of enterprises is closely related to their reliance, involvement, and management of informationization, resulting in heterogeneity in industry profitability. Thus, the profitability effect of informationization on manufacturing green growth varies. Based on the above discussion, Hypothesis 2 is proposed as follows.

Hypothesis 2: Since there are significant differences in technology level, industry size, energy consumption, pollutant emissions, and profitability among manufacturing industries, there are significant differences in the promoting effect of informationization on the green growth of manufacturing.

5.4.3 Quantitative Model Construction and Index Description

1. Quantitative model construction

In this study, we focused on analyzing the impact of informationization on green TFP in manufacturing. Considering that the manufacturing sector is a continuous and dynamic system, the first-order lag of the dependent variable was incorporated into the quantitative model and a dynamic panel data model was established as follows:

$$\ln ML_{it} = \alpha + \tau \ln ML_{i(t-1)} + \beta \ln II_{it} + \delta \ln X_{it} + \tau_{it}$$

Where i represents manufacturing industries, t represents year, α represents the constant term, and τ represents the regression coefficient of the first-order lag of the dependent variable. *ML* represents green TFP, *II* represents the level of informationization, ε represents random disturbance, and X is the control variable. Based on the related literature, R&D intensity (*RD*), environmental regulations (*ER*), marketization (*MS*), and foreign direct investment (*FDI*) were considered as other important controlled factors that impact green TFP in manufacturing. Two main kinds of estimation methods are used in the dynamic quantitative model: first-differenced generalized method of moments (first-differenced GMM) and system generalized method of moments (system GMM). Since the problem of weak instrumental variables exists in first-differenced GMM, serious errors may exist due to limited samples. System GMM can fully utilize sample information and significantly lower the error from limited samples. Therefore, system GMM was adopted to estimate the dynamic quantitative model (1) above in order to overcome inherent problems and possible errors arising from limited samples.

2. Data source and index description

(24) Since major adjustments were made to China's national economic industry classification in 2003 and 2012, the merging and separation of some manufacturing industries might result in estimation errors and inconsistency in the statistical criteria. To accommodate these concerns, the study period was set as 2003–2012 and 29 manufacturing industries were selected as study samples (waste resources and waste materials recycling and processing was not included owing to missing data). Related data were taken from the China Statistical Yearbooks, China Industrial Economy Statistical Yearbooks, China Statistical Yearbooks on Science and Technology, and China Statistical Yearbooks on Environment. In addition, missing data were supplemented by interpolation. Below is a detailed index description of the variables. Dependent variable: green total factor productivity.

(24) In order to incorporate the constraints of energy and environment into the analysis framework, this study referred to the framework constructed by Fare et al. (2007) and Shen and Gong (2011), which considered a possible production

combination comprised of both desirable and undesirable outputs and adopted the ML index to estimate the green TFP. This method allows for not only an increase in desirable output but also a reduction in environmental pollution without requiring a selection from the measurement angle or changes in equal proportion; thus, it is more suitable for measuring green TFP with energy and environmental pollution included (Wan and Zhu 2013). Input variables, desirable output variables, and undesirable output variables were involved in the calculation; related indexes and data processing are explained as follows.

① Capital input: the approach in Zhang et al. (2004) was referred to and the perpetual inventory method was adopted to audit the capital inventory of industries. ② Labor input: the average number of workers in the industry as a whole was adopted as the labor input index.③ Energy input: the total energy consumption of all industries was adopted as the energy input index. ④ Desirable output: the total industrial output of all industries was adopted as the desirable output index. In addition, 2003 is used as the base year and the producer price index is employed to perform mean-centering of the total industrial output. ⑤ Undesirable output: considering data availability and reliability, industrial wastewater emissions, industrial waste gas emissions, and solid waste incineration were selected as the three indexes for undesirable output. Core explanatory variables: Informationization level.

Due to the lack of relevant indexes, the industrial level of the manufacturing industry has seldom been used in the measurement of China's informationization level. Therefore, in order to measure the level of informationization, many areas, such as the nature of informationization, industrial information support, and applications, should be considered. Measurement primarily involves two key issues: the construction of an evaluation index system and the index weight assignment. Among research on the evaluation index system, the National Bureau Statistics has optimized the evaluation index system for assessing the informationization level in 2010 and developed an informationization integrated evaluation index system–informationization development index (II), which serves as the scientific and quantitative reference authority. Since the informationization index measures the regional informationization level of industries, Borat's information economy was referenced. Based on data availability and validity, appropriate adjustments were made to the informationization development index in the evaluation index system. Specific details of the evaluation index system are shown in Table 5.29. The industry informationization support index reflects the supply and production capabilities of the industries primarily measured in the following three areas: instrument and equipment expenditures; expenditures in technology introduction, digestion, absorption, and transformation; and R&D FTEs. The industry informationization application index reflects the application and consumption capabilities of industries primarily measured in the following three areas: sales revenues of new products, number of valid patents of the industry, and labor productivity of the entire industry.

(24) For index weight assignment, the calculation in Li et al. (2015) was referenced and the combination weighting approach was adopted to calculate index

Table 5.29 Industry informationization evaluation index system

Overall index	Serial No.	Main index	Serial No.	Sub-index	Index unit
Industry informationization evaluation index system	A	Industry informationization support index	A1	Industry instrument and equipment expenditures	Ten thousand yuan
			A2	Expenditures in technology introduction, digestion and absorption, and transformation	Ten thousand yuan
			A3	R&D FTEs	Man/year
	B	Industry informationization application index	B1	Sales revenues of new products of industry	Ten thousand yuan
			B2	No. of valid patents of industry	Item
			B3	Labor productivity of the entire industry	Yuan/person-year

Data source China Industrial Economic Statistical Yearbook and China Statistical Yearbook on Science and Technology weights. The fuzzy analytic hierarchy process (FAHP) was adopted as the subjective weighting method and the maximizing deviation method as the objective weighting method to calculate the informationization level of each manufacturing industry in China in 2003–2011. As observed from the results calculated, significant differences existed in the informationization levels among manufacturing industries. The manufacture of computers, communication, and other electronic equipment industry had the highest average informationization index at 0.8587 while the manufacture of articles for culture, education, and sport activity industry had the lowest average informationization index at only 0.4388. The trend shows an overall increase in the informationization level of manufacturing, which rose from 0.4927 in 2003 to 0.6562 in 2011 with a growth rate of 32.78%. Among the industries, the informationization level in the manufacture of special purpose machinery had the highest growth of 44.81%; although the informationization level in the manufacture of computers, communication, and other electronic equipment industry was highest, the growth rate in the observed period was the slowest with an amplification of 22.04%; all remaining industries showed a growing trend to various extents.

(24) Control variables

The first control variable is R&D intensity (*RD*). R&D intensity is an important factor that promotes R&D capabilities of industries; improved R&D capabilities impact significantly on industries' green growth (Wan and Zhu 2013). The ratio of industrial R&D expenditures to total industrial output was adopted as the proxy variable of *RD*. We expected that RD would significantly impact green growth of manufacturing.

The second control variable is environmental regulations (*ER*). According to the Porter hypothesis, appropriate environmental regulations are conducive to stimulating technology innovations, energy saving, and emissions reduction in industries and thus, is conducive for green growth in manufacturing. The proportion of industrial wastewater and waste gas treatment expenditures to total industrial output was adopted as the proxy variable for environmental regulations (Shen 2012). We expected that environmental regulations would significantly impact green growth in manufacturing.

The third control variable is the level of marketization (*MS*). Generally, the higher the level of marketization is, the more significant is the innovation spillover effect and the higher is the resource allocation, which is conducive to green growth in manufacturing. The ratio of the number of industrial workers in privately owned economic units to the total number of workers was adopted as the proxy variable for the level of marketization. We expected that the level of marketization would significantly impact green growth in manufacturing.

The fourth control variable was foreign direct investment (*FDI*). Foreign direct investment can promote industrial technology level and energy efficiency through demonstration, competition, and technology spillover and hence, becomes conducive for green growth in manufacturing. The ratio of the output of large and medium-sized foreign-funded industrial enterprises to total industrial output was the proxy variable for *FDI*. We expected that FDI would significantly impact green growth in manufacturing.

5.4.4 Impact Analysis of Informationization on Green Growth in Manufacturing

1. Estimation results and analysis of green total factor productivity in manufacturing

MAXDEA software was employed to estimate green TFP in manufacturing. As seen from the estimation results, the green growth index among industries varied significantly. The manufacture of computers, communication, and other electronic equipment had the highest average green growth index at 1.1755, primarily due to the high level of informationization in the industry and the high integration of informationization and industrialization, which have facilitated the full promoting effect of informationization on green growth. On the other hand, smelting and pressing of ferrous metals had the lowest average green growth index at 0.9531. Although the informationization level is relatively high in that industry, there was no integration of informationization and industrialization. Thus, the promoting effect of informationization on green growth was not significant. As the trend shows, although the overall level of green growth in manufacturing rose steadily, the rate of increase was not large, increasingly by only 1.59% to 1.0491 in 2010–2011 from 1.0327 in 2003–2004. Among the industries, the green growth index in the manufacture of articles for culture, education, and sport activity industry grew the most at 2.74%. Furthermore, the green growth index in the processing of food from agricultural products and the manufacture of non-metallic mineral products declined by different magnitudes during the observation period. All remaining industries showed an upward trend to a different extent. Thus, transformation was relatively slow in China's manufacturing sector, and this is closely related to the lower level of informationization in the industry and the lack of effective integration between informationization and industrialization.

Based on the green growth index in relation to the informationization level in cluster analysis, we classified the integration mode between informationization and green growth in manufacturing into four types. The following is derived from Table 5.30. First, industries that had a high informationization–high green growth integration mode were all high-tech industries. Significant characteristics of the top five industries were high informationization level, rapid penetration and diffusion of information technologies, and lower energy consumption and environmental pollutant emissions. Hence, the promoting effect of informationization on green growth was significant. However, these industries accounted for only 17.24% of the entire

Table 5.30 Integration modes of informationization and green growth in China's manufacturing industries

Integration mode	Industry classification	Industrial percentage
High informationization–high green growth	Manufacture of general purpose machinery; Manufacture of special purpose machinery; Transportation equipment manufacturing; Manufacture of electrical machinery and equipment; Manufacture of computers, communication,, and other electronic equipment	17.24
High informationization–low green growth	Processing of food from agricultural products; Manufacture of tobacco; Manufacture of textiles; Processing of petroleum, coking, processing of nuclear fuel; Manufacture of raw chemical materials and chemical products; Manufacture of medicines; Manufacture of non-metallic mineral products; Smelting and pressing of ferrous metals; Smelting and pressing of non-ferrous metals; Manufacture of metal products	34.48
Low informationization–high green growth	Manufacture of food; Manufacture of wine, drinks, and refined tea; Manufacture of textile wearing and apparel; Processing of timber, manufacture of wood, bamboo, rattan, palm, and straw products; Manufacture of leather, fur, feather, and related products and footware; Manufacture of rubber and plastic; Manufacture of articles for culture, education, and sport activity	24.14
Low informationization–low green growth	Manufacture of furniture; Manufacture of paper and paper products; Printing, reproduction of recording media; Manufacture of articles for culture, education, and sport activity; Manufacture of chemical fibers; Manufacture of rubber and plastic; Manufacture of measuring instruments	24.14

Data source The above data were generated by the author using SPSS software in cluster analysis

manufacturing sector. Therefore, it is crucial that high-tech industries be developed further. Second, 10 of the industries, such as processing of food from agricultural products, showed a high informationization–low green growth integration mode but accounted for 34.38% of the entire manufacturing sector. These industries were mostly engaged in monopolized industries, such as petroleum processing and smelting. Energy consumption and environmental pollutant emissions were relatively high, resulting in a lower green growth index. Although informationization was high in these industries, its promoting effect on green growth was still insufficient. Appropriate system reform and management innovation should be implemented in this industry type, and further competition should be introduced to force deepened integration between informationization and industrialization. Third, 24.14% of the manufacturing industries showed a low informationization–high green growth integration mode. The majority of these industries were traditional manufacturing industries. Although the infrastructure of informationization in these industries was relatively backward, energy consumption and environmental pollutant emissions were lower, resulting in higher industrial green growth. Therefore, further development of informationization infrastructure should be accelerated to fully impact green growth. Fourth, industries with a low informationization–low green growth integration mode were mostly engaged in the manufacture of paper and paper products, which accounted for a relatively high proportion of the entire manufacturing industry, reaching 24.14%. These industries were characterized by lagging informationization and low green growth. This industry type should increase investment in informationization to improve traditional production operation, thereby expediting the promoting effect of informationization on higher green growth. As observed, the integration between informationization and green growth in China's manufacturing industries is not ideal at present. China still has a long way to go to achieve overall deepened integration of high informationization and high green growth.

2. Impact analysis of informationization on green growth in the manufacturing sector

Stata 12.0 was used to conduct system GMM estimation of the regression model. The estimation results are shown in Table 5.31. As shown from the diagnostic test of the model, the AR (1) and AR (2) tests indicated that only first-order autocorrelation existed in the first-order differential of the residual sequence, suggesting

Table 5.31 Dynamic GMM estimation results of the impact effect of informationization on green growth in the manufacturing sector

Variable	Model (1)	Model (2)	Model (3)	Model (4)
Cons	-0.2436^{***} (−17.26)	-0.2336^{***} (−13.28)	-0.2413^{***} (−15.74)	-0.2385^{***} (−14.61)
τ	0.1134^{***} (27.54)	0.1095^{***} (25.49)	0.0989^{***} (28.39)	0.1025^{***} (29.14)
ln*II*	0.0896^{***} (5.26)	0.0917^{***} (4.86)	0.0935^{***} (4.66)	0.0964^{***} (5.07)
ln*RD*		0.2594^{***} (9.53)	0.2579^{***} (8.97)	0.2606^{***} (10.36)
ln*II***RD*				0.0714^{*} (1.83)
ln*ER*			0.0536^{***} (4.28)	0.0565^{***} (4.92)
ln*MS*			0.0935^{**} (2.15)	0.0897^{**} (2.26)
ln*FDI*			−0.0983 (−1.04)	−0.1026 (−0.95)
AR(1)	(0.0000)	(0.0000)	(0.0000)	(0.0000)
AR(2)	(0.5148)	(0.5227)	(0.5372)	(0.5305)
Hansen test	(0.9998)	(1.0000)	(0.9998)	(0.9999)

Note *, **, and *** represent having passed the 10, 5, and 1% significance tests, respectively; values inside parenthesis are the progressive statistical values of t that the first-order lag estimate in the dynamic panel model was reasonable. In addition, Hansen's over-identification test showed that the model design was reasonable and the selection of tools and variables was also reasonable and valid. After the control variables were introduced, the sign of the coefficient of the informationization variable and the significance level did not change; only the general estimation value of the parameter was different. This indicated that the regression results of the model were highly reliable. Among the regression results, since the first-order lag of the regression coefficients to the variables were positive and passed the 1% significance test, the initial preparation for and progress in informationization for manufacturing development, as a continuously dynamic system, inevitably were due to such factors as economic development, technology level, and labor resources, and will show in the current period of manufacturing development or in subsequent periods.

The informationization coefficient of each model passed the 1% significance test, indicating that informationization is significantly conducive to the green growth of the manufacturing sector. Hence, Hypothesis 1 was established. After the control variables were introduced, the promoting effect of informationization improved the manufacturing sector's green growth to a certain extent, indicating that the innovation spillover effect of informationization also relied on the combined effect from the other factors to impact manufacturing green growth to the fullest. We consider that the following reasons explain why informationization promotes green growth in the manufacturing sector at present. First, the rapid development and widespread application of information technologies have provided an efficient and smart R&D platform as well as an information platform for innovation development in manufacturing, which have effectively improved the innovation efficiency in the technological activities in the industries and their management efficiency. The technology level of industrial production equipment and labor productivity have been enhanced significantly. Thus, the substitution effect of informationization has fully utilized the traditional factors of production. Second, informationization not only has promoted internal structural optimization of heavy and light industries in manufacturing but also has accelerated the transformation and upgrading of traditional industries. Thus, energy consumption and environmental pollutant emissions effectively were reduced in high energy-consuming and high environmental pollution emissions manufacturing industries. In addition, informationization is conducive to transformation, upgrading, and updating of manufacturing equipment, providing enterprises with more advanced and greener production technologies and processes, improving the energy utilization rate and pollutant handling, and facilitating better performance of the energy-saving effect and environmental pollution emissions effect of informationization on energy consumption. Third, informationization infrastructure and development enable enterprises to acquire competitive advantage through continual technology innovations, effectively shortening the R&D and production cycles of new products, increasing information content and technical value-added of products, and enhancing efficiencies in decision making and transactions, thereby lowering transaction and information costs. All these result in a promoting effect of informationization on economic growth. In summary, informationization is significantly conducive for green growth in China's manufacturing sector, meaning that informationization has gradually become a new direct source of power for the green growth of China's manufacturing sector.

The cross-term of informationization and R&D investment intensity to a certain degree had a significantly positive impact on manufacturing green growth, indicating that R&D investment intensity significantly has strengthened the promoting effect of informationization on the manufacturing sector's green growth, which is consistent with the research results of Wan and Zhu (2013). According to theories of technological innovation, R&D and applications of industrial information technologies primarily come from independent R&D and technological introduction, the promoting effect of which relies on industrial R&D investment intensity as the foundation. When R&D investment intensity is low, independent R&D in the industries is weaker; technologies introduced cannot be digested and absorbed effectively, but rather restrict the promoting effect of informationization on manufacturing green growth to a certain extent. When R&D investment intensity is higher, independent R&D in the industries is stronger; technologies introduced domestically and from abroad can be digested and absorbed effectively, leading to a second round of innovations in production processes and technological details using informationization. Thus, the promoting effect of informationization on manufacturing green growth can be achieved to the fullest (Zhang 2015). As observed, R&D investment intensity plays an integral role in mediating the promoting effect of informationization on the green growth process in manufacturing. Increasing R&D investment intensity has become an indirect source of power in promoting green growth in China's manufacturing sector.

Based on the analysis of the control variables, the following conclusions are drawn. (1) R&D investment intensity has a significant impact on manufacturing green growth and has a significantly stronger promoting effect than that of other factors. This is because R&D investment intensity can directly enhance technological innovation in the industries and labor productivity. Thus, it is directly conducive to manufacturing green growth. The other reason is that an increase in R&D investment intensity is conducive to deepened integration of informationization and industrialization, and thus, has a better indirect role in bringing out the promoting effect of informationization on green growth in manufacturing. (2) Environmental regulations are conducive to green growth in manufacturing, which supports the Porter hypothesis to a certain extent and suggests that appropriate environmental regulations can stimulate technological innovations in industries, improve the energy utilization rate, and reduce pollution emissions; thus, it is conducive to green growth in manufacturing. (3) The level of marketization has a significantly positive impact on manufacturing green growth primarily because, when the level of marketization is higher in the industries, their resource allocation rate is higher, and enterprises spend more time and effort on product R&D and cost control in order to overcome intense market competition, which is conducive to manufacturing green growth. (4) Contrary to our expectation, foreign direct investment is insignificant for manufacturing green growth, possibly for the following two reasons. First, the investment structure of foreign direct investment in

China's manufacturing sector is still concentrated on resource-intensive and labor-intensive industries at the lower end of the value chain, and has not resulted in significant knowledge and technology spillovers. Instead, a certain degree of overcapacity, high energy consumption, and high pollutant emissions have resulted, which is not conducive to the manufacturing sector's green growth. Second, distribution of foreign direct investment in China's manufacturing industry is uneven. The majority of investment is concentrated in heavy industries with relatively smaller foreign direct investment in light industries. Therefore, the impact of foreign direct investment observed in all regression models of various industry types was insignificant.

3. Analysis of heterogeneous effect of informationization on green growth in manufacturing

The cross-term for the dummy variables of informationization and industry characteristics was introduced to measure the impact of heterogeneity of informationization on green growth in the manufacturing sector, where dum represents the dummy variables that reflect industry characteristics, defined as follows: (1) *tec* represents technological level; if industry *i* were an industry with medium to low technology, the value = 0; if industry *i* were an industry with high technology, the value = 1; (2) *avs* represents industry size; if industry *i* were an industry of medium to small size, the value = 0; if industry *i* were a large industry, the value = 1; (3) ene represents energy consumption; if industry *i* were an industry with low energy consumption, the value = 0; if industry *i* were an industry of high energy consumption, the value = 1; (4) *pol* represents environmental pollution; if industry *i* were an industry of low pollution, the value = 0; if industry *i* were an industry of high pollution, the value = 1; and (5) *pro* represents profitability; if industry *i* were an industry of medium to low profitability, the value = 0; if industry *i* were an industry of high profitability, the value = 1. Models (5)–(9) showed the regression results of the cross-terms between the dummy variables of informationization to technology level, industry size, energy consumption, environmental pollution, and profitability, respectively. Model (10) showed the regression results of all the cross-terms, with the estimation results shown in Table 5.32. In Models (5)–(10), the AR (1) and AR (2) tests showed that first-order lag taken in the dynamic panel data model for estimation was reasonable, while the Hansen test also indicated that the selection of tools and variables in the model was reasonable and valid; these results suggest that the use of dynamic GMM for the estimation was reasonable and feasible. After the cross-terms were introduced, there was no significant change in the regression results of the informationization index or any of the control variables, further verifying the robustness of the estimation results.

In the regression results, all the cross-terms passed the 1% significance test, indicating that significant heterogeneity exists in the promoting transmission mechanism of informationization on the manufacturing sector's green growth among the industries. Therefore, Hypothesis 2 was established. (1) When the coefficients were significantly positive, the promoting effect of informationization

Table 5.32 Estimation results of dynamic GMM on heterogeneous effect of informationization on manufacturing green growth

Variable	Model (5)	Model (6)	Model (7)	Model (8)	Model (9)	Model (10)
Cons	−0.2056*** (−12.73)	−0.2182*** (−11.68)	−0.2074*** (−13.17)	−0.2143*** (−12.36)	−0.2067*** (−13.15)	−0.2205*** (−12.76)
τ	0.1024*** (17.54)	0.1083*** (19.48)	0.1047*** (20.74)	0.0997*** (22.47)	0.0976*** (19.84)	0.1026*** (20.48)
ln*II*	0.0928*** (4.74)	0.0935*** (4.26)	0.0958*** (4.39)	0.0976*** (4.87)	0.0949*** (5.06)	0.0941*** (4.55)
ln*II***tec*	0.0715* (1.80)					0.0736** (2.07)
ln*II***avs*		−0.1036* (−1.76)				−0.1023* (−1.79)
ln*II***ene*			−0.0941*** (−3.65)			−0.0926*** (−3.86)
ln*II***pol*				−0.0864*** (−4.49)		−0.0837*** (−5.03)
ln*II***pro*					0.0576*** (3.85)	0.0546*** (4.11)
ln*RD*	0.2433*** (8.48)	0.2504*** (9.37)	0.2479*** (10.14)	0.2418*** (9.05)	0.2523*** (10.19)	0.2491*** (9.96)
ln*ER*	0.0593*** (4.76)	0.0567*** (3.84)	0.0607*** (4.16)	0.0595*** (3.93)	0.0612*** (4.48)	0.0596*** (4.27)
ln*MS*	0.0946* (1.81)	0.0913** (2.07)	0.0902* (1.77)	0.0914** (2.02)	0.0925** (2.19)	0.0898* (1.84)
ln*FDI*	−0.0983 (−0.75)	−0.1021 (−0.68)	−0.1016 (−1.05)	−0.0977 (−0.94)	0.1023 (−0.87)	−0.1046 (−0.98)
AR(1)	(0.0000)	(0.0000)	(0.0000)	(0.0000)	(0.0000)	(0.0000)
AR(2)	(0.5784)	(0.5536)	(0.5385)	(0.5636)	(0.5396)	(0.5537)
Hansen test	(1.0000)	(0.9997)	(0.9998)	(1.0000)	(0.9996)	(0.9998)

Note *, **, and *** represent having passed the significance test at 10, 5, and 1%, respectively; values inside parenthesis are the progressive statistical values of t

on green growth in high-tech industries was higher than that in low-tech industries. In other words, as the industrial technology level increases, the promoting effect of informationization on China's green growth in the manufacturing sector continues to improve. A possible reason is that as the industrial technology level improves, industries' reliance on informationization continues to strengthen, the speed of technology penetration and diffusion in the industries accelerates significantly; and the integration level between informationization and industrialization increases (Wang and Pan 2015). Thus, the promoting effect of informationization on green growth in the manufacturing sector becomes even more significant. (2) When the coefficients were significantly negative, the promoting effect of informationization on green growth in medium- to small-sized industries was higher than in large-sized industries. In other words, as the average industrial size increases, the promoting effect of informationization on green growth in China's manufacturing continues to drop, indicating that the characteristic of industry size significantly impacts the promoting effect of informationization on green growth in manufacturing. A possible reason is that as the size of the industry increases, its innovation risks and dimensions increase as well, leading to a gradual weakening of the promoting effect of informationization on technical innovations in the industry (Han et al. 2014). Thus, the promoting effect of informationization on green growth in the manufacturing sector continues to drop. (3) When the coefficients were significantly positive, the promoting effect of informationization on green growth in low energy-consuming industries was higher than that in high energy-consuming industries. As industrial energy consumption increases, the promoting effect of informationization on green growth in China's manufacturing sector continues to drop. This is possibly because more high energy-consuming industries are engaged in the resource-intensive industry type, and technology penetration tends to be slower in these industries, making integration between informationization and industrialization lower and thereby weakening the promoting effect of informationization on green growth in these industries. (4) When the coefficients were significantly negative, the promoting effect of informationization on green growth in medium- to low-pollution industries was higher than that in high pollution industries. In other words, as the industrial pollution level increases, the promoting effect of informationization on green growth in China's manufacturing sector continues to drop. A possible reason is that when the industrial pollution level is lower, industries' reliance on informationization is relatively higher, and technologies are more conducive for green industries or improvement in the technological innovation capabilities of industries (Jin 2014). Thus, the promoting effect of informationization on these industries' green growth becomes more significant. (5) When the coefficients were significantly positive, the promoting effect of informationization on the green growth of industries with high profitability was higher than that on industries with medium to low profitability. In other words, as industry profitability increases, the promoting effect of informationization on green growth in China's manufacturing sector continues to strengthen. This is possibly because as industry profitability increases, the need for industrial informationization gradually increases as well; industries will increase their informationization input and R&D investment intensity. Thus, the promoting effect of informationization on green growth in the manufacturing sector becomes more significant.

5.4.5 Conclusions and Implications

We first analyzed the transmission mechanism and heterogeneous transmission of informationization on promoting green growth in China's manufacturing sector. Two theoretical hypotheses were proposed based on theoretical analysis. Then, based on the industrial informationization level index and green growth index, panel data of 29 manufacturing industries in China were employed to empirically verify the impact of informationization and its heterogeneity among different manufacturing industries. The research conclusions reveal the following. (1) Informationization is significantly conducive for green growth in China's manufacturing sector and has become a new source of power in promoting green growth in manufacturing. R&D investment intensity plays an important role in bringing out the promoting effect of informationization on the green growth process in the manufacturing sector. (2) The promoting effect of informationization on the manufacturing sector's green growth is significantly heterogeneous among industries. Under conditions in which the technology level is higher, average industry size is smaller, energy consumption is lower, pollution is lower, and profitability of the industry is stronger, the promoting effect of informationization on manufacturing green growth is higher. Based on the above conclusions, the following implications can be drawn.

First, China should accelerate its informationization infrastructure and promote deepened integration between informationization and industrialization in order to take full advantage of the promoting effect of informationization on green growth in manufacturing. On the one hand, China should gradually implement an "Internet +informationization" plan with a target date of 2025. It should actively develop related technologies, including high-tech information technologies, such as the Internet, Internet of things, big data, cloud computing, and smart manufacturing. China should adopt an "Internet + manufacturing industry" plan to change product features; it should employ an "Internet of Things + manufacturing industry" plan to integrate production processes; it should adopt a "big data + manufacturing industry" plan to reform management styles; it should adopt a "cloud computing +manufacturing industry" plan to strengthen systems coordination; and it should adopt a "smart manufacturing+manufacturing" plan to innovate manufacturing modes in order to gradually improve coverage and penetration of information technologies in the manufacturing industry. In addition, China should focus on promoting manufacturing transformation and upgrading so as to accelerate deepened integration between informationization and industrialization. It should take advantage of the innovation R&D design mode in informationization to continue improving independent innovation capabilities in industries; it should utilize informationization to accelerate the construction of modern production systems and to promote smart production equipment and automated production processes; it should make use of informationization to establish modern management systems to improve enterprise management information systems and collaborative integration; and it should adopt informationization to promote green development in industries to enhance energy consumption efficiency and safety in production.

The second main implication is that China should focus on various industries that have different integration modes between informationization and green growth in manufacturing industry in order to develop and implement new industrial development strategies that are appropriate for these industries. For high informationization–high green growth industries, key support should be provided for the development of financial and tax policies so as to establish a sound financial investment policy system and improvement of talent, policies, and intellectual property protection, such that clustering development of high-tech industries can be accelerated to fully bring out its innovation spillover effect and enable these industries to play leading roles.

For high informationization–low green growth industries, China should implement appropriate institutional reform and management innovation to continue raising the utilization rate in information resources; it should gradually dismantle industry monopolies and introduce competition to further deepen integration of informationization and industrialization; and it should adopt informationization to transform traditional production and related processes to make better use of the promoting effect of informationization on green growth in the manufacturing sector. For low informationization–high green growth industries, China should increase investment in informationization to speed up informationization infrastructure and development. It should focus on improving industrial digital controls to promote resource allocation efficiency and management decision making, so as to fully encourage the advantages of industries' green growth. For low informationization–low green growth industries, China should increase R&D investment intensity for the integration of informationization and industrialization; and it should adopt informationization to improve traditional production operation modes so as to gradually promote production process automation and intelligence levels, such that industrial management efficiency and energy consumption efficiency will be enhanced and industries' green growth will be improved as soon as possible.

The third main implication is that China should pay more attention to the heterogeneous impact of industry characteristics on the promoting effect of informationization on green growth in the manufacturing sector. The research conclusions show that the promoting effect of informationization is higher for green growth in high-tech industries, small industries, low energy-consuming industries, low environmental pollution industries, and industries with high profitability, and vice versa. Thus, we recommend the implementation of informationization infrastructure and measures to promote green growth in these industries, taking into account industry characteristics. On the one hand, the promoting effect of informationization is lower for the green growth of medium- and low-tech industries, which are mostly found in traditional industries that focus on industrial structural adjustment and optimization. This requires active implementation of advanced technologies, such as e-commerce, supply chain management, and production operation management, to further improve product design capabilities, enhance product quality and technological value-added, and make use of informationization in order to continue to promote industrial optimization and upgrading. On the other hand, the promoting effect of informationization tends to be lower for green growth of high energy-consuming and high pollution emissions industries, which are mostly engaged in capital-intensive industries in which energy savings and pollution emissions reduction are key issues. Therefore, utilization of informationization is necessary in order to accelerate the establishment of a green low-carbon industry system, to push this industry type to become more high-end and refined, and to develop it more intensively, and to accelerate the promotion of new energy-saving and emissions-reduction production and technologies. China should focus on supporting recycling production and green manufacturing projects, and continue to promote the transformation of its industrial development model from an extensive mode to an intensive mode.

5.5 Impact of Industrial Clustering on Technological Innovation in China's Manufacturing Sector

5.5.1 Introduction

China is the world leader in manufacturing. Since the period of reform and opening, the manufacturing sector has been the dominant force driving China's economic growth, is the basis upon which China's people depend for survival and development, and is an important foundation for China's socioeconomic development. The manufacturing sector has allowed China to rapidly develop the capacity to produce a large volume of products for global export, and has provided China with successive economic miracles (Li et al. 2012). With the increasing pace of global economic integration, ever-increasing costs of raw materials, growing shortages of the means of production, and gradual weakening of ecological carrying capacities, development models that rely solely on inexpensive labor and abundant resources are no longer suitable for current development demands. China's manufacturing sector is now gradually shifting toward more advanced technological production, industrial upgrading, and innovation-driven modes of development (Li et al. 2013a, b). This gradual shift from "factor-driven" to "innovation-driven" production is giving way to a shift from a situation in which products are "made in China" to one in which they are "created in China"; the key in this process lies in the improvement of China's capacity for technological innovation. Li et al. (2014) proposed that technological innovation plays a major role in the transformation and upgrading of the manufacturing sector, and that only by fully utilizing science and technology and relying on technological innovation, is it possible to transform the current "made in China" production mode to one in which products are "created in China." Following this mode of thinking, China's 18th National Congress clearly proposed the implementation of an innovation-driven development strategy. It is evident that technological innovation has become the main engine driving economic development. Thus, there is an urgent need for China to promote technological innovation as a means to drive long-term economic development. Nevertheless, in stark contrast to this necessity, China's manufacturing sector has always been situated at the low end of the technological spectrum; its capacity for technological innovation is as yet relatively weak, and there is still a sizable gap between it and the manufacturing sectors of the world's developed nations (Yang et al. 2014).

At the same time, with the increasing pace of regional economic integration, and the gradual rise in levels of regional industrial specialization and marketization, the phenomenon of clustering in the manufacturing sector within China is increasingly apparent (Fan 2004). The appearance of industrial clusters is gradually becoming an underlying trend in the geographic structure of development in China's manufacturing sector. Recently in China, especially in the eastern coastal regions, several successful models of industrial clusters have been formed. Examples of these include Beijing's Zhongguancun, Shanghai's Zhangjiang Hi-Tech Park, Tianjin's

Huayuan IndustrialPark, Shenzhen's Nanshan High-Tech Industrial Park, and Suzhou's International Science Park.

Marshall (1890), Arrow (1962), and Romer (1986) proposed that the clustering of neighboring firms within an industry is beneficial for inter-firm knowledge spillovers and technology spillovers, and thereby promotes firms' technological innovation; thus, industrial clustering is helpful in promoting technological innovation. The following uncertainties must be considered. Can China can draw upon industrial clustering to improve its capacity for technological innovation, and thereby continue to drive long-term economic growth? What are the effects industrial clustering on technological innovation? What is the extent of these effects? What are the mechanisms of action? Are there differences across regions or industries in the effects of industrial clustering on technological innovation? What is the real significance of the conclusions to these questions? We study the impact of industrial clustering on innovation in China's manufacturing sector in order to answer each of these questions.

5.5.2 Literature Review

Academic research related to the effect of industrial clustering on technological innovation can be divided roughly into two groups: theoretical research and empirical research.

1. Current state of international research

(1) Theoretical research

Marshall (1890) investigated the factors driving industrial clustering from the external economic perspectives of labor pooling, investment in intermediate goods, and technology spillover. He proposed that externalities that generally exist in industrial clustering and economic activity are closely linked, while at the same time, the technology spillovers of industrial clustering (technological externalities) play a definite role in promoting a firm's technological innovation activities. Thereafter, academics conducted a large amount of research on the impact of industrial clustering on technological innovation. Arrow (1962) and Romer (1986), consistently with Marshall, proposed that the clustering of neighboring firms within an industry helps knowledge spillover and technology spillover between firms, and thereby promotes firms' technological innovation. MAR externalities (named from the initials of Marshall, Arrow, and Romer) refer to the externalities of knowledge or technology spillovers brought about by clustering within an industry. However, besides being generated within neighboring firms clustered in the same industry, the clustering of different industries in the same region can also generate knowledge and technology spillovers, and thereby promote technological innovation (Jacobs 1969). The externalities brought about by this type of inter-industry clustering are called Jacobs externalities. Continuing in this vein, Porter (1998) explored the competitive advantages of industrial clustering from the perspective of technological innovation, proposing that industrial clustering can promote technological innovation for the following four reasons: 1. for customers and other related entities engaging in close observation and face-to-face communication, clustered firms have good "windows" to knowledge innovation opportunities; 2. clustered firms can more easily unify suppliers and other partners to participate in innovation, and thereby achieve rapid innovation for necessary resources; 3. through coordination with other organizations, firms within clusters can engage in low-cost innovative experimentation; and 4. sustained competitive pressure within clusters can effectively push firms to be consistently innovative. Although Porter considered that knowledge and technology spillovers within industrial clusters are helpful for technological innovation, he differed from Marshall, Arrow, and Romer in his belief that competition within industrial clusters, not monopolies, is better for promoting technological innovation. This type of knowledge or technology spillover created by competition within industrial clusters is called a Porter spillover.

MAR externalities, Jacobs externalities, and Porter spillovers are all viewed from the perspective of knowledge or technology spillovers, and they study the mechanisms through which industrial clustering affects the development of technological innovation. Furthermore, many researchers have conducted research on the impact of industrial clustering on technological innovation from the perspectives of, among others, technology trade and cooperative innovation. For example, Nobel Prize-winning economist Paul Krugman (1991), writing from the perspective of technological trade, proposed that trade is a key factor influencing technological innovation activities. Industrial clustering has increased the technological trade within industries, thereby promoting technological innovation activities within industries. Meyer-Stamer (2002), from the perspective of cooperative innovation, proposed that innovative environments can be created through inter-firm cooperation, thereby raising the capacity for technological innovation of an industry and improving the competitive advantages of firms.

(2) Empirical research

Baptista and Swann (1998), through a study based on 248 manufacturing sector firms in the UK between 1975 and 1982, discovered that firms located within industrial cluster areas were better able to promote technological innovation compared to isolated firms located outside of industrial clusters. Feldman and Audretsch (1999), through an empirical study using the United States as an example, showed that diversity of economic activity (Jacobs externalities) promotes knowledge spillover, by clustering all types of complementary economic activities, and thus, better promotes innovation. Furthermore, the authors showed that specialization of economic activity (MAR externalities) does not play a significant role in promoting regional innovation. Kelly and Hageman (1999) discovered through their empirical study of two-digit industries that spatial clustering of industries could raise a region's level of innovation significantly. Paci and Usai (1999), using an empirical statistical study of 784 regions in Italy in 1990 and 1991, discovered that both Marshall externalities and Jacobs externalities can significantly raise the production of regional industrial innovation. Yoguel et al. (2000), through an analysis of the relationship between production networks and firm capacity for technological innovation, discovered that, through network relationships, industrial clustering can promote the economic performance and capacity for technological innovation of firms within cluster areas. Bagella and Becchetti (2002) conducted an empirical study based on Italian firms and discovered that increased levels of industrial clustering do not affect R&D expenditure, but rather, raise the quality of overall regional innovation. Greunz (2004), in an empirical study of 16 manufacturing industries in 153 European regions, discovered that both specialization externalities (MAR externalities) and diversification externalities (Jacobs externalities) can significantly affect regional capacity for technological innovation, and in high population density and high-tech industries, the importance of diversification externalities was more apparent. Research by Storper and Venables (2004) showed that industrial clustering provided opportunities for face-to-face interaction, and thus, promoted technological innovation activities. Carlino et al. (2007) used employment density (the number of employed people/km^2), to measure levels of industrial clustering; their results showed that per capita patents and employment density were positively correlated, and that where levels of industrial clustering doubled, per capita patents increased by 20%. Gilbert et al. (2008), through an empirical experiment based on the relationships between industrial clustering, knowledge spillover, and new firm performance, discovered that firms located in geographic clusters were able to extract more knowledge from their local environment, and had higher profits and innovation performance. Silvestre and Dalcol (2009) undertook an analysis of the results of an empirical study of 10 oil and natural gas firms clustered in Brazil's Campos Basin, and discovered that geographic proximity plays a significant role in promoting technological innovation activities.

2. Current Chinese research conditions

Compared to international research, Chinese studies on the impact of industrial clusters on technological innovation began relatively late, although Chinese academics have now conducted both theoretical and empirical studies on this phenomenon.

(1) Theoretical research

Cai and Nie (2003) proposed the theory that industrial clustering, through strong relationships, weak relationships, and structural holes within social networks, lowers the cost of information acquisition and trade in dense and sparse networks, and increases the rate of exchange and transmission of information and knowledge, thereby promoting technological innovation in firms and raising firms' competitiveness. Zhao Tao et al. (2005) constructed an industrial cluster innovation system from a core layer, service-support layer, and macro-environment layer, and analyzed three processes of industrial clustering innovation: the process of technological innovation itself; the innovation process of economic decision making and behavior, and the innovation process of organizational management. Huang and Chen (2006) considered that industrial clustering can promote technological innovation in the following four ways: 1. industrial clustering is, to a certain extent, capable of coordinating the conflicts between a firm's capacity for technological innovation and the scale of the firm, allowing for the maximization of the firm's capacity for technological innovation; 2. industrial clustering promotes the formation of support networks necessary for technological innovation by firms; 3. industrial clustering provides support at every step of technological innovation by firms; and 4. industrial clustering allows for more rapid dispersion of technology. Huang and Yu (2007) proposed that under the innovation mechanisms of industrial cluster network structures, compared to firms outside of industrial cluster areas, those within these clustered areas possess the performance and innovation advantages of the ability to create multiple times the number of new products, high success rates for low-risk and low-investment innovation, the ability to disperse the products of innovation rapidly, and compressed periods for achieving innovation. He Jun (2008) proposed that the externalities of industrial cluster areas, as well as the regional characteristics of regional network structures and technological innovation, similarly promote the increased development pace of technological innovation. Duan and Liang (2009) explored the relationship between knowledge spillover linkages and clustering from the aspects of innovation in knowledge production, transfer, and accumulation. They proposed that knowledge linkages are an important force in industrial clustering, and that the constant flow and configuration optimization of all types of information technology resources within clusters promotes innovation behavior in firms. Clusters are the basis for the formation of innovative industrial networks. Zhang Cui (2010), on the basis of reviewing existing studies, identified three micro-level mechanisms through which industrial clustering promotes technological innovation. These are the knowledge spillover mechanism, the unique knowledge mechanism, and the clustered firm interaction mechanism. Gong and Liu (2011) proposed that technological innovation is the basis for the materialization of industrial clustering, and directly promotes the intensification, improvement, and upgrading of industrial clusters. In addition, through such effects as technology spillover, cooperative innovation, and market sharing, industrial clusters provide the mutually reinforcing tools of networks and environments for innovation. Ma et al. (2012) proposed the theory that for industrial clustering, the coordination and cooperation of different industries, disciplines, and firms provides favorable conditions whereby cross-sectional integration can form new growth-points for technological innovation, and thus, allows different parties to increase one another's capacity for technological innovation. Li et al. (2012) proposed that industrial clustering, by generating savings in the costs of interactions, knowledge sharing in industrial chains, complementary capital, and knowledge externalities, promotes the sustained increase of technological innovation capacities. Liu (2013) proposed the theory that industrial clustering promotes technological innovation in the following three ways: 1. it promotes the formation of technological innovation systems; 2. it provides advantages for technological innovation; and 3. it promotes the dispersion of technology.

(2) Empirical research

Research conducted by Chinese academics on the impact of industrial clustering on technological innovation in the manufacturing sector has primarily focused on the following three types: empirical research based on regional manufacturing industries (excluding high-tech industries), empirical research based on manufacturing industries (not including high-tech industries), and empirical studies based on high-tech manufacturing industries.

For empirical research based on regional manufacturing industries (excluding high-tech industries), Yu (2007) discovered, through an empirical study of the Shanghai manufacturing sector, that industrial clustering can increase the rate of technology trade in an industry, as well as intra-industry and inter-industry technological cooperation and technological alliances. In addition, it increases network linkages of firms and cooperation in technological innovation, while reducing the cost of technological innovation and dispersing the risks of technological innovation, thereby allowing for increases in an industry's technological innovation activity. Zhang et al. (2007), conducted empirical research based on questionnaire surveys of 342 manufacturing sector firms in Jiangsu province regarding how positioning, division of labor, and clustering in industrial chains affects the intensity of firm innovation. He discovered that industrial clustering did not have a positive influence on micro-level firm innovation activity in China, and had not become an organic vehicle for driving cluster innovation. Chai (2008), in an empirical study based on panel data for Chinese regional manufacturing industries, discovered that industrial clustering is capable of bringing a certain level of advantage for innovation, although cluster bodies do not necessarily bring about relatively high regional capacities for innovation, and are susceptible to many influencing factors. Tao (2011), through an empirical study based on panel data for Chinese provincial manufacturing industries, discovered that regional industrial clustering can generate a positive influence on regional technological innovation. Liu et al. (2015), based on panel data for Chinese provincial manufacturing industries between 2001 and 2012, studied the effect of industrial clustering on technological innovation capacities and regional differences in this effect. The results showed that after controlling for such factors as investment in scientific personnel, technology, foreign direct investment, and institutional innovation, industrial clustering significantly promotes technological innovation. A regional comparative analysis showed clear regional differences in the effects of industrial clustering on technological innovation capacity, where industrial clustering in eastern regions yields greater benefits to technological innovation compared with western regions.

Considering empirical research based on manufacturing industries (excluding high-tech industries), Wu and Yao (2004), conducted an empirical study of the Shaoxing textile industry regarding innovation dispersion in industrial clusters; the authors discovered that industrial clustering is beneficial to the dispersion of technological innovation. Zhang and Li (2007), using an empirical study of Chinese electronics and communications equipment manufacturing industries, discovered that specialized knowledge spillovers produced by the aggregation of production behavior have a positive influence on the generation of regional innovation, and that diversified knowledge spillovers generate a negative impact on regional innovation. Lin (2008) discovered through an empirical study of segments of the manufacturing sector that institutional innovation, industrial clustering, and degree of industrial openness all have significant and positive effects on the efficiency of technological innovation. Liu et al. (2010), based on a panel data study of 28 manufacturing industries in 30 Chinese regions, discovered that the effects of industrial agglomeration on capacity for technological innovation vary by industry, such that many high-tech and traditional industry clusters can promote technological innovation, while resource-dependent industrial clusters can inhibit technological innovation. Wang (2011) discovered through an empirical study of the Chinese pharmaceutical, electronics, and communications equipment manufacturing industries that specialized knowledge spillovers within an industry (MAR externalities) clearly have a positive effect on regional technological innovation. In addition, the study revealed that intra-industry competition (Porter spillovers) is not helpful for regional technological innovation, and that the effects of diversified spillovers (Jacobs externalities) vary by industry. Peng and Xiong (2011) discovered in a panel data study of China's equipment manufacturing industry between 1998 and 2009 that significant technological spillovers exist within clustered areas of this industry. Peng and Jiang (2011), using an empirical study of 21 manufacturing industries in 30 Chinese provinces between 1999 and 2007, discovered that MAR externalities and Jacobs externalities have significant positive effects on technological innovation for regional industries. Zhang (2012) discovered through an empirical study of 20 manufacturing sector industries between 2000 and 2005 that industrial clustering has a significant positive effect on technological innovation for the sector. Luo (2012), conducting an empirical study of China's nine largest industries, discovered that industrial clustering and technological innovation exhibit a non-linear relationship; before the inflection point of this relationship, industrial clustering and technological innovation are negatively correlated, while after the inflection point, industrial clustering can promote increased capacity for technological innovation by firms. Wang et al. (2012) studied the effect of industrial clustering on technological innovation in the electronic information industry; their results showed that geographic proximity and frequent contact between firms did not promote mutual trust or technological innovation. Su and Song Sheng (2013), in an empirical analysis of panel data for 13 industries of China's manufacturing sector between 2000 and 2010, discovered that in the country as a whole, foreign direct investment, industrial diversification, and industrial specialization have a positive effect on innovation performance. Considering the analysis of grouped regression results, foreign direct investment and industrial specialization had relatively large effects on the performance of high-tech industries, while the effect of industrial diversification on low-technology industries' innovation performance was more apparent.

Finally, with regard to empirical research of high-tech industries, Li and Zhou (2006) conducted an empirical study of China's photoelectric industry in the

"Wuhan Chinese Optics Valley" and found that for industrial clustering in high-tech industries, clustered supply chains' organizational succession has a definite impact on technological innovation. Li and Xie (2007), undertaking an empirical analysis of 53 Chinese national high-tech zones, found that China's high-tech industrial clusters were not industrial clusters in the true sense of the term, and therefore, did not fully bring into play the effects of promoting technological innovation in high-tech industries. Niu (2008), based on an empirical analysis of 53 Chinese high-tech zones, concluded that industrial clustering has a prominent effect of raising innovation performance in high-tech zones. Zhang (2009) found through an empirical study of Chinese national high-tech development zones that levels of high-tech industrial clustering and capacities for technological innovation were positively correlated, meaning that as levels of industrial clustering increased, capacities for technological innovation within clustered areas became stronger. Huang et al. (2011) conducted an empirical study of the high-tech industries of 29 Chinese provincial-level regions and found that industrial clustering plays a definite role in promoting technological innovation. Liu and Shao (2011) proposed that industrial clusters indirectly increase intangible capital, raise the utilization efficiency of technological capital, and promote increased innovation performance in high-tech zones; in addition, the study used the high-tech industrial cluster area constituted by Beijing's Zhongguancun Science Park as its research object to carry out empirical experiments. Zhou and Li (2011) conducted an empirical study of technological innovation in regional high-tech industries from the perspective of industrial clustering based on statistical data related to Chinese high-tech industries between 1998 and 2006. Their results showed that factors of industrial clustering could significantly promote the capacity for technological innovation of high-tech industries within provincial regions, as well as promote inter-province knowledge spillovers. Niu et al. (2012) studied the effect of high-tech industrial clustering on regional technological innovation based on data related to 29 mainland Chinese provincial-level regions. The results showed that high-tech industrial clustering has a significant positive effect on regional technological innovation. Zheng (2012), using an empirical analysis of Shaanxi's high-tech industries, discovered that regional manufacturing industrial clustering is positively correlated with capacity for technological innovation. Chen et al. (2013) studied the effect of industrial clustering on technological innovation based on China's high-tech industry. Their results revealed that under different levels of clustering, the effects of industrial clustering on technological innovation vary; where levels of industrial clustering are relatively low, specialized clustering is beneficial to technological innovation, while diversified clustering can inhibit technological innovation. By contrast, under relatively high levels of clustering, specialized clustering is not helpful for technological innovation, while diversified clustering can promote technological innovation.

Despite the results produced in the research described above, academics have yet to form a consensus regarding the effect of industrial clustering on technological innovation. Furthermore, existing research tend to be conducted from the unitary perspective of a region or industry when investigating the effects of industrial clustering on technological innovation, but there are relatively few studies that investigate this question from the dual perspective of region and industry. In addition, the vast majority of existing research stops short of applying time-varying dynamic laws, and instead focuses on the use of static analysis, thereby overlooking the use of dynamic panel data models to conduct research on the effects of industrial clustering on technological innovation. For this reason, in order to conduct a scientific, comprehensive, and integrated analysis of the effects of industrial clustering on technological innovation, we first discuss the internal mechanisms of the impact of industrial clustering on technological innovation. Next, the effects of industrial clustering on China's manufacturing sector on technological innovation are studied from both static and dynamic perspectives, and variations in these effects by region and industry are considered.

5.5.3 Theoretical Analysis and Hypotheses

1. Theoretical analysis and hypotheses for the effects of industrial clustering on technological innovation

Our analysis of the mechanisms through which industrial clustering affects technological innovation is conducted primarily from the two perspectives listed below; following this analysis, related theoretical hypotheses are proposed.

(1) Industrial clustering is beneficial for promoting the generation of technological innovation

Within industrial clusters, many bodies, including governments, firms, universities, research institutions, and users, agglomerate in the same area. Mechanisms, including interdependence, mutual cooperation and complementation, specified division of labor, and resource sharing, interact to form MAR and Jacobs externalities, which promote technology and knowledge spillovers, thereby promoting collaborative innovation. In addition, based on the benefits of industrial clustering and collaborative innovation for the integrated and efficient allocation of innovation resources in industrial cluster areas, technological innovation is promoted, and thus, consistent increases in the technological innovation capacities within these areas are achieved. Following this, lively competition within these regions of industrial clustering causes specialization in areas of competitive advantage, leading to more efficient technological innovation, and thereby benefiting knowledge and technology spillovers. In other words, this process involves creating Porter spillovers, promoting the generation of technological innovation, and thus, promoting sustained increases in the technological innovation capacities of core industrial cluster areas.

(2) Industrial clustering is helpful in promoting the dispersion of technological innovation

The high-volume clustering and interaction of knowledge, human resources, and information within industrial cluster regions is helpful in the formation of exchange networks for knowledge, technology, and information. This is beneficial to reducing the costs associated with diffusing technological innovation between firms in an industrial cluster area, thereby increasing the pace of inter-firm dispersion of technological innovation, and promoting the sustained increase of the capacity of an industrial cluster area for technological innovation. In another regard, improvement of knowledge, technology, and information exchange networks not only can provide more platforms for mutual exchange between different firms in an industrial cluster area, but also can provide more and better channels for transmitting technological innovation between firms. This helps to increase the pace of the diffusion of technological innovation, and promotes continuous increases in the capacity for technological innovation of industrial cluster areas (Cheng and Liu 2008; Liu and Shao 2011).

Based on this theoretical analysis, Hypothesis 1 is proposed.

Hypothesis 1: Industrial clustering has the positive effect of promoting technological innovation.

It can be observed from the states of the development of manufacturing industries in different regions of China that there are significant differences between the levels of industrial clustering of the manufacturing sector in China's east region and its central and western regions; these differences might result in variations in the regional effects of industrial clustering on technological innovation. On the basis of Hypothesis 1, we propose Hypothesis 2.

Hypothesis 2: There are regional variations in the effect of industrial clustering in promoting technological innovation.

2. Theoretical analysis and hypotheses related to other factors' effect on technological innovation

We briefly analyze the mechanisms through which other factors affect technological innovation, and then propose relevant theoretical hypotheses.

Investment in technical personnel can influence regional technological innovation; generally, the more a region spends on technical personnel, the more abundant the technological labor of that region is, and the stronger its capacity for technological innovation is. Based on this analysis, Hypothesis 3 is proposed.

Hypothesis 3: Investment in technical personnel has a positive effect on technological innovation.

In addition, investment in technical personnel is an important factor influencing regional technological innovation. It reflects the technological activity and investment conditions of a region; where more is spent on technological funding, there is a larger effect of promoting the region's technological innovation. Based on this analysis, Hypothesis 4 is proposed.

Hypothesis 4: Investment in technology has a positive effect on technological innovation.

Foreign direct investment is an important factor influencing regional technological innovation. Kokko (1992) proposed that while foreign direct investment brings capital to the host country, it could also generate technology spillover effects through such channels as the demonstration-imitation effect, technical personnel flow, and the effects of competition and linkages. Once firms in the host country absorb these types of technology spillovers, they effectively promote technological advances in host country firms, thereby raising the capacity for technological innovation of these firms. Based on this analysis, Hypothesis 5 is proposed.

Hypothesis 5: Foreign direct investment has a positive effect on technological innovation.

North (1973) proposed that institutional innovation is an important factor determining technological innovation. Good institutional choices can promote technological innovation, while poor institutional choices can inhibit technological innovation. For this reason, institutional innovation is an important factor affecting regional technological innovation. Thus, we consider institutional innovation. Since China's reform and opening up, it has implemented a socialist market economic system. In one regard, this has created a good innovation environment for regional technological innovation; in another respect, it has provided lasting impetus for regional technological innovation, and greatly increased China's capacity for technological innovation. Based on this above analysis, Hypothesis 6 is proposed.

Hypothesis 6: Institutional innovation has a positive effect on technological innovation.

5.5.4 Regional Variation of the Effect of Industrial Clustering on Technological Innovation

1. Explanation of econometric model and data

(1) Establishment of econometric model

We observe from the theoretical analysis in Sect. 5.5.3 that in addition to industrial clustering, such factors as investment in technical personnel, foreign direct investment, and institutional innovation have the ability to impact technological innovation. In order to measure the effect of industrial clustering on technological innovation more accurately, it is necessary to use the abovementioned variables as control variables. Thus, our econometric model is established in the following form:

$$\ln Innovation_{it} = \alpha + \beta_1 \ln Agglo_{it} + \beta_2 \ln Pinput_{it} + \beta_3 \ln Einputit + \beta_4 \ln FDI_{it} + \beta_5 \ln Insti_{it} + \varepsilon_{it}$$

Here, Innovation is the dependent variable representing capacity for technological innovation; Agglo is an independent variable representing the level of industrial clustering; and Pinput, Einput, FDI, and Insti are control variables representing investment in technical personnel, investment in technology, foreign direct investment, and institutional innovation, respectively. Subscripts i and t express region and year, respectively; α, β and ε indicate the coefficients and residuals of constant terms and variables.

(2) Explanation of variable and index selection

Below, we briefly explain the index and variables.

Capacity for technological innovation (Innovation): Drawing on the research of Liu et al. (2010), in this study, we use the number of patents to measure a region's capacity for technological innovation.

Level of industrial clustering (*Agglo*): Borrowing from the methods of Ciccone and Hall (1996), and Fan (2006), we use the density of the manufacturing sector (people/km^2) to measure the level of industrial clustering in a region. The level of industrial clustering at time *t* in region *i* is $Agglo_{it} = M_{it}/S_{it}$, where M_{it} is the population employed in the manufacturing sector at time *t* in region *i*, and $S_{it}$is the land area of region *i* at time *t*. Generally, an increasing value of this index indicates that the level of industrial clustering of the manufacturing sector in this region is higher.

Investment in technical personnel *(Pinput)*: We use the number of full time R&D employees to measure the investment in technical personnel of regional manufacturing industries.

Investment in technology (*Einput*): We use spending on R&D to measure the investment in technology of regional manufacturing industries.

Foreign direct investment (*FDI*): Borrowing from the methods of Liu and Shao (2011), we use the annual amount of foreign capital utilized to measure regional foreign direct investment.

Institutional innovation (*Insti*): Borrowing from the methods of Liu (2010), we use non-state investment in fixed economic assets as a proportion of total regional fixed asset investment to measure regional institutional innovation.

Data are taken from the *China Statistical Yearbooks* (2012–2014), *China Statistical Yearbooks on Science and Technology* (2002–2014), and the DRCnet Macroeconomic Database. Herein, for investment in technical personnel and investment in technology, the statistical classification for 2001–2010 is "large and medium enterprises," and for 2011–2013, "industrial enterprises above a designated scale." The actual utilized foreign capital for each year is calculated using the average exchange rate between the yuan and the US dollar for that year. Where data for any year are missing, the averages of the neighboring years' values are used instead.

Table 5.33 Descriptive statistics for main variable indexes

Variable	Sample size	Average value	Standard deviation	Smallest value	Largest value
ln*Innovation*	390	8.4273	1.5730	4.2485	12.5060
ln*Agglo*	390	2.3186	1.6888	-1.6376	6.3187
ln*Pinput*	390	9.7239	1.3812	4.4427	12.9630
ln*Einput*	390	12.7180	1.6277	7.2123	16.3329
ln*FDI*	390	13.9191	1.7305	9.4491	17.0114
ln*Insti*	390	-0.4607	0.1920	-1.0745	-0.1216

Below, we briefly describe the statistical characteristics of each major variable. The statistical results are shown in Table 5.33.

2. Measurement results and analysis

(1) Selection of measurement methods

The estimation methods for panel data include pooled ordinary least squares regression, the FE model, and the RE model. In addition to these models, the FGLS model could also be used for the estimation; this method eliminates possible heteroscedasticity and serial correlation, and can achieve effective estimation results (Wooldridge 2002). In order to achieve relatively consistent conclusions, the FE, RE, and FGLS models are all used for the estimation, and the software Stata 12.0 is used for the analysis.

(2) Analysis of regression results

[1.] Overall regression analysis of the effects of industrial clustering on technological innovation

Based on the measurement methods outlined in the previous section, we use panel data from 30 Chinese provincial-level regions between 2001 and 2013 to estimate the coefficient values for each explanatory variable. Equations 1, 2, and 3 in Table 5.34 report the estimation results of the FE, RE, and FGLS models, respectively.

First, we can accept our null hypothesis,[2] since the results of the Hausman test provide a p value of 0.1177, which is greater than the necessary 0.05. Thus, between the FE and RE models, the RE model is selected. That is, according to the Hausman test, the result of Eq. 2 is superior to that of Eq. 1. Second, by comparing the estimation results of Eqs. 2 and 3, we observe that the signs of the coefficients of explanatory variables are all consistent. Because the FGLS model eliminates possible heteroscedasticity and serial correlation to a certain extent, we discuss the findings of the overall panel data regression analysis on the basis of the results of Eq. 3.

[2]The null hypothesis is that when choosing between the FE and RE models, the RE model is used.

Table 5.34 Analysis of overall regression results

Independent variable	Equation 1 (FE)	Equation 2 (RE)	Equation 3 (FGLS)
ln*Agglo*	0.4131*** (0.0934)	0.2200*** (0.0460)	0.1198*** (0.0186)
ln*Pinput*	0.0082 (0.0684)	0.0173 (0.0647)	0.0780 (0.0483)
ln*Einput*	0.7424*** (0.0579)	0.7488*** (0.0546)	0.6441*** (0.0428)
ln*FDI*	0.0367 (0.0363)	0.0326 (0.0343)	0.1136*** (0.0241)
ln*Insti*	0.0197 (0.2224)	0.1144 (0.2068)	0.0669 (0.1538)
Constant term	-2.4128*** (0.5136)	-2.2804*** (0.4914)	-2.3568*** (0.3612)
Hausman test *p* value	0.1177		
R^2	0.8753	0.8974	
OBS	390	390	390

Note Values within parentheses indicate standard errors; ***, **, and * indicate that the variable coefficient passed a significance test of 1, 5, and 10%, respectively; OBS indicates the number of observations in the sample

By observing the estimated values for explanatory variable coefficients from Eq. 3, we arrive at the following conclusions.

First, after controlling for investment in technical personnel, investment in technology, foreign direct investment, and institutional innovation, industrial clustering has a positive influence on technological innovation. Having passed the 1% significance test, we can accept Hypothesis 1. In addition, where industrial clustering levels increase by 1%, technological innovation capacities rise by 0.1198%. This shows that industrial clustering in the manufacturing industries of China's eastern, central, and western regions can be used to raise innate capacities for technological innovation. Thus, industrial clustering can promote the implementation of an "innovation-driven development strategy," as proposed by China's 18th National Congress, and help to drive China's long-term economic growth.

Second, investment in technical personnel positively impacts technological innovation, but does not pass a significance test, and Hypothesis 3 is confirmed. Furthermore, where investment in technical personnel increases by 1%, capacity for technological innovation increases by 0.0780%. This shows that investment in technical personnel plays an important role in technological innovation throughout China.

Third, investment in technology positively affects technological innovation, and passes the significance test at the 1% level; thus, Hypothesis 4 is confirmed. Furthermore, where investment in technology increases by 1%, capacity for technological innovation increases by 0.6441%; this shows that investment in technology plays an important role in technological innovation throughout China.

Fourth, foreign direct investment has a positive effect on technological innovation, passing the significance test at the 1% level; Hypothesis 5 is confirmed. Furthermore, when foreign direct investment increases by 1%, capacity for technological innovation increases by 0.1136%. This shows that at the present stage, all areas of China can use the vigorous attraction of foreign direct investment as a means to promote technological innovation.

Fifth, institutional innovation positively affects technological innovation, but does not pass a significance test; this confirms the stance of North (1973) and Hypothesis 6. Furthermore, when institutional innovation increases by 1%, capacity for technological innovation increases by 0.0669%. This shows that institutional innovation plays an important role in increasing the capacities for technological innovation of all Chinese regions, and that those areas within China hoping to achieve large strides in technological innovation must further expand economic structural reform and promote institutional innovation.

[2.] Regional regression analysis of the effect of industrial clustering on technological innovation

The above analysis proves that industrial clustering is beneficial for increasing capacities for technological innovation. Furthermore, it can be observed from the development conditions of China's regional manufacturing industries that there are large variations in the levels of industrial clustering in China's eastern regions compared to central and western regions. Thus, does this type of variation lead to regional differences in capacities for technological innovation?

We use China's eastern region as well as its central and western regions as objects of observation to study regional variations in the effect of industrial clustering on technological innovation. Drawing on the studies of Liu and Xu (2010a, b) and Yang et al. (2015), using the central and western regions as a reference group and the regional control variable "East," we again conduct an econometric analysis. The methods of measurement and analysis software are consistent with those used in the previous section. Here, East is a dummy variable for the eastern region; when the research sample comprises the eastern region, East takes a value of 1; when it comprises a different region, East takes a value of 0. Table 5.35 reports the results of the regional regression analysis, where Eqs. 4, 5, and 6 use the FE, RE, and FGLS models for estimation, respectively.

Table 5.35 Analysis of regional regression results

Independent variable	Equation 4 (FE)	Equation 5 (RE)	Equation 6 (FGLS)
ln*Agglo*	0.3017^{***} (0.1103)	0.2050^{***} (0.0658)	0.0697^{***} (0.0261)
ln*Pinput*	0.0347 (0.0696)	0.0184 (0.0648)	0.0895^{*} (0.0502)
ln*Einput*	0.7430^{***} (0.0577)	0.7488^{***} (0.0547)	0.6516^{***} (0.0437)
ln*FDI*	0.0534 (0.0373)	0.0326 (0.0343)	0.1086^{***} (0.0243)
ln*Insti*	0.0566 (0.2225)	0.1209 (0.2086)	0.0769 (0.1503)
East × ln*Agglo*	0.3363^{*} (0.1784)	0.0163 (0.0570)	0.0408^{***} (0.0147)
Constant term	-2.6274^{***} (0.5243)	-2.2839^{***} (0.4918)	-2.4204^{***} (0.3607)
Hausman test p value	0.0210		
R^2	0.8178	0.8979	
OBS	390	390	390

Note Values within parentheses indicate standard errors; ***, **, and * indicate that the variable coefficient passed a significance test of 1, 5, and 10%, respectively; OBS indicates the number of observations in the sample

First, according to the Hausman test, the resulting p value of 0.0210 is smaller than 0.05, and thus, the null hypothesis is rejected. Therefore, between the FE and RE models, we select the FE model. That is, according to the Hausman test, the result of Eq. 4 is superior to that of Eq. 5. Second, by comparing the estimation results of Eqs. 4 and 6, we know that the coefficients of explanatory variables are completely consistent. Because the FGLS model to a certain extent eliminates possible heteroscedasticity and serial correlation, we discuss the findings of the overall panel data regression analysis on the basis of the results of Eq. 6.

By observing the estimated values of variable coefficients in Eq. 6, we observe that there are clear regional differences in the effect of industrial clustering on technological innovation, and thus, Hypothesis 2 is confirmed. With the central and western regions serving as the reference group, the coefficient of the influence of industrial clustering on capacity for technological innovation is 0.0697; this passes the 1% significance test. For the eastern region, the coefficient of the influence of industrial clustering on capacity for technological innovation is 0.1105 (this coefficient is the sum of the central and western regions' reference group coefficient and the eastern region dummy variable coefficient), and passes the significance test at the 1% level. This illustrates that industrial clustering can promote technological innovation in both the eastern region and central and western regions, but there are significant regional variations. Specifically, industrial clustering's effect of promoting technological innovation is greater in the eastern region than in the central and western regions. Because levels of industrial clustering and capacities for technological innovation are clearly higher in east China than in central and western China, disparity in the capacities for technological innovation between regions will continue to increase.

(3) Endogeneity testing

Industrial clustering is helpful in the generation of technological innovation, and creates good conditions for the dispersion of technological innovation. Conversely, technological innovation is helpful in promoting the clustering of industries related to manufacturing. That is, some endogeneity issues exist in the relationship of interactive effects between industrial clustering and technological innovation. In response, the system proposed by Arellano and Bover (1995) and amended by Blundell and Bond (1998), uses the GMM method to overcome the endogeneity issues of the dependent variables in the model. Borrowing from the methods of Yang et al. (2014), we expand the model into a dynamic one by incorporating the one-period lagged value of capacity for technological innovation.[3] For our further research on whether industrial clustering significantly increases technological innovation, all panel data and measurement analysis software are consistent with the preceding text. Equations 13 and 14 in Table 5.36 report the endogeneity test

[3]Here, Eq. 5.1 becomes a dynamic autogregressive model:

LN $\text{Innovation}_{it} = \alpha + \delta\text{n}\ \mathit{Innovation}_{i,t\text{-}1} + \beta_1 \ln \mathit{Agglo}_{it} + \beta_2 \ln \mathit{Pirput}_{it} + \beta_3 \ln \mathit{Eirputit} + \beta_4 \ln \mathit{FDI}_i + \beta_5 \ln \mathit{Insti}_{it} + \varepsilon_{it}$, where In $\mathit{novation}_{i,t\text{-}1}$ is the one-period lagged value of the dependent variable (capacity for technological innovation).

Table 5.36 Regression analysis results for endogeneity testing

Independent variable	Overall endogeneity test: Eq. 13	Regional endogeneity test: Eq. 14
ln*Agglo*	0.1505*** (0.0291)	0.1425** (0.0565)
ln*Pinput*	0.1883*** (0.0370)	0.1957*** (0.0381)
ln*Einput*	0.6876*** (0.0323)	0.6798*** (0.0331)
ln*FDI*	0.0901*** (0.0264)	0.0854*** (0.0292)
ln*Insti*	0.3072 (0.0949)	0.1745 (0.1011)
East × ln*Agglo*		0.0074 (0.0545)
Constant term	-3.9087*** (0.3493)	-3.7456*** (0.3865)
P Value for each test		
AR (1)	0.044	0.047
AR (2)	0.448	0.427
Hansen test	1.000	1.000
OBS	360	360

Note Values within parentheses indicate standard errors; ***, **, and * indicate that the variable coefficient passed a significance test of 1, 5, and 10%, respectively; OBS indicates the number of observations in the sample

results of the estimations of explanatory variable coefficients for overall regression analysis results and regional regression analysis results, respectively. The method used here is the two-step system of GMM estimation.[4]

In Table 5.36, the p value of the AR (1) test[5] for Eq. 13 is 0.044; this is smaller than 0.1, and thus, the null hypothesis is rejected. This shows that the residuals of the estimation equation exhibit first-order serial correlation. The p value of the AR (2) test[6] is 0.448; this is greater than 0.1, and thus, the null hypothesis is accepted. This shows that the residuals of the estimation equation do not exhibit second-order serial correlation. Tests AR (1) and AR (2) illustrate that our model is suitable. The p value of the Hansen test[7] is 1.000; this is greater than 0.1, and thus, the null hypothesis is accepted. This indicates that the instrumental variables used in the regression equation are suitable. The estimation results of Eq. 13 show the

[4]The GMM estimation system consists of one-step system GMM estimation and two-step system GMM estimation. Drawing on the research of Liu and Yin (2008), the two-step system GMM estimation method is chosen to estimate independent variable coefficients.

[5]The null hypothesis for the AR (1) test is that estimation equation residuals do not exhibit first-order serial correlation.

[6]The null hypothesis for the AR (2) test is that estimation equation residuals do not exhibit second-order serial correlation.

[7]Li and Qi (2011) proposed that, compared to the Sargan test, the Hansen Test is better suited for use where heteroscedasticity exists. Thus, in order to eliminate possible heteroscedasticity, the Hansen Test is used in our study. The null hypothesis for the Hansen Test is that all selected instrumental variables are valid.

following. (1) After controlling for investment in technical personnel, investment in technology, foreign direct investment, and institutional innovation, the effect of industrial clustering on technological innovation is positive and passes the significance test at the 1% level, further confirming Hypothesis 1. (2) Investment in technical personnel positively affects technological innovation, and passes the significance test at the 1% level, further confirming Hypothesis 3. (3) Investment in technology positively affects technological innovation, and passes the significance test at the 1% level, thereby further confirming Hypothesis 4. (4) Foreign direct investment positively effects technological innovation, and passes the significance test at the 1% level, thereby further confirming Hypothesis 5. (5) Institutional innovation positively effects technological innovation but fails to pass a significance test; the view of North (1973) and Hypothesis 6 are further confirmed.

In Table 5.36, the p value of the AR (1) test for Eq. 14 is 0.047; this is less than 0.1, and thus, the null hypothesis is rejected. This shows that the residuals of the estimation equation exhibit first-order serial correlation. The p value of the AR (2) test is 0.427; this is greater than 0.1, and thus, the null hypothesis is accepted. This shows that the residuals of the estimation equation do not exhibit second-order serial correlation. Tests AR (1) and AR (2) illustrate that our model is suitable. The p value of the Hansen test is 1.000; this is greater than 0.1, and thus, the null hypothesis is accepted. This indicates that the instrumental variables used in the regression equation are suitable. The estimation's results of Eq. 14 show that the effect of industrial clustering on technological innovation exhibits clear regional variation, thereby further confirming Hypothesis 2. The central and western regions, which serve as the reference group, have a coefficient for the effect of industrial clustering on capacity for technological innovation of 0.1452, and pass the significance test at the 5% level. The coefficient for the effect of industrial clustering on capacity for technological innovation for the eastern region is 0.1499 (this coefficient is the sum of the coefficient for the reference group for the central and western regions and the coefficient for the eastern region dummy variable), but did not pass a significance test. This further illustrates that industrial clustering can promote technological innovation in both the eastern region, and the central and western regions, although there is clear regional variation. Specifically, the effect of industrial clustering in promoting technological innovation is greater in the eastern region than the central and western regions. This shows that the results of regional regression analysis in the previous section are reliable. From the current reality that levels of industrial clustering and capacities for technological innovation are greater in the eastern region than in the central and western regions, we again deduce that disparity between regional capacities for technological innovation will increase further.

5.5.5 Variation in the Effect of Industrial Clustering on Technological Innovation by Industry

In Sect. 5.5.4, we study the regional variation in the effect of industrial clustering on technological innovation, although this is only one facet of the impact of industrial clustering on technological innovation in China's manufacturing sector. Currently, with the continuous development of China's manufacturing sector, there are large variations in capacities for technological innovation between industries. At the same time, there are also relatively large variations in the levels of industrial clustering between different industries. Thus, what role does industrial clustering play in technological innovation for China's manufacturing sector? Are there differences in the impact of industrial clustering on technological innovation between industries? Researching these questions and properly establishing strategies for corresponding industries of the manufacturing sector further promotes technological advances in the manufacturing sector and raises the capacity for technological innovation for the manufacturing sector as a whole, and thus, has a certain level of theoretical value and practical significance for the transformation and upgrading of China's manufacturing sector.

1. Empirical observations

In order to directly reflect the relationship between industrial clustering in various industries of China's manufacturing sector and capacity for technological innovation, we must first measure the capacities for technological innovation of 20 industries of China's manufacturing sector,[8] then measure their levels of industrial clustering, and finally conduct a correlation analysis based on scatterplots of these two measurements.

(1) Industry capacity for technological innovation

The number of valid invention patents held by industries of the manufacturing sector reflects the efficacy of technological innovation activity in these industries. For this reason, this study uses the number of valid invention patents of industries

[8]According to the principles of data availability and consistency, this chapter considers only the following 20 manufacturing sector industries: processing of food from agricultural products; manufacture of foods; manufacture of wine, drinks, and refined tea; manufacture of textiles; manufacture of textile wearing and apparel; manufacture of paper and paper products; processing of petroleum, coking, processing of nuclear fuel; manufacture of raw chemical materials and chemical products; manufacture of medicines; manufacture of chemical fibers; manufacture of non-metallic mineral products; smelting and pressing of ferrous metals; smelting and pressing of non-ferrous metal; manufacture of metal products; manufacture of general purpose machinery; manufacture of special purpose machinery; transportation equipment manufacturing; manufacture of electrical machinery and equipment; computer, communication, and other electronic equipment; and manufacture of measuring instruments. Herein, the Chinese manufacture of automobiles industry and the manufacture of railway, ship, aerospace, and other transportation equipment industry are collectively referred to as transportation equipment manufacturing.

of China's manufacturing sector to measure the capacities for technological innovation of these industries. In order to directly reflect the state of distribution of capacities for technological innovation in industries of China's manufacturing sector, we assemble a distribution diagram of valid invention patents[9] by industries of China's manufacturing sector, as shown in Fig. 5.21 (for the year 2012).

We observe from Fig. 5.21 that there is great variation between capacities for technological innovation between the 20 industries of China's manufacturing sector. The five industries with the greatest capacities are computer, communication, and other electronic equipment manufacturing, manufacture of electrical machinery and equipment, manufacture of general purpose machinery, manufacture of special purpose machinery, and transportation equipment manufacturing. The sum of the valid invention patents of these industries accounts for 69.12% of China's overall manufacturing sector. By comparison, the number of valid patent applications of the manufacture of textile wearing and apparel industry, manufacture of paper and paper products industry, and manufacture of chemical fibers industry each account for less than 0.42% of the number of patents for the manufacturing sector as a whole. Thus, there is evident disparity in capacities for technological innovation between industries in China's manufacturing sector.

(2) Levels of industrial clustering

Compared to the clear variation between industries' capacity for technological innovation, the 20 industries of China's manufacturing sector in our sample here have even greater variation in their levels of industrial clustering. In order to reflect this difference across industries accurately, in this study, we borrow from the research of Wang (2014), utilizing industrial output density[10] (10,000 yuan/km^2) to measure the levels of industrial clustering in industries of China's manufacturing sector. Generally, greater densities of industrial output reflect higher levels of industrial clustering in a given industry. In order to directly reflect the state of distribution of industrial clustering in industries of China's manufacturing sector, we assemble a distribution diagram of industrial output density[11] for industries of China's manufacturing sector, as displayed in Fig. 5.22 (for the year 2012).

We observe from Fig. 5.22 that there are relatively large variations in levels of industrial clustering between the 20 industries of China's manufacturing sector. In this figure, the level of industrial clustering is highest in the manufacture of computer, communication, and other electronic equipment industry, at 724,200 yuan/km^2. The industries with the lowest levels of industrial clustering are the manufacture of measuring instruments, manufacture of chemical fibers, and manufacture of paper and paper products, none of which exceed 132,000 yuan/km^2.

---

[9]The statistical classification is "Industrial Enterprises above a Designated Scale.".

[10]Due to missing annual employment and industrial added-value data for industries of China's manufacturing sector, in this chapter, we use industrial output density to measure the levels of industrial clustering in industries of China's manufacturing sector.

[11]The statistical classification is "Industrial Enterprises above a Designated Scale.".

A comparison of Figs. 5.21 and 5.22 shows that industries with higher levels of industrial clustering tend to have comparatively greater numbers of valid invention patents, meaning that for industries with higher levels of industrial clustering, capacity for technological innovation is relatively strong. In addition, industries with lower levels of industrial clustering tend to have comparatively fewer numbers of valid invention patents, meaning that for industries with comparatively low levels of industrial clustering, capacity for technological innovation is weaker.

(3) Correlation analysis of industrial clustering and capacity for technological innovation

The above analysis shows a definite correlation between industrial clustering and capacity for technological innovation. In order to reflect the correlative relationship between industrial clustering and capacity for technological innovation more directly, we draw a scatterplot diagram of industrial clustering levels of China's manufacturing industries, and their capacities for technological innovation (see Fig. 5.22). In this figure, the logarithm of levels of industrial clustering (lnAgglo) in 20 industries of China's manufacturing sector between 2005 and 2012 are plotted along the horizontal axis, while the logarithm of capacities for technological innovation (lnInnovation) in these 20 industries between 2005 and 2012 are plotted along the vertical axis.

Figure 5.22 shows a high level of correlation between levels of industrial clustering and capacities for technological innovation for industries of China's manufacturing sector between 2005 and 2012, where an industry's capacity for technological innovation increases with its level of industrial clustering.

2. Explanation of econometric model and data

(1) Establishment of econometric model

We observe from the theoretical analysis in Sect. 5.5.3, in addition to affecting the technological innovation of an industry, technological innovation is also affected by investment in technical personnel, investment in technology, foreign direct investment, and institutional innovation. In order to more accurately measure the effect of industrial clustering on technological innovation capacities in industries of China's manufacturing sector, it is necessary to incorporate the abovementioned variables as control variables in our econometric model. Thus, we establish our econometric model in the following form:

$$\begin{aligned} \ln Innovation_{it} = {} & \alpha + \beta_1 \ln Agglo_{it} + \beta_2 \ln Pinput_{it} + \beta_3 \ln Einput_{it} \\ & + \beta_4 \ln FDI_{it} + \beta_5 \ln Insti_{it} + \varepsilon_{it} \end{aligned} \tag{5.2}$$

Here, *Innovation* is the dependent variable representing capacity for technological innovation; *Agglo* is an independent variable representing the level of industrial clustering; and *Pinput*, *Einput*, *FDI*, and *Insti* are control variables representing investment in technical personnel, investment in technology, foreign direct investment, and institutional innovation, respectively. Subscripts i and *t* express sectoral industries and year, respectively; α,β,and ε indicate the coefficients and residuals of constant terms and variables. The industries here are those of China's manufacturing sector. Because data prior to 2005 are difficult to acquire for some variables, and because the most recent available data on industries in China's manufacturing sector are for 2012, in this study, we select panel data for industries of China's manufacturing sector between 2005 and 2012.

(2) Explanation of variable selection and indexes

Below, we briefly explain the variables and indexes.

Capacity for technological innovation (*Innovation*): As previously described, we use the number of valid invention patents of industries of China's manufacturing sector to measure industries' capacity for technological innovation.

Industrial clustering level (*Agglo*): As previously described, we use the density of industrial output of industries of China's manufacturing sector to measure the level of industrial clustering in an industry.

Investment in technical personnel (*Pinput*): We use the number of full-time R&D employees to measure the investment in technical personnel of regional manufacturing industries.

Investment in technology (*Einput*): We use spending on R&D to measure the investment in technology of regional manufacturing industries.

Foreign direct investment (*FDI*): Because foreign direct investment data for industries of China's manufacturing sector are difficult to acquire, we use "foreign investment" in China's manufacturing sector industries as a proxy for the variable "*FDI*."

Institutional innovation (Insti): Borrowing from the research of Liu (2010), this chapter uses non-state investment in fixed economic assets as a proportion of an industry's total fixed asset investment to measure the industry's institutional innovation.

Data are from the *China Statistical Yearbook (2006–2013)*, *China Statistical Yearbook on Science and Technology (2006–2013)*, *China Industry Statistical Yearbook 2013*, and DRCnet Industry Statistical Database. Herein, the value of foreign investment in industries of the Chinese manufacturing sector for each year is calculated using the average yuan–US dollar exchange rate for the relevant year. Where data for any year are missing, the averages of the neighboring years' values are used instead.

Below, we briefly list the descriptive statistics for important variables; the statistical results are shown in Table 5.37.

3. Measurement results and analysis

(1) Overall regression analysis for effect of industrial clustering on technological innovation in manufacturing sector industries countrywide

Based on the measurement methods of Sect. 5.5.4, we use panel data for 20 Chinese manufacturing sector industries from 2005 to 2012 to estimate coefficients

Table 5.37 Descriptive statistics of important variable indexes

Variable	Sample size	Average value	Standard deviation	Smallest value	Largest value
ln*Innovation*	160	7.4211	1.4380	4.1589	11.3337
ln*Agglo*	160	2.8198	0.8303	0.9996	4.2825
ln*Pinput*	160	10.2957	1.1230	8.2049	12.8492
ln*Einput*	160	13.5865	1.2123	11.2828	16.1808
ln*FDI*	160	4.9945	0.8662	2.5349	7.1971
ln*Insti*	160	-0.3676	0.3575	-1.7328	-0.0165

Table 5.38 Regression analysis results for manufacturing sector industries countrywide

Independent variable	Equation 1 (FE)	Equation 2 (RE)	Equation 3 (FGLS)
ln*Agglo*	0.1978 (0.2438)	0.2699* (0.1584)	0.1778** (0.0789)
ln*Pinput*	1.0664*** (0.2418)	1.0222*** (0.1918)	1.3350*** (0.1348)
ln*Einput*	0.2575 (0.2543)	0.2425 (0.2041)	0.0044 (0.1380)
ln*FDI*	0.1719 (0.1180)	0.2337 (0.1010)	0.2718*** (0.0619)
ln*Insti*	0.5016 (0.4490)	0.5274** (0.2346)	0.5296*** (0.1112)
Constant term	-6.5715*** (1.5469)	-5.7981*** (1.1097)	-5.3198*** (0.6571)
Hausman test *p* value	0.0353		
R^2	0.8565	0.8573	
OBS	160	160	160

Note Values within parentheses indicate standard errors; ***, **, and * indicate that the variable coefficient passed a significance test of 1, 5, and 10%, respectively; OBS indicates the number of observations in the sample

for independent variables. Equations 1, 2, and 3 in Table 5.38 report the estimation results for the FE, RE, and FLS models, respectively.

First, according to the Hausman test, the *p* value here is 0.0353; this is smaller than 0.05, and thus, the null hypothesis is rejected. Thus, between the FE and RE models, the FE model is selected. That is, according to the Hausman test, the result of Eq. 1 is superior to that of Eq. 2. Next, through a comparison of the estimation results of Eqs. 1, 2, and 3, we know that the signs of all independent variable coefficients are consistent. Because the FGLS model to a certain extent eliminates possible heteroscedasticity and serial correlation, we discuss the findings of the overall panel data regression analysis of the effect of industrial clustering on capacity for technological innovation for manufacturing sector industries countrywide on the basis of the results of Eq. 3.

By observing the estimated values for explanatory variable coefficients fromEquation 3, we arrive at the following conclusions.

First, after controlling for investment in technical personnel, investment in technology, foreign direct investment, and institutional innovation, industrial clustering has a positive influence on technological innovation; having passed the

5% significance test, Hypothesis 1 is confirmed. Furthermore, when industrial clustering levels increase by 1%, technological innovation capacities increase by 0.1778%. This shows that industrial clustering in China's manufacturing sector industries can be used to raise innate capacities for technological innovation, and thereby promote the transformation and upgrading of China's manufacturing sector.

Second, investment in technical personnel positively impacts technological innovation, and passes the significance test at the 1% level, thereby confirming Hypothesis 4. Furthermore, when investment in technical personnel increases by 1%, capacity for technological innovation increases by 1.3350%, which clearly is higher than for other independent variables. This shows that investment in technical personnel plays an important role in technological innovation across China's manufacturing sector industries.

Third, investment in technology positively affects technological innovation, but fails to pass a significance test; Hypothesis 5 is confirmed. Furthermore, when investment in technology increases by 1%, capacity for technological innovation increases by 0.0044%; this shows that investment in technology plays an important role in technological innovation throughout the industries of China's manufacturing sector.

Fourth, foreign direct investment has a positive effect on capacity for technological innovation, passing the significance test at the 1% level, and confirming Hypothesis 6. Furthermore, when foreign direct investment increases by 1%, capacity for technological innovation increases by 0.2718%. This shows that at the present stage, China can use the vigorous attraction of foreign direct investment as a means of promoting technological innovation.

Fifth, institutional innovation positively affects technological innovation, and passes the significance test at the 1% level; this confirms the stance of North (1973), and Hypothesis 7. Furthermore, when institutional innovation increases by 1%, capacity for technological innovation increases by 0.5296%. This shows that institutional innovation plays a similarly important role in increasing the capacities for technological innovation in each industry of China's manufacturing sector.

(2) Regional group regression analysis of the effect of industrial clustering on technological innovation in industries of China's manufacturing sector

The previous analysis in (1) has proven the robustness of the FGLS model, and so in this section, based on panel data for industries of China's manufacturing sector in regional groups between 2005 and 2012, we use only the FGLS model to estimate the coefficients of independent variables; the results of this are shown in Table 5.39.

Table 5.39 shows there is variation in the effect of industrial clustering on technological innovation, thereby confirming Hypothesis 3. Among the 20 industries of China's manufacturing sector, industrial clustering in the following 17 industries has a definite effect of promoting capacity for technological innovation: processing of food from agricultural products, manufacture of foods, manufacture of wine, drinks, and refined tea, manufacture of textiles, manufacture of textile wearing and apparel, manufacture of paper and paper products, manufacture of raw chemical materials and chemical products, manufacture of medicines, manufacture

Table 5.39 Results of regional group regression analysis for industries of China's manufacturing sector

Industry	lnAgglo	lnPinput	lnEinput	lnFDI	lnInsti	Constant term	OBS
Processing of food from agricultural products	0.1284***	0.1610**	0.5924***	0.1814***	0.6553	-3.9591***	240
	(0.0299)	(0.0711)	(0.0700)	(0.0267)	(0.2025)	(0.4115)	
Manufacture of foods	0.0919***	0.1740**	0.5830***	0.1646***	-0.2617	-3.3304***	240
	(0.0228)	(0.0681)	(0.0651)	(0.0281)	(0.1648)	(0.4704)	
Manufacture of wine, drinks, and refined tea	0.2110***	0.3082***	0.3944***	0.1430***	0.2866	-2.1775***	240
	(0.0224)	(0.0680)	(0.0683)	(0.0228)	(0.1673)	(0.4096)	
Manufacture of textiles	0.0408**	0.1206*	0.6466***	0.1916***	0.3114	-3.9129***	240
	(0.0176)	(0.0660)	(0.0586)	(0.0297)	(0.1645)	(0.4216)	
Manufacture of textile wearing and apparel	0.0233*	0.1120	0.6675***	0.1884***	0.2559	-3.9705***	240
	(0.0139)	(0.0698)	(0.0619)	(0.0309)	(0.1688)	(0.4486)	
Manufacture of paper and paper products	0.0982***	0.1009	0.6603***	0.1317***	-0.4888	-3.2157***	240
	(0.0157)	(0.0613)	(0.0566)	(0.0280)	(0.1567)	(0.3868)	
Processing of petroleum, coking, processing of nuclear fuel	-0.0397**	0.1399**	0.6168***	0.2131***	-0.0763	-3.9491***	240
	(0.0175)	(0.0684)	(0.0660)	(0.0265)	(0.1792)	(0.4195)	
Manufacture of raw chemical materials and chemical products	0.1069***	0.2681***	0.4858***	0.1745***	-0.3842	-3.3645***	240
	(0.0196)	(0.0699)	(0.0658)	(0.0267)	(0.1558)	(0.3747)	
Manufacture of medicines	0.0990***	0.1812**	0.5845***	0.1727***	-0.3861	-3.6257***	240
	(0.0204)	(0.0713)	(0.0663)	(0.0271)	(0.1694)	(0.3890)	
Manufacture of chemical fibers	0.0750***	0.0069	0.7828***	0.1063***	-0.2831	-3.1393***	240
	(0.0093)	(0.0650)	(0.0584)	(0.0307)	(0.1855)	(0.3979)	
Manufacture of non-metallic mineral products	0.1420***	0.2466***	0.4987***	0.1536***	-0.5032	-3.1559***	240
	(0.0256)	(0.0704)	(0.0693)	(0.0271)	(0.1638)	(0.4297)	
Smelting and pressing of ferrous metals	-0.0586**	0.0523	0.7628***	0.2284***	-0.1466	-4.8952***	240

(continued)

Table 5.39 (continued)

Industry	lnAgglo	lnPinput	lnEinput	lnFDI	lnInsti	Constant term	OBS
	(0.0234)	(0.0737)	(0.0696)	(0.0265)	(0.1699)	(0.3677)	
Smelting and pressing of non-ferrous metals	-0.0391**	0.0936	0.7344***	0.2104***	-0.2156	-4.8102***	240
	(0.0183)	(0.0713)	(0.0670)	(0.0260)	(0.1747)	(0.3743)	
Manufacture of metal products	0.1054***	0.2705***	0.4803***	0.1429***	-0.2675	-2.6602***	240
	(0.0193)	(0.0710)	(0.0697)	(0.0287)	(0.1660)	(0.5047)	
Manufacture of general purpose machinery	0.0245	0.0889	0.6715***	0.2112***	0.2640	-4.1556***	240
	(0.0206)	(0.0701)	(0.0663)	(0.0299)	(0.1682)	(0.4895)	
Manufacture of special purpose machinery	0.0422**	0.0759	0.6842***	0.1996***	0.2910	-4.0887***	240
	(0.0197)	(0.0704)	(0.0668)	(0.0278)	(0.1692)	(0.4244)	
Transportation equipment manufacturing	0.0861***	0.1767**	0.6169***	0.1245***	-0.3571	-3.3635***	240
	(0.0136)	(0.0698)	(0.0640)	(0.0286)	(0.1799)	(0.4126)	
Manufacture of electrical machinery and equipment	0.1525***	0.3210***	0.3918***	0.1213***	0.1205	-1.8633***	240
	(0.0180)	(0.0636)	(0.0622)	(0.0260)	(0.1674)	(0.4455)	
Computer, communication, and other electronic equipment	0.0935***	0.2090***	0.5322***	0.0969***	0.0516	-1.9493***	240
	(0.0124)	(0.0639)	(0.0616)	(0.0291)	(0.1662)	(0.4729)	
Manufacture of measuring instruments	0.0530***	0.0609	0.7072***	0.1533***	0.1880	-3.4707***	240
	(0.0137)	(0.0672)	(0.0603)	(0.0300)	(0.1698)	(0.4488)	

Note Values within parentheses indicate standard errors; ***, **, and * indicate that the variable coefficient passed a significance test of 1, 5, and 10%, respectively; OBS indicates the number of observations in the sample of chemical fibers, manufacture of non-metallic mineral products, manufacture of metal products, manufacture of general purpose machinery, manufacture of special purpose machinery, transportation equipment manufacturing, manufacture of electrical machinery and equipment, computer, communication, and other electronic equipment manufacturing, and manufacture of measuring instruments. This shows that the usefulness of industrial clustering for the capacities for technological innovation already has fully demonstrated itself in most industries of the Chinese manufacturing sector. Such industries not only include manufacture of metal products, manufacture of general purpose machinery, manufacture of special purpose machinery, transportation equipment manufacturing, manufacture of electrical machinery and equipment, computer, communication, and other electronic equipment, and manufacture of measuring instruments within a broader category of the machinery and electronics manufacturing industry,[12] but also processing of food from agricultural products, manufacture of foods, manufacture of wine, drinks, and refined tea, manufacture of textile wearing and apparel, and manufacture of paper and paper products within a broader category of the textiles and other light industrial goods industry. This is an important discovery, as we generally believe that industrial clustering within the broader textiles and other light industrial goods industry inhibits technological innovation; our research, however, reveals that in addition to industrial clustering promoting technological innovation in the machinery and electronics manufacturing industry, industrial clustering can also promote technological innovation in the textiles and other light industrial goods industry. This discovery provides a theoretical basis for manufacturing sector development strategies suited to the industrial bases and comparative advantages of each Chinese region. Thus, in order to advance and improve the overall capacity for technological innovation of China's manufacturing sector, thereby promoting the transformation and upgrading of the sector, China's eastern region could develop its machinery and electronics manufacturing industry on the basis of its industrial base and regional advantages of capital, technology, and human resources, while the central and western regions could choose to focus on the development of its textiles

[12]Borrowing from the research of Li and Du (2004), we divide the abovementioned 20 industries of China's manufacturing sector into the three main industrial categories of textiles and other light industrial goods manufacturing; resource processing; and machinery and electronics manufacturing. Herein, the category of textiles and other light industrial goods manufacturing includes the following six industries: processing of food from agricultural products; manufacture of foods; manufacture of wine, drinks, and refined tea; manufacture of textile wearing and apparel; and manufacture of paper and paper products. The resource processing category includes the following seven manufacturing industries: processing of petroleum, coking, processing of nuclear fuel; manufacture of raw chemical materials and chemical products; manufacture of medicines; manufacture of chemical fibers; manufacture of non-metallic mineral products; smelting and pressing of ferrous metals; and smelting and pressing of non-ferrous metals. The machinery and electronic goods category includes the following seven industries: manufacture of metal products; manufacture of general purpose machinery; manufacture of special purpose machinery; transportation equipment manufacturing; manufacture of electrical machinery and equipment; computer, communication, and other electronic equipment; and manufacture of measuring instruments.

and other light industrial goods industry (Yang 2014). In addition, it is worth noting that industrial clustering has a negative effect on technological innovation in the following industries of the manufacturing sector: processing of petroleum, coking, processing of nuclear fuel; smelting and pressing of ferrous metals; and smelting and pressing of non-ferrous metals. This effect passes the significance test at the 5% level for all three sectors, showing that technological innovation is inhibited by industrial clustering for these three industries within the broader category of resource processing, which are heavily influenced by resource endowment. This effect of industrial clustering inhibiting technological innovation in the resource-processing category is primarily caused by the fact that most resource processing industries are resource-dependent industries that rely excessively on natural resources, causing diminished enthusiasm for technological innovation, and thereby causing a certain degree of capacity restriction for raising technological innovation.

In summary, there are significant variations across industries in the effects of industrial clustering on technological innovation. For most industries in the machinery and electronics category and the textiles and other light industrial goods category, clustering is helpful in promoting technological innovation, but for some industries in the resource-processing category, clustering can inhibit technological innovation to a certain extent.

(3) Endogeneity testing

In Sect. 5.5.4, we discuss the mutual influence between regional industrial clustering and technological innovation, or the issue of endogeneity. Similarly, industrial clustering within industries and technological innovation exert mutual influences on each other, and therefore, they also exhibit endogeneity. For this reason, this chapter continues to utilize the two-step system GMM estimation method to handle endogeneity for the independent variables in the model. All the panel data and measurement analysis software used for the analysis are consistent with the contents of Sect. 5.5.4. Table 5.40 reports the estimation results of independent variable coefficients from endogeneity testing of the overall regression analysis of the countrywide effect of industrial clustering on technological innovation in manufacturing sector industries, as well as from endogeneity testing of regional regression analysis for this effect.

In Table 5.40, the p values for the AR (1) test are larger than 0.1, and thus, the null hypothesis is accepted. This shows that the residuals of the estimation equation do not exhibit second-order serial correlation. The p values for the AR (2) test are larger than 0.1, and thus, the null hypothesis is again accepted; this shows that the residuals of the estimation equation do not exhibit second-order serial correlation. The AR (1) and AR (2) tests show that our model is sound, and because the p values of the Hansen test are greater than 0.1, we accept the null hypothesis. This shows that the instrumental variables used in the regression equations are appropriate, meaning there is indeed a relationship of mutual influence between industrial clustering and technological innovation, that is, endogeneity.

Table 5.40 National and regional group manufacturing industry sector endogeneity testing regression analysis results

Sector	lnAgglo	lnInput	lnEinput	lnFDI	lnInsti	Constant term	AR (1)	AR (2)	Hansen test	OBS
Countrywide endogeneity test regression analysis results for industries of the manufacturing sector										
National manufacturing industry	0.0421	0.9509^{***}	0.5960	0.4386	0.2380	-8.2662^{***}	0.186	0.143	0.995	140
	(0.2208)	(0.2999)	(0.3795)	(0.0961)	(0.2084)	(1.8080)				
Regional group endogeneity test regression analysis results for industries of the manufacturing sector										
Processing of food from agricultural products	0.0929^{*}	0.3083^{***}	0.6224^{***}	0.1792^{***}	1.1920	-5.8354^{***}	0.343	0.731	0.584	210
	(0.0541)	(0.0836)	(0.1101)	(0.0289)	(0.1510)	(0.7054)				
Manufacture of foods	0.0720^{***}	0.1700^{*}	0.5815^{***}	0.3941^{***}	-1.0543	-6.8216^{***}	0.644	0.922	0.504	210
	(0.0246)	(0.0907)	(0.0924)	(0.0559)	(0.1066)	(1.0195)				
Manufacture of wine, drinks, and refined tea	0.0770	-0.3406	0.7760^{***}	0.5672^{***}	0.8015^{***}	-6.0327^{***}	0.860	0.554	0.577	210
	(0.0714)	(0.0919)	(0.0820)	(0.0892)	(0.1626)	(1.2547)				
Manufacture of textiles	0.1182^{***}	0.3883^{***}	0.4265^{***}	0.4982^{***}	-0.0041	-7.5440^{***}	0.880	0.744	0.496	210
	(0.0182)	(0.0596)	(0.0242)	(0.0448)	(0.1257)	(0.4462)				
Manufacture of textile wearing and apparel	0.1613^{***}	0.1106^{***}	0.5635^{***}	0.1194^{***}	0.7231^{***}	-1.3459^{***}	0.485	0.628	0.514	210
	(0.0113)	(0.0255)	(0.0141)	(0.0286)	(0.1089)	(0.4288)				
Manufacture of paper and paper products	0.0415^{***}	0.3498^{***}	0.5874^{***}	0.1463^{***}	-0.0351	-4.6314^{***}	0.386	0.569	0.501	210
	(0.0075)	(0.0497)	(0.0391)	(0.0257)	(0.1389)	(0.4256)				
Processing of petroleum, coking, processing of nuclear fuel	- 0.0697^{***}	-0.0624	0.6308^{***}	0.6333^{***}	0.0078	-7.7402^{***}	0.797	0.709	0.521	210
	(0.0131)	(0.0291)	(0.0202)	(0.0263)	(0.0575)	(0.3272)				
Manufacture of raw chemical materials and chemical products	0.3683^{***}	0.2682^{***}	0.4796^{***}	0.0915^{***}	-1.1261	-3.3782^{***}	0.244	0.808	0.596	210
	(0.0412)	(0.0465)	(0.0421)	(0.0336)	(0.1427)	(0.3744)				
Manufacture of medicines	0.1465^{***}	-0.0413	0.8027^{***}	0.3411^{***}	-2.0834	-7.4982^{***}	0.542	0.955	0.735	210
	(0.0556)	(0.0754)	(0.0489)	(0.0541)	(0.1629)	(0.5578)				
Manufacture of chemical fibers	0.0761^{***}	0.9213^{***}	0.1167	-0.1723	2.2950^{***}	1.4317^{***}	0.759	0.989	0.618	210
	(0.0151)	(0.1312)	(0.1020)	(0.0598)	(0.2003)	(0.5376)				

(continued)

Table 5.40 (continued)

Sector	lnAgglo	lnInput	lnEinput	lnFDI	lnInsti	Constant term	AR (1)	AR (2)	Hansen test	OBS
Manufacture of non-metallic mineral products	0.1820***	0.6533***	0.4060***	0.0167	-2.2497	-4.7963***	0.298	0.908	0.586	210
	(0.0316)	(0.0840)	(0.0600)	(0.0314)	(0.3476)	(0.6697)				
Smelting and pressing of ferrous metals	-0.3135***	0.1403***	0.4581***	0.1766***	0.5551***	-2.1923***	0.291	0.742	0.504	210
	(0.0170)	(0.0407)	(0.0518)	(0.0318)	(0.0890)	(0.1922)				
Smelting and pressing of non-ferrous metals	-0.0754***	-0.0096	0.9035***	0.4579***	-3.1537	-10.6050***	0.688	0.606	0.520	210
	(0.0158)	(0.0413)	(0.0455)	(0.0420)	(0.2972)	(0.5883)				
Manufacture of metal products	0.1395***	0.2256***	0.6197***	0.1598***	-1.0918	-4.6973***	0.388	0.647	0.520	210
	(0.0232)	(0.0722)	(0.0887)	(0.0389)	(0.1714)	(0.3641)				
Manufacture of general purpose machinery	0.3507***	-0.3872	0.9068***	0.0022	0.4936***	-0.1670	0.247	0.607	0.524	210
	(0.0194)	(0.0717)	(0.0710)	(0.0561)	(0.1838)	(0.5591)				
Manufacture of special purpose machinery	0.0675***	0.4963***	0.4262***	0.0652***	0.7221***	-2.6061***	0.337	0.981	0.505	210
	(0.0177)	(0.0170)	(0.0314)	(0.0151)	(0.1184)	(0.4074)				
Transportation equipment manufacturing	0.3325***	0.5314***	0.4571***	-0.3225	-1.4853	0.3625	0.599	0.928	0.540	210
	(0.0178)	(0.0410)	(0.0350)	(0.0299)	(0.1647)	(0.4389)				
Manufacture of electrical machinery and equipment	0.1564***	0.3199***	0.6304***	0.1180***	-2.6590	-5.9406***	0.443	0.962	0.549	210
	(0.0184)	(0.0560)	(0.0571)	(0.0187)	(0.1143)	(0.4927)				
Computer, communication, and other electronic equipment	0.1797***	0.1721**	0.7331***	-0.0790	-0.8323	-2.2638***	0.535	0.557	0.505	210
	(0.0095)	(0.0682)	(0.0506)	(0.0290)	(0.1708)	(0.5014)				
Manufacture of measuring instruments	0.0872***	0.6198***	0.0482**	0.3041***	0.9693***	-2.0582***	0.830	0.550	0.606	210
	(0.0102)	(0.0302)	(0.0221)	(0.0333)	(0.1693)	(0.5006)				

Note Values within parentheses indicate standard errors; ***, **, and * indicate that the variable coefficient passed a significance test of 1, 5, and 10%, respectively; values under AR (1), AR (2), and the Hansen test are *p* values for the corresponding test; OBS indicates the number of observations in the sample

By observing the results of the endogeneity testing regression analysis, we conclude as follows.

First, after controlling for investment in technical personnel, investment in technology, foreign direct investment, and institutional innovation, industrial clustering has a positive influence on technological innovation, although fails to pass a significance test; Hypothesis 1 is further confirmed.

Second, investment in technical personnel positively affects technological innovation, passing a significance test at the 1% level. This further confirms Hypothesis 4.

Third, investment in technology positively affects technological innovation, but this effect does not pass a significance test, and Hypothesis 5 is further verified.

Fourth, foreign direct investment has a positive effect on capacity for technological innovation, but does not pass a significance test, and verifies Hypothesis 6.

Fifth, institutional innovation positively affects technological innovation, but does not pass a significance test, and it confirms the view of North (1973) and Hypothesis 7. This shows that the overall regression analysis results for the countrywide effect of industrial clustering in manufacturing sector industries on technological innovation are reliable.

By observing the endogeneity test regression analysis results for regional groups of manufacturing sector industries, we know there are variations in the effects of industrial clustering on technological innovation, thereby further verifying Hypothesis 3. Among the 20 industries of China's manufacturing sector, industrial clustering in the following 17 industries has a definite effect of promoting capacity for technological innovation: processing of food from agricultural products; manufacture of foods; manufacture of wine, drinks, and refined tea; manufacture of textiles; manufacture of textile wearing and apparel; manufacture of paper and paper products; manufacture of raw chemical materials and chemical products; manufacture of medicines; manufacture of chemical fibers; manufacture of non-metallic mineral products; manufacture of metal products; manufacture of general purpose machinery; manufacture of special purpose machinery; transportation equipment manufacturing; manufacture of electrical machinery and equipment; computer, communication, and other electronic equipment; and manufacture of measuring instruments. In addition, industrial clustering has a definite negative impact on capacities for technological innovation for the following three manufacturing industries: processing of petroleum, coking, processing of nuclear fuel; smelting and pressing of ferrous metals; and smelting and pressing of non-ferrous metals. This goes a step further in showing the reliability of the regional group regression analysis results for the effect of industrial clustering on technological innovation for industries of the Chinese manufacturing sector.

In summary, we further conclude that there are variations across industries in the effects of industrial clustering on technological innovation, with the promotion of technological innovation in most industries within the machinery and electronics manufacturing category and the textiles and other light industrial goods manufacturing category benefiting from industrial clustering. Nonetheless, for some industries in the resource-processing category, technological innovation is inhibited to a certain extent by industrial clustering.

5.5.6 Conclusions and Policy Recommendations

1. Research conclusions

The research in this Sect. 5.5 leads us to the following primary conclusions.

Theoretical analysis shows that industrial clustering promotes technological innovation primarily through the following two mechanisms. First, the advantages of specialization and diversification in industrial cluster areas benefit knowledge and technology spillovers, thereby promoting the production of technological innovation. Second, the grouping and flow of large volumes of knowledge, human resources, and information within industrial cluster areas are beneficial to the formation of knowledge, technology, and information communication networks, and thus, promote the dispersion of technological innovation. In addition, the effects of industrial clustering on technological innovation exhibit clear variations across industries and across regions. Furthermore, investment in technical personnel, investment in technology, foreign direct investment, and institutional innovation exert positive influences on technological innovation. Our study of regional variation in the effects of industrial clustering on technological innovation show that, after controlling for regional investment in technical personnel, regional investment in technology, regional foreign direct investment, and regional institutional innovation, regional industrial clustering significantly promotes technological innovation. The regional comparative analysis showed clear regional differences in the effects of industrial clustering on technological innovation, whereby industrial clustering has a larger impact on promoting technological innovation in the east than central and western regions. The combination of higher levels of industrial clustering and higher capacities for technological innovation that currently exist in the east compared to central and western regions further increases the disparity of technological innovation capacity across regions.

Our study of variations in the impact of industrial clustering on technological innovation across industries revealed that after controlling for an industry's investment in technical personnel, investment in technology, foreign direct investment, and institutional innovation, industrial clustering significantly promotes technological innovation. Our comparative analysis by industry revealed clear differences in the impact of industrial clustering on technological innovation across industries, with technological innovation in most industries of the machinery and electronics manufacturing category and the textiles and light industrial goods manufacturing category benefiting from industrial clustering; meanwhile, technological innovation in some sectors of the resource processing category is inhibited to a certain extent by industrial clustering.

2. Policy recommendations

Here, we summarize the research in this section, and propose policy recommendations for achieving increased capacities for technological innovation within China's manufacturing sector.

(1) Industrial cluster development

We observe that industrial clustering is helpful in promoting technological innovation. Therefore, according to the industry base, and resource endowment advantages of each locality in China's eastern, central, and western regions, industries in which an area holds comparative advantages should be cultivated, and clustered industrial development should be encouraged, thereby raising levels of industrial clustering. This is especially the case in the central and western regions, where levels of industrial clustering are lower; here, it is particularly advisable to cultivate industries in which regional comparative advantages are held, and to encourage clustered industrial development, thereby promoting increased levels of industrial clustering. Clustered industrial development can be encouraged in these regions in the following two ways. First, by combining a region's local industrial base and labor force conditions to cultivate clustered industrial development, levels of industrial clustering are raised. Second, in relatively developed coastal regions, such as the Yangtze River Delta, the Pearl River Delta, Beijing, Tianjin, and Hebei, industrial transformation and upgrading provide an opportunity to actively guide the marginal industries of these areas in westward migration, and thereby gradually promote the clustered industrial development of the central and western regions. Through these two paths, levels of industrial clustering in China's central and western regions can be raised gradually, and local capacities for technological innovation can be increased.

(2) Sensible dispersion of industrial clusters across regions

At present, the Yangtze River Delta, Pearl River Delta, Beijing, Tianjin, and Hebei gradually have formed the core of China's manufacturing sector, with central and western regions forming the periphery of the industrial clustering model; these regions have clearly lower capacities for technological innovation than the eastern region. We observe from the analysis that the effect of industrial clustering in promoting technological innovation is greater in the eastern region than in the central and western regions. Thus, China's industrial planning departments and government departments should develop a variety of preferential policies to actively guide a portion of the manufacturing sector in the eastern region to migrate toward central and western regions, thereby promoting regional distribution of industrial clusters between regions, and advancing the rapid increase of levels of industrial clustering in China's central and western regions. In this way, disparity in capacities for technological innovation between China's eastern region and its central and western regions can be diminished.

(3) Role of talent and capital in promoting technological innovation

We observe that investment in technical personnel and investment in technology are both helpful in promoting technological innovation. At present, levels of investment in technical personnel and technology in China are clearly insufficient compared to developed countries, leaving significant room for increases. Thus, all Chinese regions should increase investment intensity in technical personnel and technology significantly. Through organic linkages between industry associations, firms, and universities, and by various methods, such as running universities like associations and firms, the integration of production, teaching, and research, human resources, and developmental research should be cultivated. This will raise levels of technical human capital and increase investment in technology, drawing on the usefulness of human resources and capital in promoting technological innovation. In addition, to address the issues of fewer technical personnel and insufficient investment in technology, it is necessary to increase intensity of investment in technical personnel, and conscientiously implement a strategy to invigorate China through science and education. On the one hand, by adhering to an education-based orientation and vigorously developing schools, the scientific and cultural value of labor will be increased; on the other hand, by introducing talent and increasing remuneration for technical personnel, incentive policies can be improved and developed, thereby attracting more talent.

(4) Technology spillover effect of foreign investment

We observe from the analysis in Sect. 5.5 that foreign direct investment is capable of significantly promoting technological innovation. Since the period of reform and opening, development practices in China's eastern coastal regions of Yangtze River Delta, Pearl River Delta, Beijing, Tianjin, and Hebei have also proven that foreign direct investment is beneficial to promoting technological progress. Thus, localities in China's eastern, central, and western regions should vigorously attract foreign investment, further expand the scale of utilization of foreign capital, and focus on the attraction of advanced technology and management experience. Through the demonstration-imitation effect, a greater number of technology spillovers can be attained, thus promoting technological progress, and increasing the rate at which China's eastern, central, and western regions raise their capacities for technological innovation.

(5) Accelerate the adjustment and optimization of the structure of the manufacturing sector

We observe that industrial clustering in industries within the machinery and electronics category and textiles and other light industrial goods category is helpful in promoting technological innovation, and that for a portion of industries within the resource-processing category, clustering inhibits technological innovation to a certain extent. Thus, China's industrial planning departments and governmental departments should develop appropriate strategies for increasing the pace of adjustment and optimization of the structure of the manufacturing sector, by appropriately increasing the concentration of firms from the machinery and electronics category, and textiles and other light industrial goods category, and by reducing the concentration of firms in the resource-processing category. Advances in technology in China's manufacturing sector should be actively promoted, along with overall capacity for technological innovation; in this way, the transformation and upgrading of China's manufacturing sector can be promoted.

(6) Implementing a complementary industrial development strategy

We observe that in addition to promoting technological innovation in the machinery and electronics-manufacturing category, industrial clustering can also promote technological innovation in the textiles and other light industrial goods manufacturing category. Thus, based on the manufacturing sector industrial base, resource endowment, and local characteristics of each area of China's eastern, central, and western regions, development strategies that utilize regional comparative advantages suited to a region's own manufacturing sector should be implemented. Specifically, in China's eastern region, advantages in labor, technology, capital, and superior industrial bases can be used to vigorously develop industries within the machinery and electronics-manufacturing category. In the central and western regions, advantages, including abundant natural resources and labor, can be used to develop industries within the textiles and other light industrial goods category. Of course, central and western localities with better conditions can develop industries within the machinery and electronics-manufacturing category according to their own advantages.

5.6 Summary

This chapter first analyzed China's manufacturing sector based on the following four dimensions: capacity for economic innovation, capacity for technological innovation, resource consumption, and capacity for environmental protection. Next, using the deviation maximization method, a new industry model was assessed, after which the effect of technological innovation on the economic growth of China's manufacturing sector was studied, along with energy efficiency in the manufacturing sector and its influencing factors, analysis of the effect of informatization on green growth in the manufacturing sector, and the effect of industrial clustering on technological innovation in China's manufacturing sector. Finally, a summary of the entire chapter was conducted.

In assessing the new industry model, levels of transition to the new industry model were believed to be higher in such industries as the manufacture of computers, communication, and other electronic equipment; manufacture of electrical machinery and equipment; manufacturing of automobiles; manufacture of general purpose machinery; and manufacture of special purpose machinery. Thus, governments can vigorously promote their development. Furthermore, levels of transition to the new industry model were lower in such industries as smelting and pressing of ferrous metals; manufacture of paper and paper products; and manufacture of non-metal mineral products. The development of this type of industry must be controlled by governments.

The results of our research on the effect of technological innovation on economic growth in China's manufacturing sector showed that independent R&D and technological advances have a positive influence on the economic growth of China's manufacturing sector. This means that independent R&D and technological advances both can promote growth in China's manufacturing sector. In addition, investment in physical capital and labor as well as degree of openness have positive effects on China's manufacturing sector. Furthermore, foreign direct investment has a negative effect on growth of the manufacturing sector. Thus, we propose the following policy recommendations: increase independent R&D and investment in technological advances; increase the technological content of investment in physical capital; raise the quality of laborer technical ability; and balance openness levels. Through these actions, economic stability and sustainable development of the manufacturing sector can be achieved in a meaningful way.

In our study of energy efficiency in China's manufacturing sector and its influencing factors, our results showed the following. 1. Between 2003 and 2012, energy efficiency in China's manufacturing sector increased steadily, with an energy-saving potential of 30–35%. Because of considered undesired outputs, measurements for industrial energy efficiency are underestimated, which is in line with expectations, and realistic. 2. Corporate ownership structure, technological progress, energy consumption structure, openness, industry scale, and firm scale all have positive effects on energy efficiency in the manufacturing sector. 3. Four types of industry were derived from cluster analysis, and the influence of five independent variables had different sizes and directions for different types of industry. Thus, we propose the following policy recommendations: First, high-efficiency and high-polluting industries, on the one hand, should have their trade structures optimized, lowering the proportion of exports of high resource-consumption products, and raising the proportion of low resource-consumption export products. On the other hand, by developing energy-saving technologies, increasing technological research intensity, introducing advanced technology production equipment, and lowering the energy consumption of high resource-consumption product processes, energy efficiency can be improved from the root causes of inefficiency. Second, China should focus on the development of high-energy efficiency and low-polluting industries, reducing the proportion of state-owned enterprises, and optimizing the structures of this type of firm. China should increase the intensity of technological R&D investment, while expanding the scales of industries and firms. Third, the development of low-efficiency and high-polluting industries should be restricted. Restricting internal support for R&D expenditure, reducing investment in technology, and optimizing resource structures for this class of industry, while also restricting exports and reducing the proportion of high resource-consuming products for export, can improve energy efficiency. Fourth, for low-efficiency and low-polluting industries, the proportion of state-owned enterprises should be increased, and national policies should be implemented actively. Expenditure on technological R&D should be increased, and advanced energy equipment should be incorporated into production processes, thereby promoting energy savings and improving energy efficiency. Energy expenditure structures should be reasonably planned, reducing the proportion of spending on coal, and actively developing clean energies. Export trade should be encouraged, along with the scalar expansion of this class of industry.

In our study of the effect of informatization on green growth in the manufacturing sector, the results show the following. 1. Informatization is significantly helpful for green growth in the manufacturing sector, and has become the new primary driver of green growth in the sector. Intensity of R&D investment plays an important role in the process of informatization driving green growth in the manufacturing sector. 2. There is significant variation between industries in the effect of informatization on promoting green growth in the manufacturing sector; this effect is greater where technology is more advanced, average firm scale is smaller, energy consumption is lower, levels of pollution are lower, and profitability is stronger. Thus, we propose the following policy recommendations. First, China should increase the pace of establishing informatization, promote the deep integration of informatization and industrialization, and bring the role of informatization in promoting green growth in the manufacturing sector into play fully. Second, for industries with different modes of integrating informatization and green growth of the manufacturing sector, we recommend the targeted establishment and implementation of a new model of industrial development strategies. Third, China should pay greater attention to the variation that specific types of industries bring to the effect of informatization promoting green growth in the manufacturing sector.

In our study of the effect of industrial clustering on technological innovation in China's manufacturing sector, the results showed that industrial clustering promotes technological innovation primarily through the following two mechanisms. 1. The advantages of specialization and diversification in industrial clustering areas are helpful for knowledge and technology spillovers, and thereby promote the generation of technological innovation. 2. The agglomeration and flow of large volumes of knowledge, talent, and information in industrial clustering areas is beneficial for the flow of knowledge and technology, and for the formation of information networks, thereby promoting the dispersion of the results of technological innovation. In addition, the effect of industrial clustering on technological innovation exhibits clear regional and industry variations. Furthermore, investment in technical personnel, investment in technology, foreign direct investment, and institutional innovation all have positive effects on technological innovation. Thus, we propose the following policy recommendations. First, China should encourage clustered industrial development. Second, China should advance reasonable distribution of industrial clusters between regions. Third, China should bring into play the role of talent and funding in promoting technological innovation. Fourth, it should bring into play the technological spillover effect of foreign investment. Fifth, China should accelerate the adjustment and optimization of the manufacturing sector's structure. Sixth, China should implement a complementary development strategy for its manufacturing sector.

References

Acemoglu, D., D. Autor, and D. Dorn. 2014. Return to the Solow Paradox? IT, Productivity, and Employment in U.S. Manufacturing. Prepared for the AEA Papers and Proceedings 104 (5):394–399.

Ahmed, E.M. 2010. Information and communications technology effects on East Asian productivity. *Journal of the Knowledge Economy* 1 (3): 191–201.

Alma, M.S., and G.S. Elina. 2014. Impact of information and telecommunication technologies development on labour productivity. *Procedia Social and Behavioral Sciences* 110 (1): 1271–1282.

Arundel, Anthony. 2001. The relative effectiveness of patents and secrecy for appropriation. *Research Policy* 30 (4): 611–624.

Arellano, M., and O. Bover. 1995. Another look at the instrumental variable estimation of error-components models. *Journal of Econometrics* 68 (1): 29–51.

Arrow, K.J. 1962. The economic implications of learning by doing. *The Review of Economic Studies* 29 (3): 155–173.

Blundell, R., and S. Bond. 1998. Initial conditions and moment restrictions in dynamic panel data models. *Journal of Econometrics* 87 (1): 115–143.

Badescu, M., and C. Garces-Ayerbe. 2009. The impact of information technologies on firm productivity: Empirical evidence from Spain. *Technovation* 29 (2): 122–129.

Baptista, R., and P. Swann. 1998. Do firms in clusters innovate more? *Research Policy* 27 (5): 525–540.

Bagella, M., and L. Becchetti. 2002. The, "geographical agglomeration-private R&D expenditure" effect: Empirical evidence on Italian data? *Economics of Innovation and New Technology* 11 (3): 233–247.

Basu, S., and J. Fernald. 2007. Information and communications technology as a general purpose technology: Evidence from US industry data. *German Economic Review* 8 (2): 146–173.

Birol, B., and J.H. Keppler. 2000. Prices, technology development and the rebound effect. *Energy Policy* 28: 457–469.

Brülhart, M., and N.A. Mathys. 2008. Sectoral agglomeration economies in a panel of European regions. *Regional Science and Urban Economics* 38 (4): 348–362.

Carlino, G.A., S. Chatterjee, and R.M. Hunt. 2007. Urban density and the rate of invention. *Journal of Urban Economics* 61 (3): 389–419.

Chien, T., and J.L. Hu. 2007. Renewable energy and macroeconomic efficiency of OECD and non-OECD economies. *Energy Policy* 35 (7): 3606–3615.

Chou, Y.C., H. Chuang, and B. Shao. 2014. The impacts of information technology on total factor productivity: A look at externalities and innovations. *International Journal of Production Economics* 158 (2): 290–299.

Ciccone, A., and R.E. Hall. 1996. *Productivity and the Density of Economic Activity*. National Bureau of Economic Research.

Dahl, C.M., H.C. Kongsted, and A. Sorensen. 2011. ICT and productivity growth in the 1990s: Panel data evidence on Europe. *Empirical Economics* 40 (1): 141–164.

Dekle, R., and J. Eaton. 1999. Agglomeration and land rents: Evidence from the prefectures. *Journal of Urban Economics* 46 (2): 200–214.

Farrell, M.J. 1957. The measurement of productive efficiency. *Journal of the Royal Statistical Society Series A* 120 (3): 253–290.

Fare, R., S. Grosskopf, and C.A. Pasurka. 2007. Environmental production functions and environmental directional distance functions. *Energy* 32 (7): 1055–1066.

Feldman, M.P., and D.B. Audretsch. 1999. Innovation in cities: Science-based diversity, specialization and localized competition. *European Economic Review* 43 (2): 409–429.

Fisher-Vanden, K., G.H. Jefferson, and Q. Liu Tao. 2004. What is driving China's decline in energy intensity? *Resource and Energy Economics* 26 (1): 77–97.

Glaeser, E.L., H.D. Kallal, J.A. Scheinkman, and A. Shleifer. 1991. *Growth in Cities*. National Bureau of Economic Research.

Gilbert, B.A., P.P. McDougall, and D.B. Audretsch. 2008. Clusters, knowledge spillovers and new venture performance: An empirical examination. *Journal of Business Venturing* 23 (4): 405–422.

Greunz, L. 2004. Industrial structure and innovation—Evidence from European regions. *Journal of Evolutionary Economics* 14 (5): 563–592.

Huang, J.P. 1993. Industry energy use and structure change: A case study of the People's Republic of China. *Energy Economics* 15(2):131–13.

Inklaar, R., O. Mahony, and M.P. Timmer. 2005. ICT and Europe's productivity performance: Industry level growth account comparisons with the United States. *Review of Income and Wealth* 51 (4): 505–536.

Jacobs, J. 1969. *The economy of cities*. New York: Random House.

Hu, Jin-Li, and Shih-Chuan Wang. 2006. Total factor energy efficiency of regions in China. *Energy Policy* 34: 3206–3217.

Jorgenson, D.W., M.S. Ho, J.D. Samuels, and Stiroh, K.J. 2008. Industry origins of the American productivity resurgence. *Interdisciplinary Information Sciences* 14(1):43–59.

Kelly, M., and A. Hageman. 1999. Marshallian externalities in innovation. *Journal of Economic Growth* 4 (1): 39–54.

Kennedy, C, and A.P. Thirlwall. 1972. Surveys in applied economics: technical progress. *The Economic Journal* 82(325):11–72.

Khazzoom, J.D. 1980. Economic implications of mandated efficiency in standards for household appliances. *The Energy Journal* 1:4(4):21–40.

Krugman, P.R. 1991. *Geography and trade*. MIT Press, 55–76.

Kokko, A. 1992. Foreign direct investment, host country characteristics and spillovers.

Lan Sue Wing, E., Blind Knut, and Jungmittag Andre. 2008. The impact of patents and standards on macroeconomic growth: A panel approach covering four countries and 12 sectors. *Journal of Productivity Analysis* 1:51–60.

Lawrence, J Lau. 1999. The source of East Asian economic growth. In *The Political Economy of Contemporary Development into the 21st Century*, eds. Gustav Rains, Sheng-Cheng Hu, and Yun-Peng Chu, 45–75. Edward Elgar Press.

Lovric, Ljiljana. 2012. Information-communication technology impact on labor productivity growth of EU developing countries. *Social Science Electronic Publishing* 30 (2): 223–245.

Lovell, C.A.K. 1993. Production frontier and productive effieiency? In *The Measurement of Productive Efficiency*, ed. H.O. Fried, C.A.K. Lovell, and S.S. Scbmidi, 3–67. NewYork: Oxford University Press.

Meyer-Stamer, J. 2002. Clustering and the creation of an innovation-oriented environment for industrial competitiveness: Beware of overly optimistic expectations. *International Small Business Journal* 20(3).

North, D.C. 1973. *The rise of the western world: A new economic history*. Cambridge: Cambridge University Press.

Cai, Chunquan, Shi Ouyan, and Shen Changhong. 2010. Ouyan surgical treatment of a patient with human tail and multiple abnormalities of the spinal cord and column.

Patterson, M.G. 1996. What is energy efficiency? Concepts, indicators and methodological issues. *Energy Policy* 24 (5): 377–390.

Porter, M.E. 1998. Clusters and the new economics of competition. *Watertown. Harvard Business Review* 76 (6): 75–90.

Paci, R., and S. Usai. 1999. The role of specialisation and diversity externalities in the agglomeration of innovative activities 2(2):235–268.
Roach, S. 1998. America's technology dilemma: A profile of the information economy. *Zeitschrift Angewandte Mathematik Und Mechanik* 78(Supplement S3):973–974.
Paul, M. Romer. 1986. Increasing returns and long-run growth. *Journal of Political Economy* 1002–1037.
Schreyer, P. 2001. The OECD productivity manual: A guide to the measurement of industry-level and aggregate productivity. *International Productivity Monitor* 2 (2): 37–51.
Shadbegian, R.J., and W.B. Gray. 2005. Pollution abatement expenditures and plant-level productivity: A production function approach. *Ecological Economics* 54 (2): 196–208.
Shahiduzzaman, M., and K. Alam. 2014. Information technology and its changing roles to economic growth and productivity in Australia. *Telecommunications Policy* 38 (3): 125–135.
Silvestre, B.S., and P.R.T. Dalcol. 2009. Geographical proximity and innovation: Evidences from the Campos Basin oil & gas industrial agglomeration—Brazil. *Technovation* 29 (8): 546–561.
Sinton, J.E., and D.G. Fridley. 2000. What goes up: Recent trends in China's energy consumption. *Energy Policy* 28:671–687.
Solow, R. 1987. *We'd better watch out*. New York: Times Book Review.
Stephen, D.O., and E.S. Daniel. 2000. The resurgence of growth in the late 1990s: Is information technology the story? *Journal of Economic Perspectives* 14(4):3–22.
Stiroh, K., and M. Botsch. 2007. Information technology and productivity growth in the 2000s. *German Economic Review* 8(2):255–280.
Storper, M., and A.J. Venables. 2004. Buzz: Face-to-face contact and the urban economy. *Journal of Economic Geography* 4(4):351–370.
Vaelav, S. 1990. *China's Energy*, in Report Prepared for the U. S. Congress, Washington, D. C: Office of Technology Assessment.
Wang, Y.C. 2015. How ICT penetration influences productivity growth: Evidence from 17 OECD countries. *Economic Development Quarterly* 29 (1): 79–92.
Watanabe, M., and K. Tanaka. 2007. Efficiency analysis of Chinese industry: A directional distance function approach. *Energy Policy* 35:6323–6331.
Wooldridge, J.M. 2002. *Econometric analysis of cross section and panel data*. MIT Press.
Yoguel, G., and A. Marin. 2000. Production Networks: Linkages, Innovation Processes and Social Management Technologies: A Methodological Approach Applied to the Volkswagen Case in Argentina. Department of Industrial Economics and Strategy, Copenhagen Business School.
Bai, Xuejie, and Ying Song. 2009. Environment regulation, technology innovation and efficiency improvement of Chinese thermal power industry. *China Industrial Economics* 8: 68–77.
Chen, Demin, and Rui Zhang. 2012. The influence of environment regulation on total factor energy efficiency in China—An empirical experiment based on panel data. *Economic Science* 4: 49–65.
Chen, Guanju. 2014. Research on total factor energy efficiency of Chinese manufacture industry and its influence factors based on SFA. *China Soft Science* 1: 180–192.
Chen, Shiyi. 2011. Green industrial revolution in China: A perspective from the change of environmental total factor productivity. *Economic Research Journal* 11: 21–34.
Chu, Deyi, and Tongbin Zhang. 2013. Independent R&D, technology import and high-tech industry growth. *Science Research Management* 11: 53–60.
Diao, Xinke, and Anbao Tang. 2012. Study on the impact of energy price variation on energy efficiency. *China Mining Magazine* 6: 37–41.
Ding, Rijia. 2008. FDI, human capital, R&D, and Chinese energy efficiency. *Finance & Trade Economics* 9: 95–99.
Dong, Li. 2008. Energy intensity trend and its influential factors in China. *Industrial Economics Research* 1: 8–17.
Dou, Lichen, and Cui Zhao. 2005. The influence of scientific tech investment to economy growth. *Journal of Yunnan Finance and Trade Institute* 3: 9–12.

Fan, Jianyong, Meng Feng, and Fangwen Li. 2014. Industrial clustering and firm total factor production efficiency. *The Journal of World Economy* 5: 51–73.

Fan, Jianyong. 2006. Industrial agglomeration and difference of regional labor productivity: Chinese evidence with international comparison. *Economic Research Journal* 11: 72–81.

Guan, Xin, Xiaoyong Qiao, and Qingguo Meng. 2012. Empirical research on the impact of informatization development to scientific & technological progress: Based on the panel data of 31 provinces in China. *Studies in Science of Science* 30(7):1020–1030.

Guo, Baomei. 2006. Impact of technological progress on China's industrial structure. *Journal of Shanxi Finance and Economics University* 2:37–39.

Guo, Ju E., Jian Chai, and Ximin Xi. 2008. Effect of the primary energy consumption structure change to the energy use per unit of GDP. *China Pollulation. Resources and Environment* 04:38–42.

Informatization Assessment Group of the National Bureau of Statistics Research Institute. 2011. Research report on optimization of informatization development indices. *Management World* 12: 1–11.

Han, Zhiyong, Yiming Wei, and Ying Fan. 2004. Research on change features of Chinese energy intensity and economic structure. *Application of Statistics and Management* 1: 1–6.

Han, Xianfeng, Ning Hui, and Wenfei Song. 2014. Can informatization improve the technology innovation efficiency of Chinese industrial sectors. *China Industrial Economy* 12: 70–82.

Hang, Leiming, and Meiceng Tu. 2006. The impacts of energy prices on energy intensity. *The Journal of Quantitative & Technical Economics* 12: 93–100.

He, Jie. 2000. A more precise quantification of the impact of FDI on spillover effects in China's industrial sector. *The Journal of World Economy* 12: 29–36.

Hu, Qiuyang. 2014. Major industries selection based on the rebound effects and energy efficiency policy. *Economic Research Journal* 2: 128–140.

Hu, Shaolin. 2009. *Influencing factors for Chinese energy efficiency*. Shanghai Jiao Tong University.

Hu, Xiaojuan, and Han Shao. 2010. Environmental effects of foreign investment in Chinese industrial sectors—An empirical study based on industrial panel data. *World Economy Study* 7: 55 61.

Jin, Bei. 2014. The mission and value of industry the theoretical logic of industrial transformation and upgrading in China. *China Industrial Economy* 9:51–64.

Jiang, Jinhe. 2004. Energy efficiency improvement and economic structure adjustment. *The Journal of Quantitative & Technical Economics* 10: 16–23.

Jiang, Lei, and Minhe Ji. 2011. Spatial difference analysis of regional energy pressures in China: Based on STIRPAT model. *Finance and Economics* 4: 64–70.

Jiang, Xiujuan, and Feng Shao. 2010. T-type correlation degree analysis for China's technological investment and economic growth. *Science & Technology Progress and Policy* 11: 4–6.

Huang, Zhilin, and Peiguo Yu. 2007. The effect of recent technological advances on China's economic development—A panel data model analysis. *Science and Technology Management Research* 5:74–77.

Liang, Xi, Hengyuan Zhu, and Guisheng Wu. 2006. Relationship between innovation activity and economic growth in China——A preliminary research based on co-integration theory. *Journal of Tsinghua University (Philosophy and Social Sciences)* 6.

Li, Bin, and Lin Liu. 2009. The contribution of informationzation to economic growth—Based on the statistics of Hunan province. *Economic Geography* 29 (10): 1685–1690.

Li, Bin, and Lejun Huang. 2009. Empirical analysis of the contribution of scientific and technological progress to China's economic growth. *Science & Technology and Economy* 3:65–58.

Li, Kai, and Shaozhou Qi. 2011. Trade openness, economic growth and carbon dioxide emission in China. *Economic Research Journal* 11:60–72.

Li, Juan, Chengliang Zhu, and Liren An. 2013. The fluctuation and regional difference of low carbon transition performance in China. *China Soft Science* 5:167–182.

Li, Lianshui, Zhonghua Cheng, and Jun Liu. 2015. The "new pattern" of Chinese manufacturing industry and its evaluation research. *China Industrial Economy* 2: 63–75.

Li, Lianshui, and Zhanyuan Du. 2004. *Research study on the development of China's manufacturing sector, 2014*. Beijing: Science Press.

Li, Lianshui, Caihong Zhou, and Jun Liu. 2012. *Research study on the development of China's manufacturing sector, 2012*. Beijing: Science Press.

Li, Lianshui, Caihong Zhou, and Jun Liu. 2013. *Research study on the development of China's manufacturing sector, 2013*. Beijing: Science Press.

Li, Lianshui, Haochang Yang, and Jun Liu. 2014. Evaluation study on comprehensive developing ability of regional manufacturing industry in China based on the empirical analysis of manufacturing industry in east, middle and west China. *China Soft Science* 2: 121–129.

Li, Lianshui, and Yong Zhou. 2006. Can technological advances improve high-energy efficiency? An empirical experiment based on China's industrial sector. *Management World* 10: 82–89.

Li, Lianshui, and Yong Zhou. 2005. Empirical analysis of the status of the "new industrial model" in China's manufacturing sector—An assessment based on the manufacturing sectors of 30 Chinese regions. *Management World* 6: 76–81.

Li, Jianping, and Shuyu Xie. 2007. Analysis of economic growth based on technological advances. *Mathematics in Economics* 1:50–53.

Li, Shixiang, and Jinhua Cheng. 2008. Study on the energy efficiency of China and its determinants. *Statistical Research* 25 (10): 28–36.

Li, W., and X. Jin. 2011. Efficiency of Environmental Technological of Chinese Manufacturing Industry (2):33–38.

Li, Zicheng, and Zhesong Jin. 2011. An analysis of the relativity between opening rate and economic growth in Yunnan Province inquiry into economic issues 2:148–153.

Liao, Xiaoling. 2011. *Research on determinants of energy efficiency in China*. Hubei University.

Lin, Boqiang, and Kerui Du. 2013. The energy effect of factor market distortion in China. *Economic Research Journal* (9):125–136.

Liu, Chang, Xianli Kong, and Tiemei GAO. 2008. Empirical analysis of changes in China's industrial sector energy consumption intensity and influential factors. *Resources Science* 09: 1291–1298.

Liu, Jun, and Jun Shao. 2011. Technology gap and FDI spillover effect: An empirical analysis based on quantile regression. *International Business* 3: 82–89.

Liu, Jun, and Kangning Xu. 2010. Industrial agglomeration, industrialization level and regional disparity: An empirical study based on provincial panel data in China. *Finance & Economics* 10:65–72.

Liu, Jun, and Kang Ning Xu. 2010. Industrial agglomeration, economic growth and regional disparity—An empirical study based on provincial panel data in China. *China Soft Science* 7:91–102.

Liu, Xiu Yan, and Xing Min Yin. 2008. Spatial externalities and regional wage differences: A dynamic panel-data study. *China Economic Quarterly* 8 (1): 77–98.

Long, Ruyin, and Zhonggui Li. 2009. An empirical study on the relationship between technology progress and energy intensity. *East China Economic Management* 4:36–39.

Lu, Fangyuan, and Dandan Jin. 2011. An empirical analysis on the effect of R&D input to economic growth based on panel data. *China Industrial Economics* 3: 150–157.

Niu, Yongze, and Maohui Sun. 2013. Technological progress and economic growth in underdeveloped areas—A case study of Gansu Province. *Journal of Jiangxi Agricultural University(Social Sciences Edition)* (3):334–337.

Qi, Zhixin, Wen ying Chen, and Zong-xin WU. 2007. Effect of light-heavy industry structure changes on energy consumption. *China Industrial Economy* 02:35–42.

Qiu, Ling, Yuming Shen, Wangbing Ren, and Tingting Yan. 2008. Analysis on regional disparity and its influential factors of energy utilization efficiency in China. *Journal of Natural Resources* 23 (05): 920–928.

Qin, Teng, Xi Chen, and Tieying Zhang. 2014. Research on the threshold effect of technological progress on economic growth. *Commercial Times* 33: 39–41.

Qu, Xiaoe. 2011. Chinese provincial industrial energy efficiency and energy saving potential: Empirical and simulation-based DEA. *Economic Management Journal* 33 (07): 16–24.

Shao, Shuai, Lili Yang, and Tao Huang. 2013. The oretical model and experience from China of energy rebound effect. *Economic Research Journal* 12: 96–109.

Shen, Lisheng. 2007. The changes of China's foreign trade structure is harmful to energy-saving and consumption-reducing. *Management World* 10: 43–50.

Shi, Dan. 2002. The improvement of energy consumption efficiency in China's economic growth. *Economic Research Journal* 9: 49–56.

Shi, Dan. 2006. Regional differences in China's energy efficiency and conservation potentials. *China Industrial Economy* 10: 49–58.

Shi, Li, and Pei Hu. 2008. Estimation of contribution to China's TFP growth from information technology: 1980–2003. *Forecasting* 27(3):1–7.

Shen, Keting, and Jianjian Gong. 2011. Environmental pollution, technical progress, and productivity growth of energy-intensive industries in China—Empirical study based on ETFP. *China Industrial Economics* 12: 25–34.

Shen, Neng. 2012. Environmental efficiency, industrial heterogeneity and intensity of optimal regulation—Nonlinear test based on industrial panel-data. *China Industrial Economics* 3: 56–68.

Sun, Guangsheng, Yi Huang, Haifeng Tian, and Fengping Wang. 2012a. Total factor productivity, inputs substitution and regional energy efficiency disparity. *Economic Research Journal* 9: 99–112.

Sun, Linlin, Haitao Zheng, and Ruoen Ren. 2012b. The contribution of information to economic growth in China: Evidence from panel data of the industry. *The Journal of World Economy* 2: 3–25.

Tan, Zhongfu, and Jinliang Zhang. 2010. Demonstrational analysis of the relationship between energy efficiency and its influential factor in China. *Population, Resources and Environment* 20 (4): 43–49.

Tang, Anbao, and Xingmin Li. 2014. Study on the effect of energy price and technological progress on energy efficiency in China. *Statistics & Decision* 15: 98–101.

Wan, Lun Lai, and Meng Yi Tong. 2010. Analysis of the factors affecting the China energy intensity under environmental regulation. *Journal of Shanxi Finance and Economics University* 2: 6–7.

Wan, Lun Lai, and Qin Zhu. 2013. The impact of R&D investment on the industrial green total factor productivity growth–empirical data from China's industrial 1999–2010. *Economic Perspectives* 9: 20–26.

Wang, B., and D.Y. Yu. 2004. Layer-by-layer analysis on economic structures effect on informatization in China—On the basis of empirical research on econometric model. *China Industrial Economy* 7: 21–28.

Wang, Fang, and Maomao Pan. 2015. Industrial convergence, performance improvement and manufacturing industry growth: An empirical study based on panel data during 1998 to 2011. *Studies in Science of Science* 33 (4): 530–538.

Wang, Guangdong, Rensun Ye, and Lei Wang. 2008. The effect of technical progress on employment: The regional differences and the policy choice. *China Soft Science* 11: 151–160.

Wang, Keliang, Baochen Yang, and Li Yang. 2010. China's Provincial total-factor energy efficiency considering environmental effects. *Journal of Management Science* 23 (6): 100–111.

Wang, Qingyi. 2005. China's energy efficiency and international comparison Energy Conservation and Environmental Protection 6:10–13.

Wang, Qingyi. 2003. China's energy efficiency and international comparison Energy Conservation and Environmental Protection 8:6–7.

Wang, Qunwei, and Dequn Zhou. 2008. Improved model for evaluating rebound effect of energy resource and its empirical research. *Chinese Journal of Management* 5 (5): 688–691.

Wang, Yuqian. 2008. Factor analysis method and its application of the change of energy consumption intensity. *Quantitative & Technica Economics* 8: 43–52.

Wang, Zhigang, Liutang Gong, and Yuyu Chen. 2006. China's regional differences in technical efficiency and the decomposition of total factor productivity growth (1978–2003). *Social Sciences in China* 2: 55–66.

Wang, Song, and Xiaoming Yang. 2007. An empirical study on the contribution of scientific and technological progress to the economic growth of China. *Prices Monthly* 2: 42–45.

Chapter 6
Development of China's Manufacturing Sector: Firm Research

Zhou Feixue and Ji Liangyu

This chapter studies and analyzes the current development conditions of listed firms, considering such aspects as scale, efficiency, growth, and innovativeness. The chapter confirms the overall orientation of the growth of manufacturing sector firms, industries, and regions, assesses the most respectable listed manufacturing sector firms, and establishes a model for firm development, and thereby, the influence of listed firms in the manufacturing sector can be increased, the formation of economic, technological, and social momentum that benefits the sustainable development of firms in the manufacturing sector can be promoted continuously.

6.1 Basic Assessment of Listed Firm Development in China's Manufacturing Sector

This section deals primarily with the assessment of conditions, including the distribution of the number of listed firms in China's manufacturing sector, scale of firm development and efficiency, as well as firm growth and innovativeness in 2014. Through this assessment, we come to understand the basic conditions of listed firm development in China's manufacturing sector.

Data for this section were compiled primarily from annual reports of listed firms for 2014, as well as from the patent search system of China's State Intellectual Property Office. In total, sample data were collected for 1656 listed manufacturing sector firms, which were listed for the first time prior to December 31, 2014. In this section, industry classifications for listed manufacturing sector firms are determined in accordance with the *Guidelines for the Industry Classification of Listed Companies* (2012 revision) published by the China Securities Regulatory Commission. Here, the primary considerations are as follows. First, firms listed prior to 2014 operating in the manufacturing sector and that fit our research goals were selected for analysis. Second, the industrial classification regulations in the

L. Li and Z. Du (eds.), *A Research Report on the Development of China's Manufacturing Sector (2016)*, Current Chinese Economic Report Series,
DOI 10.1007/978-981-10-4445-8_6

Guidelines for the Industry Classification of Listed Companies (2012 revision) were adjusted periodically in order to provide an objective basis for the industrial analysis of manufacturing sector listed firms.

6.1.1 Development Distribution of Number of Listed Firms

The numerical distribution of listed Chinese manufacturing sector firms can reflect current regional and industry distribution characteristics of listed manufacturing firms. It is, to a certain extent, an important basis for measuring the development levels of firms within different regions and industries of the manufacturing sector.

1. Regional distribution of Chinese manufacturing sector firms

Considering the regional distribution of listed Chinese manufacturing sector firms, the top four provinces with the greatest number of firms are Guangdong, Zhejiang, Jiangsu, and Shandong, each of which has more than 100 firms. Among these provinces, Guangdong has the greatest number of listed manufacturing sector firms, with 249, or 15.04% of China's total. Zhejiang has 199 such firms, or 12.02% of the national total. Jiangsu has 194 such firms, or 11.71% of the national total. Shandong has 117 such firms, or 7.07% of China's total. Together, these four provinces contain 759 listed manufacturing sector firms, 45.83% of China's total. Following these top 4 provinces, other regions placed in the top 10 include Shanghai, Beijing, Sichuan, Anhui, Hubei, and Henan. Together, these 10 provincial-level regions include 1176 such firms, or 71.01% of China's total. Of China's 31 provincial-level regions (including directly controlled municipalities and autonomous regions), Tibet has the fewest number of listed manufacturing sector firms, with only 5, or 0.3% of the national total; Qinghai has only 8 listed firms, or 0.48% of the national total. Hainan has only 7 firms, or 0.42% of the national total; and Ningxia has 10 such firms, or 0.6% of the national total. The other bottom 10 provincial-level regions (i.e., those with the fewest listed manufacturing sector firms) are Tianjin (21), Heilongjiang (19), Inner Mongolia (18), Yunnan (17), Gansu (17), and Guizhou (16) (see Table 6.1).

Considering the distribution of listed firms across China's eastern, central, and western regions, The 11 provincial-level regions of Guangdong, Zhejiang, Jiangsu, Shanghai, Shandong, Beijing, Fujian, Hebei, Liaoning, Tianjin, and Hainan together account for 1109 listed manufacturing sector firms, which is more than two-thirds of the national total, or 66.97%. The eight central provinces of Anhui, Hubei, Henan, Hunan, Jiangxi, Jilin, Shandong, and Heilongjiang together account for 303 such firms, or 18.3% of the national total. The 12 western provincial-level regions of Sichuan, Inner Mongolia, Yunnan, Chongqing, Guizhou, Guangxi, Xinjiang, Shaanxi Gansu, Ningxia, Qinghai, and Tibet together account for 244 such firms, or 14.73% of the national total.

Table 6.1 Provincial distribution of 1656 listed firms in China's manufacturing sector, 2014

Rank	Province	Number of firms	Percentage of total (%)	Cumulative proportion (%)
1	Guangdong	249	15.04	15.04
2	Zhejiang	199	12.02	27.05
3	Jiangsu	194	11.71	38.77
4	Shandong	117	7.07	45.83
5	Shanghai	99	5.98	51.81
6	Beijing	91	5.50	57.31
7	Sichuan	61	3.68	60.99
8	Anhui	57	3.44	64.43
9	Hubei	56	3.38	67.81
10	Henan	53	3.20	71.01
11	Fujian	51	3.08	74.09
12	Hunan	47	2.84	76.93
13	Liaoning	44	2.66	79.59
14	Hebei	37	2.23	81.82
15	Jilin	26	1.57	83.39
16	Jiangxi	24	1.45	84.84
17	Shaanxi	25	1.51	86.35
18	Xinjiang	23	1.39	87.74
19	Chongqing	22	1.33	89.07
20	Guangxi	22	1.33	90.40
21	Shanxi	21	1.27	91.67
22	Tianjin	21	1.27	92.93
24	Heilongjiang	19	1.15	94.08
23	Inner Mongolia	18	1.09	95.17
25	Yunnan	17	1.03	96.20
27	Gansu	17	1.03	97.22
26	Guizhou	16	0.97	98.19
28	Ningxia	10	0.60	98.79
29	Hainan	7	0.42	99.21
30	Qinghai	8	0.48	99.70
31	Tibet	5	0.30	100.00
Total		1656	100	–

Data source Calculations are from information related to 2014 annual reports sourced from the Shanghai Stock Exchange (www.sse.com.cn) and the Shenzhen Stock Exchange (www.szse.com.cn)

Considering regional totals for the distribution of listed firms, China's central and eastern regions have a definitive advantage, especially the six provincial-level regions of Guangdong, Zhejiang, Jiangsu, Shandong, Shanghai, and Beijing, where the proportion of listed manufacturing sector firms is close to 60%.

2. Industry distribution for China's manufacturing sector firms

According to the *Guidelines for the Industry Classification of Listed Companies* (2012 revision), listed firms in the manufacturing sector can be divided into 29 industries as follows: manufacture of computers, communications, and other electronic equipment; manufacture of raw chemical materials and chemical products; manufacture of electrical machinery and equipment; manufacture of medicines; manufacture of special purpose machinery; manufacture of general purpose machinery; manufacture of automobiles; manufacture of non-metallic mineral products; smelting and pressing of non-ferrous metals; manufacture of rubber and plastic; manufacture of textiles; manufacture of metal products; agricultural and sideline food processing; manufacture of wine, drinks, and refined tea; manufacture of railway, ship, aerospace, and other transportation equipment; smelting and pressing of ferrous metals; manufacture of textile wearing and apparel; manufacture of paper and paper products; manufacture of measuring instruments; manufacture of foods; manufacture of chemical fibers; processing of petroleum, coking, processing of nuclear fuel; other manufactures; manufacture of articles for culture, education, and sport activity; processing of timber, manufacture of wood, bamboo, rattan, palm, and straw products; printing, reproduction of recording media; manufacture of leather, fur, feather, and related products and footware; manufacture of furniture; and utilization of waste resources.

With regard to industry distribution, each of these 29 industries contains listed firms. The manufacture of computers, communications, and other electronic equipment industry has the most listed firms (218 firms, or 13.16% of all industries of the manufacturing sector), followed by the manufacturing of raw chemical materials and chemical products industry (180 firms, or 10.87%) and the manufacture of electrical machinery and equipment industry (170 firms, or 10.27%). These three most prominent industries have a combined total 568 listed manufacturing firms, or 34.3% of the manufacturing sector total; this is more than one-third of the total. Adding manufacture of medicines, manufacture of special purpose machinery, and manufacture of general purpose machinery, these six industries account for 956 listed firms, or nearly 60% of the industry total. Thus, these six industry types have an advantage in the numerical distribution of listed manufacturing sector firms.

The three industries with the lowest numbers of listed manufacturing firms are manufacture of leather, fur, feather, and related products and footware (six firms, or 0.36% of the total), manufacture of furniture (five firms, or 0.3% of the total), and utilization of waste resources (one firm, or 0.06% of the total); together, these three industries account for less than 1% of the total number of listed manufacturing sector firms. The number of listed firms in these three industries represents an important benchmark for similar manufacturing sector firms (see Table 6.2).

3. Regional firm distribution of China's manufacturing sector

Tables 6.3, 6.4 and 6.5 reflect the distribution of China's 1656 manufacturing sector firms across provinces, autonomous regions, and direct controlled

Table 6.2 Industry distribution of 1656 firms in China's manufacturing sector in 2014

Rank	Industry	Number of firms	Percentage of total (%)	Cumulative proportion (%)
1	Manufacture of computers, communications, and other electronic equipment	218	13.16	13.16
2	Manufacture of raw chemical materials and chemical products	180	10.87	24.03
3	Manufacture of electrical machinery and equipment	170	10.27	34.30
4	Manufacture of medicines	147	8.88	43.18
5	Manufacture of special purpose machinery	140	8.45	51.63
6	Manufacture of general purpose machinery	101	6.10	57.73
7	Manufacture of automobiles	87	5.25	62.98
8	Manufacture of non-metallic mineral products	77	4.65	67.63
9	Smelting and pressing of non-ferrous metals	55	3.32	70.95
10	Manufacture of rubber and plastic	51	3.08	74.03
11	Manufacture of textiles	40	2.42	76.45
12	Manufacture of metal products	40	2.42	78.86
13	Processing of food from agricultural products	38	2.29	81.16
14	Manufacture of wine, drinks, and refined tea	36	2.17	83.33
15	Manufacture of railway, ship, aerospace, and other transportation equipment	34	2.05	85.39
16	Smelting and pressing of ferrous metals	31	1.87	87.26
17	Manufacture of textile wearing and apparel	30	1.81	89.07
18	Manufacture of paper and paper products	27	1.63	90.70
19	Manufacture of measuring instruments	28	1.69	92.39
20	Manufacture of foods	28	1.69	94.08
21	Manufacture of chemical fibers	24	1.45	95.53
22	Processing of petroleum, coking, processing of nuclear fuel	19	1.15	96.68
23	Other manufactures	17	1.03	97.71

(continued)

Table 6.2 (continued)

Rank	Industry	Number of firms	Percentage of total (%)	Cumulative proportion (%)
24	Manufacture of articles for culture, education, and sport activity	10	0.60	98.31
25	Processing of timber, manufacture of wood, bamboo, rattan, palm, and straw products	9	0.54	98.85
26	Printing, reproduction of recording media	7	0.42	99.28
27	Manufacture of leather, fur, feather, and related products and footware	6	0.36	99.64
28	Manufacture of furniture	5	0.30	99.94
29	Utilization of waste resources	1	0.06	100.00
Total		1656	100.00	–

Data source Calculations are from information related to 2014 annual reports sourced from the Shanghai Stock Exchange (www.sse.com.cn) and Shenzhen Stock Exchange (www.szse.com.cn)

Table 6.3 Regional industry distribution of 1656 listed Chinese manufacturing sector firms in 2014 (Industry Group 1) (*unit* firm)

Region	Province	Manufacture of computers, communications, and other electronic equipment	Manufacture of raw chemical materials and chemical products	Manufacture of electrical machinery and equipment	Manufacture of medicines	Manufacture of special purpose machinery	Manufacture of general purpose machinery	Manufacture of automobiles	Manufacture of non-metallic mineral products	Smelting and pressing of non-ferrous metals	Manufacture of rubber and plastic	Manufacture of textiles	Manufacture of metal products	Processing of food from agricultural products
Eastern region	Guangdong	75	16	39	16	17	6	6	6	3	8	2	5	5
	Zhejiang	22	16	22	16	12	22	15	4	4	8	8	7	2
	Jiangsu	25	21	29	7	19	18	8	6	6	9	7	10	0
	Shandong	6	22	10	9	11	9	9	5	3	4	5	4	6
	Shanghai	9	11	12	8	6	6	9	5	1	5	4	1	1
	Beijing	26	3	4	10	13	1	2	7	3	0	0	2	2
	Fujian	9	2	5	1	5	3	3	3	2	1	2	0	1
	Liaoning	1	4	5	2	6	7	3	0	3	0	0	1	2
	Hebei	2	7	2	3	5	1	2	1	0	2	1	2	1
	Tianjin	3	0	5	6	4	0	2	0	0	1	0	0	0
	Hainan	0	0	0	3	0	0	1	1	1	0	1	0	0
	Industry Total	178	102	133	81	98	73	60	38	26	38	30	32	20
Central region	Anhui	3	10	7	2	3	5	5	4	3	6	2	2	1
	Henan	3	3	5	4	6	2	4	7	5	1	1	1	2
	Hubei	13	10	4	6	6	3	2	3	0	1	1	0	0
	Hunan	1	4	3	6	7	3	1	3	2	1	2	1	4
	Jiangxi	1	4	2	3	1	1	1	1	3	0	0	0	2
	Jilin	1	1	2	7	1	0	4	2	2	0	0	0	0
	Heilongjiang	0	1	3	4	2	0	1	0	0	1	0	0	1
	Shanxi	1	3	0	4	1	0	0	2	2	0	0	0	0
	Industry Total	23	36	26	36	27	14	18	22	17	10	6	4	10

(continued)

Table 6.3 (continued)

Region	Province	Manufacture of computers, communications, and other electronic equipment	Manufacture of raw chemical materials and chemical products	Manufacture of electrical machinery and equipment	Manufacture of medicines	Manufacture of special purpose machinery	Manufacture of general purpose machinery	Manufacture of automobiles	Manufacture of non-metallic mineral products	Smelting and pressing of non-ferrous metals	Manufacture of rubber and plastic	Manufacture of textiles	Manufacture of metal products	Processing of food from agricultural products
Western region	Guangxi	0	6	1	4	1	0	2	0	0	0	0	0	1
	Guizhou	2	3	1	3	0	1	1	1	0	1	0	1	0
	Gansu	1	1	1	2	2	0	0	3	0	1	1	0	0
	Inner Mongolia	0	5	0	4	1	0	0	1	1	0	1	0	1
	Ningxia	0	2	0	0	0	1	0	2	1	0	1	1	0
	Shaanxi	4	1	4	1	5	1	0	2	2	0	0	0	0
	Sichuan	9	15	1	2	5	7	2	2	1	1	1	2	2
	Xinjiang	1	3	1	0	1	1	0	4	0	0	0	0	4
	Qinghai	0	1	1	2	0	1	0	1	0	0	0	0	0
	Tibet	0	0	0	2	0	0	0	0	1	0	0	0	0
	Chongqing	0	3	1	7	0	0	3	0	0	0	0	0	0
	Yunnan	0	2	0	3	0	2	1	1	6	0	0	0	0
	Industry total	17	42	11	30	15	14	9	17	12	3	4	4	8

Table 6.4 Regional industry distribution of 1656 listed Chinese manufacturing sector firms in 2014 (Industry Group 2) (*unit* firm)

Region	Province	Manufacture of wine, drinks, and refined tea	Manufacture of railway, ship, aerospace, and other transportation equipment	Smelting and pressing of ferrous metals	Manufacture of textile wearing and apparel	Manufacture of paper and paper products	Manufacture of measuring instruments	Manufacture of foods	Manufacture of chemical fibers	Processing of petroleum, coking, processing of nuclear fuel	Other manufactures	Manufacture of articles for culture, education, and sport activity	Timber processing and wood manufacturing, bamboo, rattan, palm, and straw products	Printing, reproduction of recording media
Eastern region	Guangdong	3	3	1	4	9	4	8	2	1	1	7	1	2
	Zhejiang	2	1	1	9	3	5	2	6	0	6	2	1	0
	Jiangsu	3	5	2	5	0	3	1	7	0	1	0	2	0
	Shandong	2	0	1	1	5	0	1	2	0	0	0	0	1
	Shanghai	1	4	2	2	2	3	2	0	1	2	1	0	1
	Beijing	2	6	1	3	0	3	1	0	0	1	0	0	1
	Fujian	1	0	1	3	4	0	1	0	0	1	0	0	1
	Liaoning	0	0	4	1	1	0	0	0	2	1	0	1	0
	Hebei	2	0	1	0	0	2	0	1	1	0	0	0	0
	Tianjin	0	0	0	0	0	0	0	0	0	0	0	0	0
	Hainan	0	0	0	0	0	0	0	0	0	0	0	0	0
	Industry Total	16	19	14	28	24	20	16	18	5	13	10	5	6
Central region	Anhui	2	0	1	0	0	0	0	1	0	0	0	0	0
	Henan	0	0	1	0	1	2	2	2	0	0	0	0	0
	Hubei	0	0	2	1	0	0	1	1	1	1	1	0	0
	Hunan	1	1	2	0	1	1	2	0	1	0	0	0	0
	Jiangxi	0	1	2	0	0	2	0	0	0	0	0	0	0
	Jilin	1	0	0	0	1	1	0	1	1	0	0	1	0
	Heilongjiang	0	2	0	0	1	0	0	0	2	1	0	0	0
	Shanxi	1	1	2	0	0	0	0	0	4	0	0	0	0
	Industry Total	5	5	10	1	4	6	5	5	9	2	1	1	0

(continued)

Table 6.4 (continued)

Region	Province	Manufacture of wine, drinks, and refined tea	Manufacture of railway, ship, aerospace, and other transportation equipment	Smelting and pressing of ferrous metals	Manufacture of textile wearing and apparel	Manufacture of paper and paper products	Manufacture of measuring instruments	Manufacture of foods	Manufacture of chemical fibers	Processing of petroleum, coking, processing of nuclear fuel	Other manufactures	Manufacture of articles for culture, education, and sport activity	Timber processing and wood manufacturing, bamboo, rattan, palm, and straw products	Printing, reproduction of recording media
Western region	Guangxi	0	0	1	0	1	1	2	1	0	0	0	1	0
	Guizhou	1	1	0	0	0	0	0	0	0	0	0	0	0
	Gansu	3	0	1	0	0	0	0	0	0	1	0	0	0
	Inner Mongolia	0	1	2	0	0	0	1	0	0	0	0	0	0
	Ningxia	1	0	0	0	0	0	0	0	0	0	0	0	0
	Shaanxi	0	2	0	0	0	0	0	0	1	1	0	0	1
	Sichuan	4	2	0	1	1	0	0	0	1	0	0	2	0
	Xinjiang	3	0	1	0	0	0	2	0	1	0	0	0	0
	Qinghai	1	0	1	0	0	0	0	0	0	0	0	0	0
	Tibet	1	0	0	0	0	0	1	0	0	0	0	0	0
	Chongqing	1	4	1	0	0	1	1	0	0	0	0	0	0
	Yunnan	0	0	0	0	0	0	0	0	2	0	0	0	0
	Industry Total	15	10	7	1	2	2	7	1	5	2	0	3	1

Data source Calculations are from information related to 2014 annual reports sourced from the Shanghai Stock Exchange (www.sse.com.cn) and Shenzhen Stock Exchange (www.szse.com.cn)

Table 6.5 Regional industry distribution of 1656 Chinese manufacturing sector listed firms in 2014 (Industry Group 3) (*unit* firm)

Region	Province	Manufacture of leather, fur, feather, and related products and footware	Manufacture of furniture	Utilization of waste resources	Industry total
Eastern region	Guangdong	0	2	1	249
	Zhejiang	1	2	0	199
	Jiangsu	0	0	0	194
	Shandong	1	0	0	117
	Shanghai	0	0	0	99
	Beijing	0	0	0	91
	Fujian	2	0	0	51
	Liaoning	0	0	0	44
	Hebei	1	0	0	37
	Tianjin	0	0	0	21
	Hainan	0	0	0	7
	Industry Total	5	4	1	1109
Central region	Anhui	0	0	0	57
	Henan	1	0	0	53
	Hubei	0	0	0	56
	Hunan	0	0	0	47
	Jiangxi	0	0	0	24
	Jilin	0	0	0	26
	Heilongjiang	0	0	0	19
	Shanxi	0	0	0	21
	Industry Total	1	0	0	303
Western region	Guangxi	0	0	0	22
	Guizhou	0	0	0	16
	Gansu	0	0	0	17
	Inner Mongolia	0	0	0	18
	Ningxia	0	0	0	10
	Shaanxi	0	0	0	25
	Sichuan	0	0	0	61
	Xinjiang	0	1	0	23
	Qinghai	0	0	0	8
	Tibet	0	0	0	5
	Chongqing	0	0	0	22
	Yunnan	0	0	0	17
	Industry Total	0	1	0	244

municipalities in the three major economic belts of China's eastern, central, and western regions. At the provincial level, the industry distribution of listed manufacturing sector firms is consistent with the industry distribution within the tables. Compared to other provinces, Guangdong has the highest number of listed firms in the manufacture of computers, communications, and other electronic equipment (75 firms), manufacture of electrical machinery and equipment (39 firms), and manufacture of foods (8 firms). For Jiangsu province, industries with the highest numbers of listed firms compared to other provinces are manufacture of special purpose machinery (19 firms), manufacture of general purpose machinery (18 firms), and manufacture of metal products (10 firms). For Zhejiang province, industries with the highest numbers of listed firms compared to other provinces are the manufacture of general purpose machinery (19 firms), manufacture of automobiles (15 firms), and manufacture of textile wearing and apparel (9 firms).

In terms of numbers of listed firms, the manufacturing sector in China's eastern region possesses a clear advantage, with more than 60% of all listed firms. For the manufacture of computers, communications, and other electronic equipment, manufacture of electrical machinery and equipment, manufacture of special purpose machinery, and manufacture of general purpose machinery, the proportion is more than 70%.

6.1.2 Scale Assessment of Listed Chinese Manufacturing Sector Firms

Firm scale is an important economic indicator of firms reaching certain levels of production and operations. Firms with a foothold in competitive market environments must be of a certain scale; a large body of research points out that the resource endowment conditions of firm scale are closely related to such factors as growth (Yu et al. 2013) and innovation (Wang and An 2014a). In designing the index for firm scale, most of the abovementioned research utilizes the firm-scale classification criteria of China's Bureau of Statistics, that is, such indexes as total capital, total income from sales (core operating revenue (Liu and Cheng 2014)), and total number of employees (Li et al. 2014). In the case that only a single index is used to represent firm scale, most of these studies elected to use the total capital index. Capital index (He and Cheng 2009) to all capital owned or controlled by an economic body and that brings economic profit; it reflects firm scale from the resources it possesses and its production factors. In the previous *Research Report on the Development of China's Manufacturing Sector*, total capital was used as the basis for differentiating between listed firms of very large scale (5 billion yuan or more), large scale (500 million yuan or more) medium scale (50 million yuan or more), and small scale (up to 50 million yuan). Income from sales refers primarily to the basic income generated by the primary operations of a firm in its typical state. Specifically, when considering industrial firms, income from sales refers to income from product sales, while for manufacturing sector firms, it refers to income from such sources as product sales and unfinished product sales, and the provision of industrial labor services. These form the foundation for accumulation and development of firms; a higher index indicates that the firm's economic base is stronger, its capacity for sustainable growth is higher, and its growth potential is larger. The index can objectively reflect the scale of a firm's operations and its competitiveness in the market, and of the indexes currently used in China, it can be supplied with relatively complete data. In addition, it is easy to use, and due to varying labor structures of listed firms in different industries and relatively large variations in employee quality, the index for number of employees is not considered very well balanced (Ren et al. 2010). Thus, here, we use total capital and core operating revenue to assess the scale of firms in the manufacturing sector.

1. Overall scale structure for firms in China's manufacturing sector

In 2014, China's 1656 listed manufacturing sector firms earned total core operating revenue of 9.36 trillion yuan. The largest individual firm had core operating revenue of 626.712 billion yuan, while that of the smallest was 8.386 million yuan. The average was 5.65 billion yuan, a 4.15% increase from the previous year, which shows that in 2014, there was a clear expansion in the scale of listed Chinese manufacturing sector firms compared to the previous year.

With regard to the revenue scale of industries, the manufacture of computers, communications, and other electronic equipment industry was the largest in 2014; its 218 listed firms had combined core operating revenue of 879.91 billion yuan, or 9.4% of the total. The second largest industry by revenue in 2014 was manufacture of electrical machinery and equipment; its 170 listed firms had combined core operating revenue of 835.6 billion yuan, or 8.93% of the total. The third largest industry by revenue in 2014 was manufacture of raw chemical materials and chemical products; its 180 listed firms had a combined core operating revenue of 644.384 billion yuan, or 6.89% of the total. The smallest industry by revenue was utilization of waste resources; its one listed firm had core operating revenue of 3.909 billion yuan, or only 0.04% of the total for all industries. Thus, we can observe that the scale structure of industries and industry distribution are essentially consistent.

With regard to core operating revenue scale by regional distribution, Shanghai ranks first (its 99 listed manufacturing sector firms had combined core operating revenue of 1.446319 trillion yuan), Guangdong second (249 firms had revenue of 1.310122 trillion yuan), and Beijing third (91 firms had revenue of 905.783 billion yuan). The fourth, fifth, and sixth ranked regions are Shandong province, Jiangsu province, and Zhejiang province, respectively. Ranked by average scale of core operating income, Shanghai, Jiangxi province, Yunnan province, Beijing, and Hebei province comprise the top five, while the five lowest ranked regions are Zhejiang province, Tibet, Fujian province, Tianjin, and Ningxia province.

With regard to revenue scale by region, the 1109 listed manufacturing sector firms in eastern China had combined core operating revenue of 6.256716 trillion yuan, or 66.87% of the total core operating income of China's listed manufacturing sector firms. The 303 listed manufacturing sector firms in China's central region had combined core operating revenue of 1.723887 trillion yuan, or just 18.42% of China's total. The 244 listed manufacturing sector firms in China's western region had combined core operating revenue of 1.376244 trillion yuan, or only 14.71% of the national total. Thus, we observe that the overall scale of listed manufacturing sector firms is largest in China's east, followed by the central region; the scale is smallest in China's west.

2. Scale (total assets) distribution of China's manufacturing sector firms

The National Bureau of Statistics of China uses the two indexes of sales revenue and total assets to differentiate between four classes of firms: very large enterprises (more than 5 billion yuan), large enterprises (more than 500 million yuan), medium enterprises (more than 50 million yuan), and small enterprises (up to 50 million yuan). We use both total assets and core operating revenue as standards for calculating the scale of China's 1656 listed manufacturing sector firms for 2014.

Table 6.6 calculates the scale of listed manufacturing sector firms on the basis of total assets. From this, we observe that the numerical scale of China's listed manufacturing sector firms shows a trend of uninterrupted expansion. This is especially so for very large listed manufacturing firms, which have exhibited distinct growth. The total asset scale of very large firms is more than three-fourths of the total scale, or 79.92%, and the proportion of very large firms is clearly increasing, with a proportional increase of 1.64%. In 2013, the average scale of very large enterprises was 21.851 billion yuan, which grew to 22.507 billion yuan in 2014. However, the proportion of the number of large enterprises decreased by 1.89% while the proportion of the number of very large enterprises increased by 2.61%. In 2013, the average total asset scale of 1608 manufacturing sector firms was 5.368 billion yuan. In 2014, the average total asset scale of 1656 manufacturing sector firms increased to 6.616 billion yuan. The increase in average scale experienced in 2014 further illustrates that listed firms in China's manufacturing sector are experiencing a trend of overall scalar growth.

Table 6.6 Scalar comparison of 1656 listed Chinese manufacturing sector firms, 2013–2014 (total assets)

Scale type	Number of firms		Proportion of total (%)		Scale (100 million yuan)		Proportion of total (%)	
	2013	2014	2013	2014	2013	2014	2013	2014
Very large	395	450	24.56	27.17	86312.1	101282.6	78.28%	79.92
Large	1153	1156	71.70	69.81	23,754	25270.75	21.54%	19.94
Medium	58	49	3.61	2.96	2008.87	172.30	0.18%	0.14
Small	2	1	0.12	0.06	0.1	0.32	0	0
Total	1608	1656	100%	100%	110267.07	126,726	100%	100

Data Source: *Research Report on the Development of China's Manufacturing Sector* (2013) and 2014 annual reports of listed firms

3. Scale distribution (core operating revenue) of Chinese manufacturing sector firms

Table 6.7 ranks the scale of listed manufacturing sector firms according to core operating revenue. In 2013, there were 286 listed manufacturing sector firms with at least 5 billion yuan in core operating revenue; this grew to 303 firms in 2014, a 0.5% increase in the proportion of total firms with such revenues. In 2013, the scale of listed manufacturing sector firms with core operating revenues greater than 5 billion yuan comprised 78.7% of the scale of all firms; in 2014, this figure declined to 78.62%, a decrease of 0.08%. In 2014, the combined core operating revenue of these firms grew by 494.243 billion yuan from the previous year. We observe from the data in Table 6.7 that the proportion of the scale of all firms comprised of firms with between 500 million yuan and 5 billion yuan in core operating revenue was 20.4%, an increase of 0.23% from the previous year. The average scale of these firms grew by 47 million yuan, while the number of firms with core operating revenues between 50 million yuan and 500 million yuan declined from 313 to 288 in the same period.

Considering core operating revenue, the overall scale of China's 1656 listed manufacturing firms in 2014 rose by 7.32% compared to the previous year, while the number of firms rose by 2.99% in the same period. On a macro level, an increasing trend is exhibited. On a micro level, for very large and large-scale firms, there is a trend of increasing scale for both the number of firms and the core operating revenues of these firms. For medium- and small-scale firms, however, there is a trend of decreasing numbers of such firms, and decreasing core operating revenues. This illustrates that there is the growth trend is concentrated in relatively large-scale listed manufacturing sector firms.

Table 6.7 Comparing the scale of 1608 Chinese listed manufacturing sector firms (core operating revenue)

Scale type	Number of firms		Proportion of total (%)		Scale (100 million yuan)		Proportion of total (%)	
	2013	2014	2013	2014	2013	2014	2013	2014
5 billion yuan and above	286	303	17.80	18.30	68,617	73559.43	78.70	78.62
500 million–5 billion yuan	993	1050	61.79	63.41	17585.8	19086.32	20.17	20.40
50 million–500 million yuan	313	288	19.48	17.39	978.45	919.42	1.12	0.98
Up to 50 million yuan	15	15	0.03	0.91	4.07	3.30	0.00	0.00
Total	1607	1656	100	100	87185.32	93568.47	100	100

Data Source: *Research Report on the Development of China's Manufacturing Sector* (2013) and 2014 annual reports of listed firms

6.1.3 Assessment of Efficiency for Listed Chinese Manufacturing Sector Firms

The goal of a firm's operations is to generate maximum revenue and to increase economic efficiency; that is, a basic guarantee of the continued development of firm operations, and an indicator of strengthened competitiveness. Because the purpose of a firm's organization of production activities is to generate profit, profitability is the most important factor for measuring a firm's economic efficiency and primarily reflects the profitability of a firm's operations. Strong profitability is the basis of a firm's rapid future development, and its generation of better economic efficiency. Profits are a central question for all parties within and outside of a firm. Profits allow investors to receive returns on their investments, allow creditors to receive payments on principal and interest, and are a central indicator of the results of firm operators and the efficiency of firm managers. Furthermore, profits are an important guarantee for continuously improving collective benefits and facilities for workers. For these reasons, profitability analysis is of crucial importance.

In designing indexes, profitability indexes include return on equity (ROE), return on assets (ROA), core operating profit margin, and earnings before interest and tax. In this study, overall ROA measures a firm's returns from assets, reflecting the usage efficiency of assets (Cao 2011). ROE includes the influences of a firm's operating efficiency and capital structure, and measures the rate of return on shareholder investment. For this reason, some literature points out that ROE is the central index for reflecting profitability (Yuan 2013), although some literature also points out that in order to conform to the provisions of regulatory authorities, listed Chinese firms engage in a large amount of manipulation of their ROE indexes (He and Cheng 2009). In selecting an operating profit margin index, the literature posits that for firms focused on product management, the leading products and services of these firms can be considered the core profit target of their competitiveness (Qian et al. 2009). Core profit should comprise the main part of financial performance within a given period. The core profit margin index is a comprehensive representation of a firm's core profitability, and might be more appropriate for the analysis of a firm's core profitability development trends. Thus, in this study, we continue to use the *Research Report on the Development of China's Manufacturing Sector* (2014), as well as the indexes of ROE, ROA, and core operating profit margin, to assess the profitability of listed manufacturing sector firms. Here, core operating profit margin = (core operating revenue—prime operating costs—core business taxes and surcharges)/core operating revenue.

Next, we analyze the profitability of China's listed manufacturing sector firms in 2014 according to the three indexes described above.

1. Sector analysis

According to industry types, the profitability of Chinese manufacturing sector firms in 2014 is listed in Table 6.8.

In 2014, the average core operating profit margin of 29 manufacturing sector industries was 22.32%; the 147 listed firms in the manufacture of medicines

Table 6.8 Industry profitability index

Industry	Number of firms	Core operating profit margin (%)	Return on assets (%)	Return on equity (%)
Manufacture of computers, communications, and other electronic equipment	218	26.71	4.25	4.21
Manufacture of raw chemical materials and chemical products	176	19.60	3.48	4.47
Manufacture of electrical machinery and equipment	169	25.06	4.15	7.66
Manufacture of medicines	146	47.77	7.67	12.09
Manufacture of special purpose machinery	139	29.69	3.90	6.05
Manufacture of general purpose machinery	101	24.40	3.75	6.05
Manufacture of automobiles	86	20.59	4.67	6.78
Manufacture of non-metallic mineral products	75	−25.38	3.67	7.18
Smelting and pressing of non-ferrous metals	54	11.14	1.63	1.87
Manufacture of rubber and plastic	50	19.60	4.20	−38.75
Manufacture of textiles	39	17.36	2.85	4.91
Manufacture of metal products	40	19.93	3.47	5.72
Processing of food from agricultural products	38	15.33	3.87	4.96
Manufacture of wine, drinks, and refined tea	36	43.09	4.85	6.55
Manufacture of railway, ship, aerospace, and other transportation equipment	34	17.29	2.33	−1.94
Smelting and pressing of ferrous metals	30	7.90	−0.33	−3.81
Manufacture of textile wearing and apparel	30	36.13	4.76	7.64
Manufacture of paper and paper products	25	16.37	−0.16	−5.80
Manufacture of measuring instruments	28	41.05	6.64	7.93
Manufacture of foods	28	33.80	5.72	8.81
Manufacture of chemical fibers	24	10.09	−0.48	−6.94
Processing of petroleum, coking, processing of nuclear fuel	19	6.97	−1.25	−19.43
Other manufactures	17	22.47	7.11	12.12
Manufacture of articles for culture, education, and sport activity	10	25.75	5.38	7.64
Timber processing and wood manufacturing, bamboo, rattan, palm, and straw products	9	25.81	5.38	9.65

(continued)

Table 6.8 (continued)

Industry	Number of firms	Core operating profit margin (%)	Return on assets (%)	Return on equity (%)
Printing, reproduction of recording media	7	31.11	7.97	10.80
Manufacture of leather, fur, feather, and related products and footware	6	22.67	3.73	4.91
Manufacture of furniture	5	36.63	7.47	10.80
Utilization of waste resources	1	18.38	2.68	6.10
Total Average Value	–	22.32	3.91	3.04

Note Values in this table are derived from calculations and manipulation of data in the 2014 annual reports of listed firms; herein, 13 of 1656 listed manufacturing sector firms had unknown ROE, while 3 firms had unknown core business taxes and surcharges. Therefore, this table considered data from only 1640 firms for its sample

industry had the highest average core operating profit margin of 47.77%. The 75 listed firms in the non-metallic mineral products industry had the lowest average core operating profit margin, at −25.38%. The following 15 industries of the manufacturing sector had core operating profit margins above the industry average, showing that the listed firms in these industries have higher core operating profit margins than the sector average: manufacture of computers, communications, and other electronic equipment; manufacture of electrical machinery and equipment; manufacture of medicines; manufacture of special purpose machinery; manufacture of general purpose machinery; manufacture of wine, drinks, and refined tea; manufacture of textile wearing and apparel; manufacture of measuring instruments; manufacture of foods; other manufactures; manufacture of articles for culture, education, and sport activity; timber processing and wood manufacturing, bamboo, rattan, palm, and straw products; printing, reproduction of recording media; manufacture of leather, fur, feather, and related products and footware; and manufacture of furniture.

In 2014, the average ROA of 29 manufacturing sector industries was 3.91%. The 7 listed firms in the printing, reproduction of recording media industry had the highest average ROA of 7.97% while the 19 listed firms in the processing of petroleum, coking, processing of nuclear fuel industries had the lowest average ROA, of −1.25%. Within the manufacturing sector, the following 15 industries had ROA rates higher than the manufacturing sector average, showing that the listed firms in these industries have higher than average asset utilization efficiency rates, more effective asset operations than average, as well as better than average levels of cost control: manufacture of computers, communications, and other electronic equipment; manufacture of electrical machinery and equipment; manufacture of medicines; manufacture of automobiles; manufacture of rubber and plastic; manufacture of wine, drinks, and refined tea; manufacture of textile wearing and apparel; manufacture of measuring instruments; manufacture of foods; other manufactures; manufacture of articles for culture, education, and sport activity; timber processing and wood manufacturing, bamboo, rattan, palm, and straw products; printing, reproduction of recording media; manufacture of leather, fur, feather, and related products and footware; and manufacture of furniture.

In 2014, the average ROE of the 29 industries of the manufacturing sector was 3.04%. The 17 listed firms in other manufacturing industries had the highest average ROE, at 12.12% while the 50 listed firms in the manufacture of rubber and plastic industry had the lowest average ROE, at −38.75%. In the manufacturing sector, the following 21 industries had ROE higher than the sector average, showing that the listed firms in these industries generated higher investment returns than the sector average: manufacture of computers, communications, and other electronic equipment; manufacture of raw chemical materials and chemical products; manufacture of electrical machinery and equipment; manufacture of medicines; manufacture of special purpose machinery; manufacture of general purpose machinery; manufacture of automobiles; manufacture of non-metallic mineral products; manufacture of metal products; processing of food from agricultural products; manufacture of wine, drinks, and refined tea; manufacture of textile wearing and apparel; manufacture of measuring instruments; manufacture of foods; other manufactures; manufacture of articles for culture, education, and sport activity; timber processing and wood manufacturing, bamboo, rattan, palm, and straw products; printing, reproduction of recording media; manufacture of leather, fur, feather, and related products and footware; manufacture of furniture; and utilization of waste resources.

2. Regional analysis

The profitability in 2014 of firms in China's manufacturing sector by regional classification is presented in Table 6.9.

In 2014, the average core operating profit margin for the manufacturing sector in China's 31 provincial-level divisions was 24.61%. The 5 listed firms in Tibet had the highest average core operating profit margins, at 38.07%, while the 18 listed firms in Inner Mongolia had the lowest average core operating profit margins, at −1.84%. Of the 31 provincial-level divisions, the following 15 regions had higher core operating profit margins than the industry average, showing that the listed manufacturing sector firms in these regions were able to attain better profitability from their core operations than the national average: Beijing, Fujian, Gansu, Guangdong, Guizhou, Hainan, Hubei, Hunan, Jilin, Qinghai, Shanxi, Shanghai, Tianjin, Tibet, and Zhejiang.

In 2014, the average ROA for China's 31 provincial-level divisions was 3.41%. The five listed firms in Tibet had the highest ROA, at 8.17%, while Ningxia's nine listed firms had the lowest ROA, at only 0.1%. Of all provincial-level regions, the following 16 divisions had ROA rates higher than the industrial average, showing that the listed firms in these regions had relatively high input–output levels, more effective asset operations than the national average, as well as better cost control: Anhui, Beijing, Gansu, Guangdong, Guizhou, Hebei, Hubei, Hunan, Jiangsu, Jiangxi, Jilin, Shandong, Shanghai, Tianjin, Tibet, and Zhejiang.

Table 6.9 Regional profitability index

Region	Number of firms	Core operating profit margin (%)	Return on assets (%)	Return on equity (%)
Anhui	56	22.40	3.42	5.81
Beijing	91	31.88	5.20	7.75
Fujian	51	24.76	2.01	−1.71
Gansu	17	27.72	4.31	8.90
Guangdong	249	27.95	4.87	7.28
Guangxi	20	20.86	2.99	4.98
Guizhou	15	32.66	5.42	7.88
Hainan	7	28.96	2.60	3.85
Hebei	37	23.82	3.56	−53.63
Henan	53	24.37	3.37	5.26
Heilongjiang	19	23.25	2.36	−3.70
Hubei	55	24.72	4.12	7.39
Hunan	47	28.36	3.57	4.35
Jilin	26	33.74	4.32	6.19
Jiangsu	193	23.93	4.28	5.58
Jiangxi	23	21.77	4.21	7.62
Liaoning	42	21.48	2.60	2.49
Inner Mongolia	18	−1.84	2.76	4.37
Ningxia	9	17.58	0.10	0.54
Qinghai	8	29.17	2.16	18.63
Shandong	117	22.62	4.35	5.99
Shanxi	21	26.65	1.28	0.43
Shaanxi	24	21.09	0.79	−0.33
Shanghai	97	26.34	5.66	6.62
Sichuan	58	24.10	2.28	−0.20
Tianjin	21	27.82	3.50	4.94
Tibet	5	38.07	8.17	14.95
Xinjiang	23	21.47	2.19	0.03
Yunnan	17	16.30	0.87	−6.99
Zhejiang	199	26.96	5.16	7.96
Chongqing	22	23.86	3.22	−1.76
Total average value	–	24.61	3.41	2.63

Note Values in this table are derived from calculations and manipulation of data in the 2014 annual reports of listed firms; herein, 13 of 1656 listed manufacturing sector firms had unknown ROE, while 3 firms had unknown core business taxes and surcharges. Therefore, this table's sample comprises data from 1640 firms

In 2014, the average ROE for China's 31 provincial-level divisions was 2.63%. The 8 listed firms in Qinghai had the highest average ROE, at 18.63%, while the 37 listed firms in Hebei had the lowest average ROE, at −53.63%. Of all regions, the following 20 divisions had ROE levels higher than the industrial average, showing that the listed manufacturing sector firms in these areas generated higher investment returns than the industrial average: Anhui, Beijing, Gansu, Guangdong, Guangxi, Guizhou, Hainan, Henan, Hubei, Hunan, Inner Mongolia, Jiangsu, Jiangxi, Jilin, Qinghai, Shandong, Shanghai, Tianjin, Tibet, and Zhejiang.

6.1.4 Growth Assessment for China's Listed Manufacturing Sector Firms

Assessing the growth of China's listed manufacturing sector firms means judging whether firms have strong future development potential, and making a judgment on the individual growth conditions and potential of different firms. Final growth targets should be attributable to a firm's financial growth. There are many financial indicators used to assess the growth capacity of manufacturing sector firms, such as sales revenue growth, capital accumulation rate, total asset growth rate, fixed asset turnover ratio, 3-year average profit growth rate, 3-year average capital growth rate, and Tobin' s Q ratio (Zhang and Meng 2011). In most literature, indexes for firm growth potential are measured using total asset growth and sales revenue growth; for example, for high-tech firms, the means of authenticating assessments of firm growth primarily depend on sales revenue growth and total assets growth for the prior 3 years, but given that many listed manufacturing sector firms have large growth in non-core operating revenue, this leads to difficulty in measuring firm growth through growth of operating revenue and total assets (Tang et al. 2008). Thus, we use sales revenue growth rates (i.e., the 3-year average growth rate of core operating revenue) as an index for the growth ability for listed manufacturing sector firms, and the following formula for 3 successive years to measure the 3-year average core operating revenue growth rate: (current year core operating revenue—previous year core operating revenue)/previous year core operating revenue. These indexes can show a firm's or product's potential and forecast for growth or change; they reflect the trends and stability of a firm's core operations growth, and embody the conditions of a firm's continuous development and development capacity. A higher index illustrates that the firm has better momentum for continuous core operating growth, stronger capacity for market expansion, and good firm growth potential. Next, using the index of 3-year average core operating revenue growth, we measure the 2014 growth potential for listed manufacturing sector firms from the perspective of sector and region.

1. Sector analysis

According to sector divisions, the 2014 growth potential for the Chinese manufacturing sector is shown in Table 6.10 and Fig. 6.1.

By collecting data for the core operating revenues of each firm between 2011 and 2014, the 3-year average core operating revenue growth rate calculated for

Table 6.10 2014 Average industry growth potential for listed firms in China's manufacturing sector

Rank	Industry	Number of firms	Industry average growth capacity index
1	Manufacture of non-metallic mineral products	74	0.8456
2	Utilization of waste resources	1	0.7077
3	Other manufactures	17	0.2688
4	Manufacture of railway, ship, aerospace, and other transportation equipment	32	0.2272
5	Manufacture of computers, communications, and other electronic equipment	214	0.2252
6	Manufacture of raw chemical materials and chemical products	170	0.1863
7	Manufacture of electrical machinery and equipment	169	0.1768
8	Smelting and pressing of non-ferrous metals	55	0.1665
9	Manufacture of furniture	5	0.1486
10	Manufacture of articles for culture, education, and sport activity	10	0.1484
11	Manufacture of measuring instruments	28	0.1438
12	Manufacture of foods	28	0.1413
13	Manufacture of textile wearing and apparel	30	0.1291
14	Printing, reproduction of recording media	7	0.125
15	Manufacture of rubber and plastic	48	0.1216
16	Manufacture of automobiles	85	0.1098
17	Manufacture of general purpose machinery	99	0.1007
18	Manufacture of special purpose machinery	136	0.0872
19	Manufacture of chemical fibers	22	0.0832
20	Manufacture of metal products	40	0.0791
21	Manufacture of leather, fur, feather, and related products and footware	6	0.0786
22	Processing of food from agricultural products	38	0.0752
23	Timber processing and wood manufacturing, bamboo, rattan, palm, and straw products	9	0.0729
24	Manufacture of paper and paper products	23	0.0513
25	Manufacture of medicines	145	0.0405
26	Manufacture of wine, drinks, and refined tea	32	0.0254
27	Manufacture of textiles	38	0.0161
28	Smelting and pressing of ferrous metals	30	−0.0218
29	Processing of petroleum, coking, processing of nuclear fuel	18	−0.0373
	Overall average value		0.1599

Note Manipulations and calculations are based on information from 2014 annual reports from the Shanghai Stock Exchange (www.sse.com.cn) and Shenzhen Stock Exchange (www.szse.com.cn). In addition, of 1656 listed manufacturing sector firms in 2014, 47 firms had stocks beginning with * ST\ST\S; thus, these firms received special treatment and had unstable operations. These listed firms were first eliminated, and the analysis for the growth potential of listed manufacturing sector firms utilized data from 1609 firms

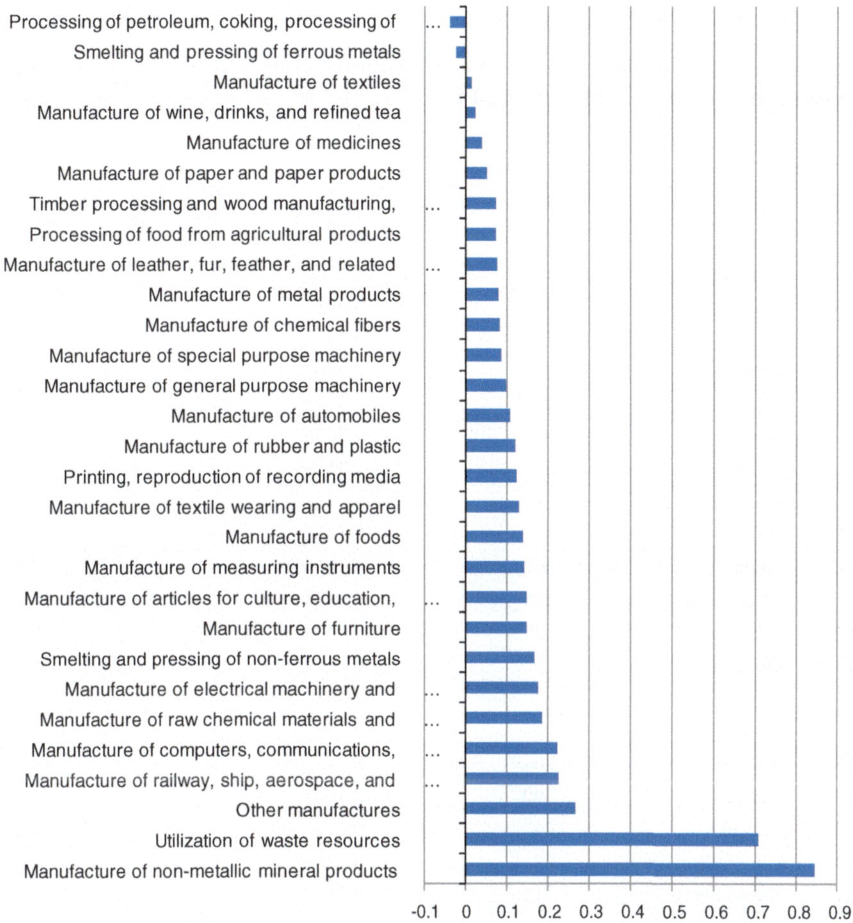

Fig. 6.1 Average growth capacity index by industry

1609 listed manufacturing sector firms is 0.1599. This reflects that the overall manufacturing sector possesses steady growth capacity. Herein, the non-metallic mineral products industry and the utilization of waste resources industry had comparatively high average revenue growth rates, both of which were greater than 0.7. This shows that these industries enjoyed relatively quick sales revenue growth and strong market expansion capacity in recent years. Two industries had negative average revenue growth rates: smelting and prossing of ferrous metals; and processing of petroleum, coking, processing of nuclear fuel. This shows that in recent years, these two industries suffered consistently declining growth capacities, and have poor outlooks for future development. In these industries, it is necessary to increase the pace of transformation and upgrading to the new industrial model.

2. Regional analysis

The 2014 growth potential for Chinese manufacturing sector firms by regional divisions is shown in Table 6.11 and Fig. 6.2.

Table 6.11 Average regional growth potential

Rank	Region	Number of firms	Average regional growth capacity index
1	Gansu	16	0.9561
2	Heilongjiang	18	0.5406
3	Shanghai	95	0.521
4	Hainan	7	0.3873
5	Jilin	26	0.3467
6	Qinghai	8	0.3123
7	Beijing	89	0.2676
8	Hubei	55	0.2579
9	Guangxi	21	0.2111
10	Hebei	36	0.1982
11	Hunan	46	0.1866
12	Guangdong	245	0.1814
13	Yunnan	17	0.1633
14	Guizhou	15	0.1628
15	Tibet	5	0.161
16	Xinjiang	23	0.15
17	Tianjin	20	0.148
18	Jiangsu	190	0.1404
19	Shaanxi	23	0.138
20	Chongqing	22	0.1361
21	Shandong	115	0.1115
22	Anhui	56	0.1091
23	Zhejiang	198	0.1063
24	Jiangxi	24	0.0962
25	Sichuan	55	0.0903
26	Fujian	49	0.0824
27	Henan	52	0.0656
28	Inner Mongolia	16	0.0594
29	Liaoning	41	0.0422
30	Ningxia	8	−0.0037
31	Shanxi	18	−0.0161
	Overall average value	–	0.2035

Note Manipulations and calculations are based on information from 2014 annual reports. In addition, of 1656 listed manufacturing sector firms in 2014, 47 firms had stocks beginning with * ST\ST\S; because these firms received special treatment and had unstable operations, they were first eliminated, and the analysis for the growth potential of listed manufacturing sector firms utilized data from 1609 firms

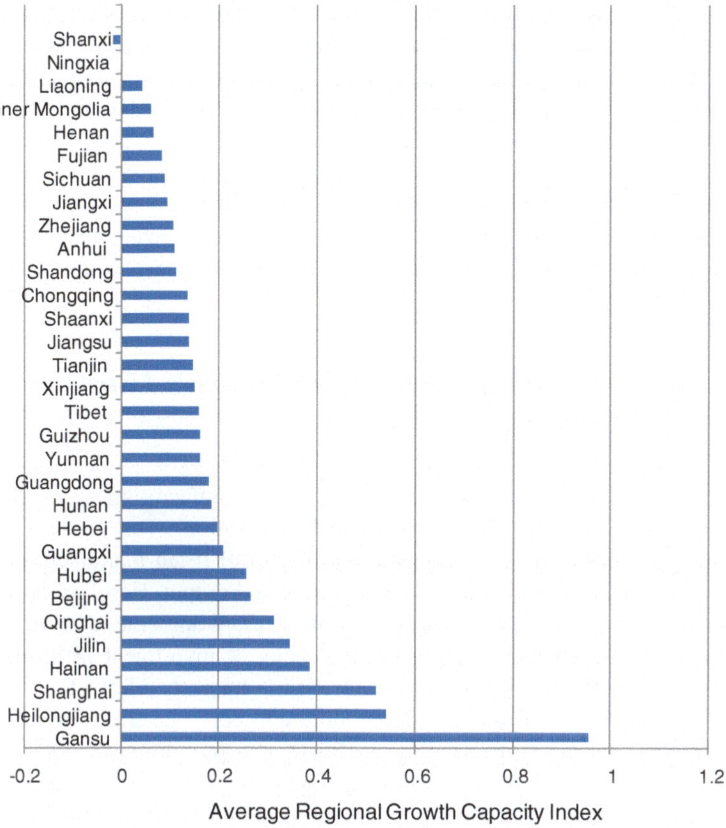

Fig. 6.2 Average regional growth capacity index

In 2014, the average growth capacity index for the manufacturing sector in all of China's 31 provincial-level divisions was 0.2035. Most provincial-level divisions had reasonable average revenue growth rates, reflecting stability and reliability of operations, and strong growth capacity for listed manufacturing sector firms in these regions. The three divisions of Gansu, Heilongjiang, and Shanghai had relatively high revenue growth rates, all greater than 0.5; this shows that the listed manufacturing sector firms in these regions have good growth potential. At the same time, Jiangxi, Sichuan, Fujian, Henan, Inner Mongolia, and Liaoning all had relatively small average growth capacity index values; this shows that the growth capacities of listed manufacturing sector firms in these regions are relatively weak. Furthermore, Ningxia and Shanxi both had negative values for this index, from which we observe that listed manufacturing sector firms here face growth constraints.

When viewing the combined growth potential of individual listed manufacturing sector firms, we discovered that three listed manufacturing firms of Shanghai

Zhongji Investment Holding Co., Ltd. (600634.SH), Gansu Shangfeng Cement Co., Ltd. (000672.SZ), and Harbin Electric Corporation Jiamusi Electric Machine Co., Ltd. (000922.SZ) had growth capacity indexes of 34.67, 12.03, and 7.66, respectively; raising the average growth potential of their respective industries (e.g., non-metallic mineral products) and regions (e.g., Gansu and Heilongjiang). Furthermore, the 3-year average core operating revenue growth rate of Jiangsu province was 0.1404, lower than the national average; however, in the national provincial-level division ranking, this was only 18th, showing that the average growth potential of listed manufacturing sector firms in Gansu is not very high.

6.1.5 Assessment of Innovativeness of China's Listed Manufacturing Sector Firms

In assessing the innovativeness of China's listed manufacturing sector firms, we investigate firm capacity for technological innovation. The primary indexes reflecting the capacity for technological innovation of listed manufacturing sector firms are investment in technological innovation, and output. Here, we draw on the methods of a great deal of literature (He et al. 2013; Wang et al. 2014b), utilizing intensity of R&D expenditure to assess the index of investment in technological innovation for listed manufacturing sector firms. Intensity of R&D expenditure refers to a comparison of total R&D spending and core operating revenue, and is used to reflect the value of spending on R&D as a proportion of core operating revenue. It is measured by the following equation: intensity of R&D expenditure = firm R&D expenditure/firm core operating revenue. Also utilized are indexes related to patents, as a measure of technological innovation output. This reflects the R&D capacities of firms and the degree of importance placed on innovation, as well as being an important index measuring a firm's development potential. This is an important factor for preserving a firm's market share, generating profits, and maintaining status within the industry; studies show that the generation of patents has the strongest effect on long-term industrial growth (Zhu et al. 2002). Patents containing technological plans are primarily classified as invention and utility model patents; because invention patents must pass the novelty and innovation tests by examiners, while utility model patents do not, invention patents have comparatively high technological content. Thus, in designing our patent index, this section uses the number of valid patent licenses divided by core operating revenue (in units of 100 million yuan) as an important index for assessing the output of technological innovation.

Below, we analyze the innovativeness of China's listed manufacturing sector firms in 2014 according to the two indexes of technological innovation input and output described above, from the perspectives of sector and region.

1. Level of investment in technological innovation

The index of intensity of R&D expenditure below is used to reflect the levels of investment in technological innovation of China's listed manufacturing sector firms.

(1) Sector analysis

According to sector divisions, the intensity of R&D expenditure for China's manufacturing sector is shown in Table 6.12 and Fig. 6.3.

In 2014, the average intensity of R&D expenditure for the 29 industries of the manufacturing sector was 2.96%, wherein the 214 listed firms in the manufacture of computers, communications, and other electronic equipment industry had the highest average intensity of R&D expenditure, at 7.69%. The 17 listed firms in processing of petroleum, coking, processing of nuclear fuel had the lowest average intensity of R&D expenditure, at 0.46%.

(2) Regional analysis

According to regional divisions, the intensity of R&D expenditure of listed Chinese manufacturing sector firms is shown in Table 6.13 and Fig. 6.4.

In 2014, the average intensity of R&D expenditure for the manufacturing sector in China's 31 provincial-level divisions was 3.54%. The average intensity of R&D expenditure for the 23 listed manufacturing sector firms in Shaanxi province was the highest, at 11.15% while the 8 listed firms in Qinghai province had the lowest average intensity of R&D expenditure, at 0.86%. Of all regions, the following 15 provincial-level divisions had average R&D expenditure intensity higher than the industrial average, showing that the listed manufacturing sector firms in these regions have relatively high levels of technological investment, and larger development potential: Shaanxi, Beijing, Hubei, Guangdong, Tianjin, Hainan, Fujian, Shanghai, Jilin, Jiangsu, Zhejiang, Henan, Hunan, Tibet, and Yunnan.

When viewing the combined conditions of investment in technological innovation for individual listed manufacturing sector firms, we observe that the levels of investment for firms, such as Caihong Group Corporation (600707.SH), were comparatively prominent, with Caihong Group's R&D expenditure intensity reaching 1.69%, and raising the average level of technological investment of its industry (i.e., manufacture of computers, communications, and other electronic equipment), and its region (i.e., Shaanxi). Furthermore, considering overall R&D expenditure intensity, the R&D expenditure intensity of Jiangsu province was 3.81%, ranking it 10th, and exceeding the national average. In addition, the overall volume of R&D expenditure for Jiangsu province was among the highest in China, illustrating that the listed firms in Jiangsu's manufacturing sector generally place great importance on R&D for technological innovation; Jiangsu's overall levels of technological R&D expenditure was among the highest in the country.

Table 6.12 Intensity of R&D expenditure for 1609 listed Chinese manufacturing sector firms in 2014

Rank	Industry	Number of firms	Intensity of R&D expenditure
1	Manufacture of computers, communications, and other electronic equipment	214	0.0769
2	Manufacture of measuring instruments	28	0.0675
3	Manufacture of chemical fibers	22	0.0553
4	Manufacture of special purpose machinery	136	0.0447
5	Manufacture of electrical machinery and equipment	169	0.0439
6	Manufacture of general purpose machinery	99	0.0408
7	Manufacture of articles for culture, education, and sport activity	10	0.036
8	Manufacture of railway, ship, aerospace, and other transportation equipment	32	0.0356
9	Utilization of waste resources	1	0.0331
10	Manufacture of leather, fur, feather, and related products and footware	6	0.032
11	Smelting and pressing of ferrous metals	30	0.031
12	Manufacture of paper and paper products	23	0.0293
13	Manufacture of automobiles	85	0.029
14	Manufacture of rubber and plastic	48	0.0279
15	Printing, reproduction of recording media	7	0.0269
16	Manufacture of non-metallic mineral products	74	0.0268
17	Manufacture of raw chemical materials and chemical products	170	0.0251
18	Manufacture of foods	5	0.0228
19	Manufacture of furniture	28	0.0228
20	Manufacture of medicines	145	0.022
21	Processing of food from agricultural products	38	0.02
22	Manufacture of metal products	40	0.0183
23	Manufacture of textile wearing and apparel	55	0.0179
24	Smelting and pressing of non-ferrous metals	30	0.0179
25	Other manufactures	17	0.016
26	Manufacture of textiles	9	0.0117
27	Manufacture of wine, drinks, and refined tea	32	0.0117
28		38	0.0117
29	Timber processing and wood manufacturing, bamboo, rattan, palm, and straw products	18	0.0046
	Overall average value		0.0296

Note Manipulations and calculations are based on information from 2014 annual reports from the Shanghai Stock Exchange (www.sse.com.cn), and Shenzhen Stock Exchange (www.szse.com.cn). In addition, of 1656 listed manufacturing sector firms in 2014, 47 firms had stocks beginning with * ST\ST\S; thus, these firms received special treatment and had unstable operations. These listed firms were first eliminated, and the analysis for the growth potential of listed manufacturing sector firms utilized data from 1609 firms

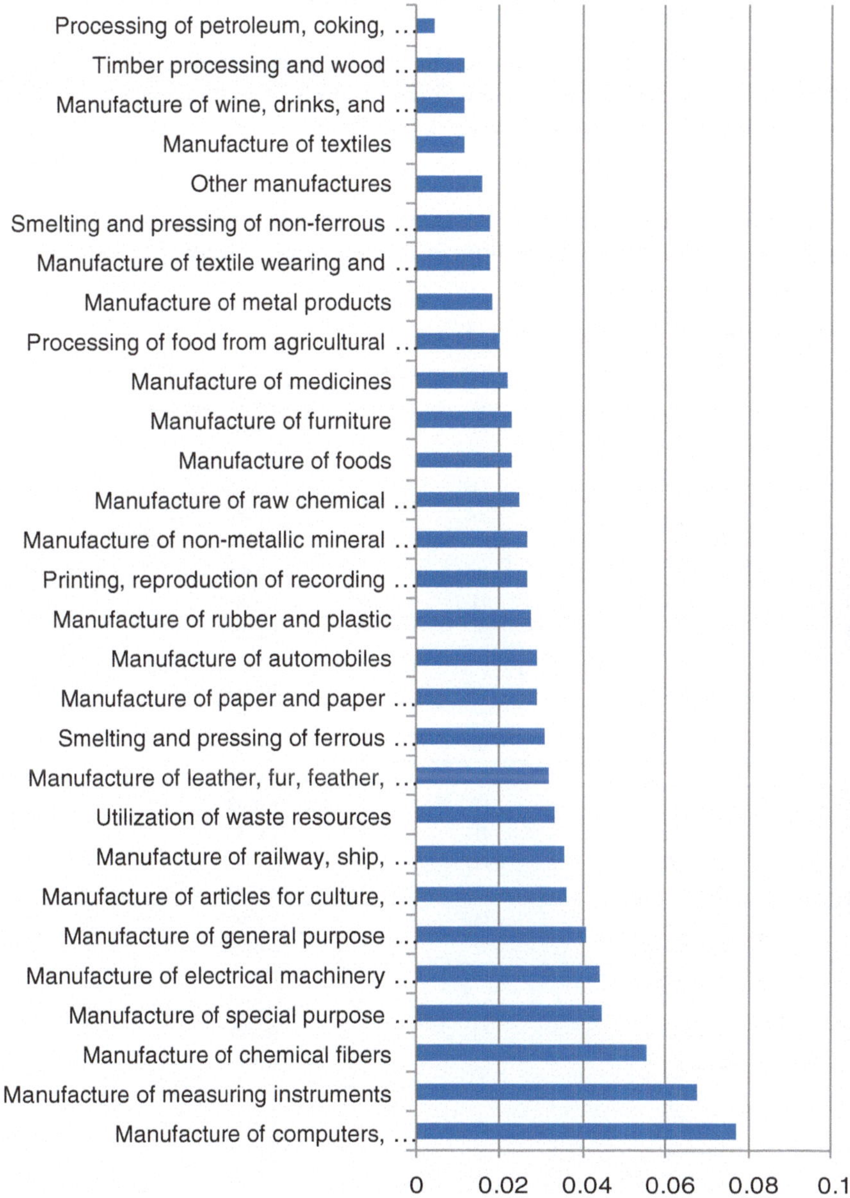

Fig. 6.3 Intensity of R&D expenditure by industry

2. Levels of technological innovation output

Next, the number of valid invention patents for each 100 million yuan in core operating revenue is used as an index to reflect levels of technological innovation output by listed Chinese manufacturing sector firms.

(1) Industry analysis

According to industry divisions, the levels of technological innovation output by listed Chinese manufacturing sector firms in 2014 are shown in Table 6.14 and Fig. 6.5.

Table 6.13 Intensity of R&D expenditure by region

Rank	Region	Number of firms	R&D expenditure intensity
1	Shaanxi	23	0.1115
2	Beijing	89	0.0669
3	Hubei	55	0.0635
4	Guangdong	245	0.0475
5	Tianjin	20	0.0458
6	Hainan	7	0.0414
7	Fujian	49	0.0409
8	Shanghai	95	0.0385
9	Jilin	26	0.0385
10	Jiangsu	190	0.0381
11	Zhejiang	198	0.0377
12	Henan	52	0.0374
13	Hunan	46	0.0366
14	Tibet	5	0.0362
15	Yunnan	17	0.0358
16	Liaoning	41	0.0353
17	Shandong	115	0.0343
18	Anhui	56	0.0307
19	Guizhou	15	0.0306
20	Chongqing	22	0.0303
21	Heilongjiang	18	0.0292
22	Hebei	36	0.0283
23	Sichuan	55	0.0275
24	Jiangxi	24	0.0236
25	Xinjiang	23	0.0184
26	Gansu	16	0.0182
27	Guangxi	21	0.0175
28	Inner Mongolia	16	0.0169
29	Ningxia	8	0.0166
30	Shanxi	18	0.0161
31	Qinghai	8	0.0086
	Total average value	–	0.0354

Note Manipulations and calculations are based on information from 2014 annual reports from the Shanghai Stock Exchange (www.sse.com.cn) and Shenzhen Stock Exchange (www.szse.com.cn). In addition, of 1656 listed manufacturing sector firms in 2014, 47 firms had stocks beginning with * ST\ST\S; thus, these firms received special treatment and had unstable operations. These listed firms were first eliminated, and this analysis of listed manufacturing sector firms utilized data from only 1609 firms

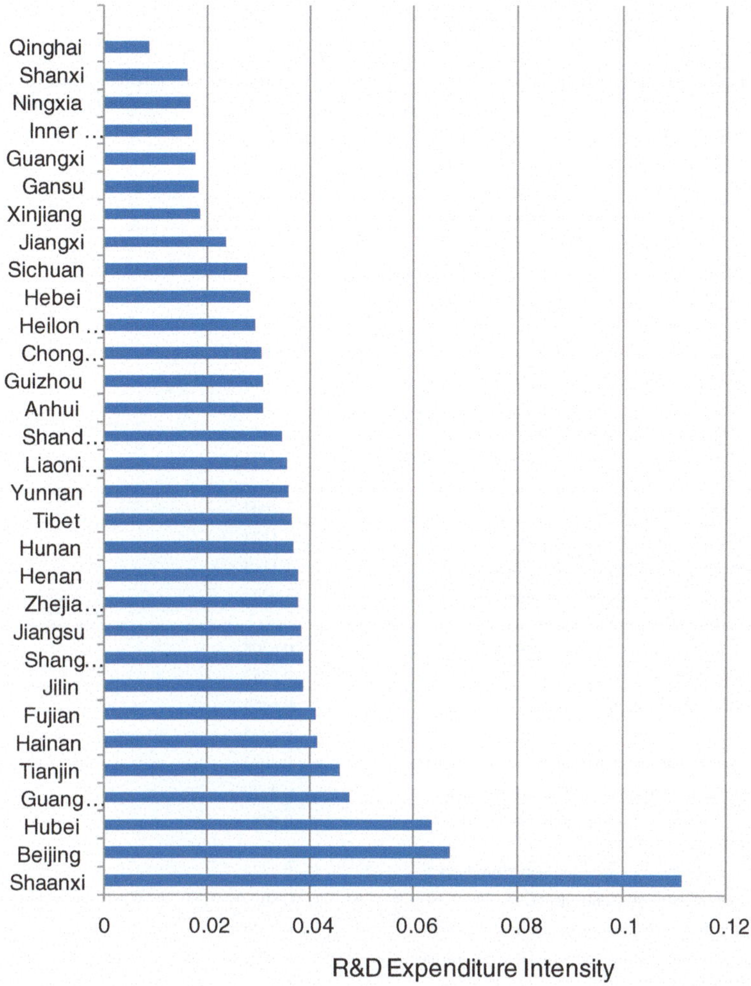

Fig. 6.4 Intensity of R&D expenditure by region

Made in China 2025 proposes another index for measuring innovativeness, namely, the number of valid invention patents per 100 million yuan in core operating revenue. Calculations reveal that the average technological innovation output of these 29 manufacturing sector industries in 2014 was 1.2836 patents per 100 million yuan. Herein, the industry with the highest average output of technological innovation was the manufacture of non-metallic mineral products, whose 73 listed firms had an average output of 4.9864 patents per 100 million yuan. The 17 listed firms in the processing of petroleum, coking, processing of nuclear fuel industry had the lowest average output, with 0.142 patents per 100 million yuan. The following 10 industries had higher than average numbers of invention patents per 100 million yuan in core operating revenue when compared with the entire

Table 6.14 Technological innovation output by 1609 listed Chinese manufacturing sector firms in 2014

Rank	Industry	Number of firms	Number of valid invention patents/core operating revenue (patents/100 million yuan)
1	Manufacture of non-metallic mineral products	74	4.9864
2	Manufacture of computers, communications, and other electronic equipment	214	4.1441
3	Manufacture of measuring instruments	28	3.7279
4	Manufacture of special purpose machinery	136	2.4751
5	Manufacture of electrical machinery and equipment	169	2.3935
6	Other manufactures	17	1.9829
7	Utilization of waste resources	1	1.8164
8	Manufacture of general purpose machinery	99	1.7615
9	Manufacture of railway, ship, aerospace, and other transportation equipment	32	1.6396
10	Manufacturing of raw chemical materials and chemical products	170	1.3271
11	Manufacture of foods	28	1.2002
12	Manufacture of metal products	40	1.1133
13	Manufacture of rubber and plastic	48	1.1114
14	Manufacture of leather, fur, feather, and related products and footware	6	1.024
15	Processing of food from agricultural products	38	0.8308
16	Manufacture of medicines	145	0.7352
17	Manufacture of automobiles	85	0.6854
18	Manufacture of textile wearing and apparel	30	0.5935
19	Manufacture of textiles	38	0.4873
20	Printing, reproduction of recording media	7	0.4539
21	Manufacture of articles for culture, education, and sport activity	10	0.4387
22	Manufacture of paper and paper products	23	0.3825
23	Manufacture of chemical fibers	22	0.363
24	Smelting and pressing of non-ferrous metals	55	0.3445
25	Timber processing and wood manufacturing, bamboo, rattan, palm, and straw products	9	0.3364
26	Manufacture of wine, drinks, and refined tea	32	0.2953
27	Manufacture of furniture	5	0.2601
28	Smelting and pressing of ferrous metals	30	0.1727
29	Processing of petroleum, coking, processing of nuclear fuel	18	0.142
	Total average value		1.2836

Note Manipulations and calculations are based on information from 2014 annual reports from the Shanghai Stock Exchange (www.sse.com.cn) and the Shenzhen Stock Exchange (www.szse.com.cn) as well as information collected from China's State Intellectual Property Office (www.pss-system.gov.cn). In addition, of 1656 listed manufacturing sector firms in 2014, 47 firms had stocks beginning with * ST\ST\S; thus, these firms received special treatment and had unstable operations. These listed firms were first eliminated, and thus, the analysis of listed manufacturing sector firms utilized data from only 1609 firms

sector, showing that the listed firms in these manufacturing industries have relatively high capacities for innovation. The manufacture of non-metallic mineral products, manufacture of computers, communications, and other electronic

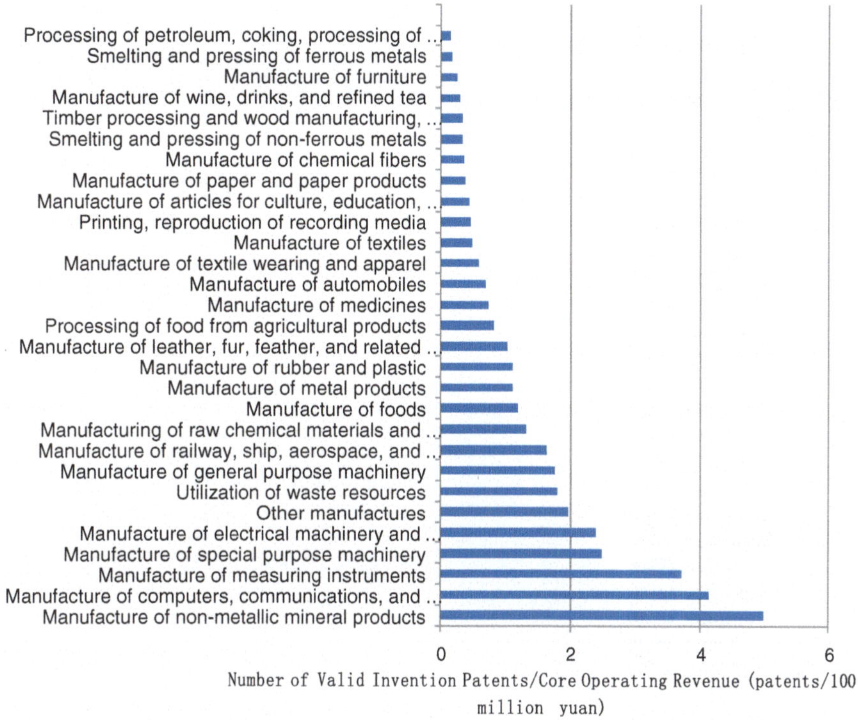

Fig. 6.5 Technological innovation output by industry

equipment, manufacture of measuring instruments, manufacture of special purpose machinery, manufacture of electrical machinery and equipment, other manufactures, utilization of waste resources, manufacture of general purpose machinery, manufacture of railway, ship, aerospace, and other transportation equipment, and manufacture of raw chemical materials and chemical products. The other 19 industries had lower than average levels of innovation output, showing that they lag significantly behind other industries in terms of numbers of invention patents; most industries have relatively low levels of technological innovation output (Fig. 6.5).

2. Regional analysis

According to regional distinctions, technological innovation output for China's manufacturing industry in 2014 is shown in Table 6.15.

In 2014, the average level of technological innovation output for the manufacturing sector of China's 31 provincial-level divisions was 2.1857 patents/100 million yuan. The average level of technological innovation output was highest for the 17 listed firms in Yunnan province, at 19.6332 patents/100 million yuan. The average level of technological innovation output for the following eight provincial-level divisions was higher than the average, showing that the capacity for technological innovation output of listed firms in these eight regions is comparatively strong: Yunnan, Shaanxi,

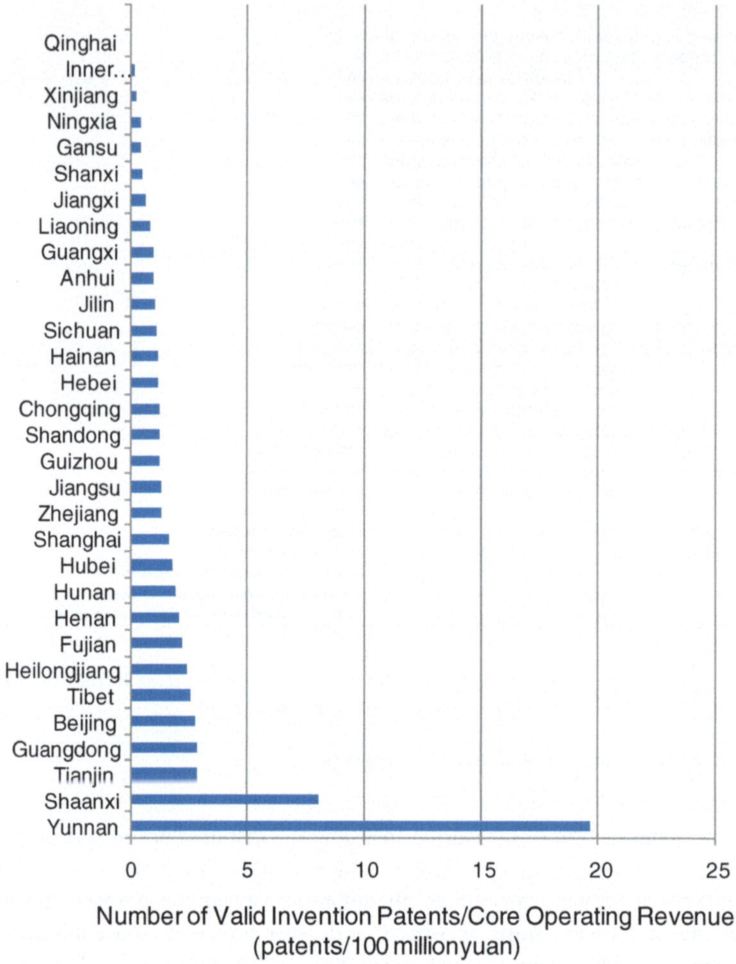

Fig. 6.6 Technological innovation output by region

Tianjin, Guangdong, Beijing, Tibet, Heilongjiang, and Fujian. For regions with very low levels of technological innovation output, such as Inner Mongolia and Qinghai, the capacity for technological innovation output is somewhat weak (Fig. 6.6).

For most of China's 31 provincial-level divisions, the average level of innovation output was comparatively reasonable; however, the levels of technological innovation output for Tianjin, Guangdong, and Beijing were relatively high, showing that the listed manufacturing sector firms in these three regions generally have stronger capacities for innovation output. Considering the combined conditions of innovation output capacities for individual listed manufacturing sector firms, the levels of innovation output by listed firms, such as Yunnan Bowin Technology Industry Co. Ltd. (600883.SH), and Caihong Group Corporation (600707.SH), stand out, with

Table 6.15 Technological innovation output by region

Rank	Region	Number of firms	Number of valid invention patents/core operating revenue (patents/100 million yuan)
1	Yunnan	17	19.6332
2	Shaanxi	23	8.0888
3	Tianjin	20	2.9042
4	Guangdong	245	2.8767
5	Beijing	89	2.8233
6	Tibet	5	2.5975
7	Heilongjiang	18	2.4393
8	Fujian	49	2.2791
9	Henan	52	2.1065
10	Hunan	46	1.9702
11	Hubei	55	1.8043
12	Shanghai	95	1.6767
13	Zhejiang	198	1.3461
14	Jiangsu	190	1.3278
15	Guizhou	15	1.2571
16	Shandong	115	1.2546
17	Chongqing	22	1.2518
18	Hebei	36	1.24
19	Hainan	7	1.1867
20	Sichuan	55	1.1593
21	Jilin	26	1.0894
22	Anhui	56	1.013
23	Guangxi	21	1.0062
24	Liaoning	41	0.874
25	Jiangxi	24	0.6906
26	Shanxi	18	0.5319
27	Gansu	16	0.4332
28	Ningxia	8	0.421
29	Xinjiang	23	0.2327
30	Inner Mongolia	16	0.1962
31	Qinghai	8	0.0442
	Total average value	–	2.1857

Note Manipulations and calculations are based on information from 2014 annual reports from the Shanghai Stock Exchange (www.sse.com.cn) and the Shenzhen Stock Exchange (www.szse.com.cn) as well as information collected from China's State Intellectual Property Office (www.pss-system.gov.cn). In addition, of 1656 listed manufacturing sector firms in 2014, 47 firms had stocks beginning with * ST\ST\S; thus, these firms received special treatment and had unstable operations. These listed firms were first eliminated, and thus, the analysis of listed manufacturing sector firms utilized data from only 1609 firms levels of invention patents/core operating revenue reaching 323.3575 patents/100 million yuan and 164.2040 patents/100 million yuan, respectively. Thus, these firms increase the average innovation output levels for their respective industries (i.e., manufacture of non-metallic mineral products; manufacture of computers, communications, and other electronic equipment), and their respective regions (i.e., Yunnan province and Shaanxi province). Furthermore, the listed manufacturing firms in Jiangsu province had an average level of technological innovation output lower than the national average (1.3278 patents/100 million yuan), ranking 14th in the country. This shows that in 2014, the listed manufacturing sector firms in Jiangsu province had a moderate capacity for innovation output.

6.2 Selecting the Most Respectable Listed Manufacturing Sector Firms

Each annual research report on the development of China's manufacturing sector (2004–2014) systematically analyzes and selects the most respectable listed manufacturing sector firms. For the connoted definition of "most respectable," the index system and selection process steps have been designed to evaluate and analyze the work of selecting such firms for 11 consecutive years. The design emphasizes that the essence of respectability lies in aspects of firms that include sense of social responsibility, corporate image, capacity for innovation, quality of management, and corporate performance. In addition, it reflects the recognition of value in each interest group in society, highlights the development direction of listed manufacturing sector firms toward the new industry model, and utilizes many indexes, stages, and steps in its assessment system. Furthermore, it uses a comprehensive assessment method integrating qualitative and quantitative aspects. In this way, it objectively and pragmatically selects the most respectable listed manufacturing sector firms, possessing positive theory and practical significance. This section continues to uphold this objective and pragmatic stance, on the basis of the theory and practice of previous reports' selection analysis, by selecting once again the most respectable listed manufacturing sector firms.

6.2.1 Basis for Selecting the Most Respectable Listed Manufacturing Sector Firms

The first selection in 2004 clearly stated that the connotation of "most respectable" is in line with the features of the new industry model. Overall efficiency is achieved, growth potential is large, both of which have features of the technological economy, such as knowledge and technology concentration; listed manufacturing sector firms that satisfy each societal concern are most respectable. Because the basic conditions of being listed require these manufacturing sector firms to be of a certain scale, maintain profitability, and comply with laws and regulations upon entering the market, they already possess a certain reputation, and receive basic recognition and respect. At the same time, the information of current industrial development conditions, demands for industrial development planning, and standards for relevant industry divisions reflects that the new industry model characteristics, such as capacity for economic innovation and technological innovation, have become the key factors in selecting the most respectable listed manufacturing sector firms.

1. Conditions of industrial development

With the economic development of industrial technology, high-tech development trends, such as informatization, smart technology, transitions toward networking, as well as green and high-end development, now comprise the key value orientation for manufacturing sector development. Some firms consistently produce innovations, such as smart machine tools, high-end robotics, radio frequency identification, sensors, 3D printing, and industrial automation. They have formed new "high-end manufacturing," "smart manufacturing," and "internet manufacturing" development formats that echo the links between R&D, production, and sales in the manufacturing sector. At the same time, the expanded application of these high-tech tools and their development modes include important development opportunities for all types of firms within the traditional manufacturing sector, at any stage of development, to transform and upgrade, thereby comprehensively raising the development level of firms and expanding their space for further development. In this way, it is possible to differentiate between types of firms with different levels of development, especially for listed manufacturing sector firms most closely integrated with industrial development trends. Thus, the selection of the most respectable listed manufacturing sector firms must adapt to new developments and changes, distinguish between firm development benchmarks of different industries, and thereby, promote faster and improved development of listed manufacturing sector firms.

2. Industrial development planning

Current industrial development plans have already established the mid- to long-term development focal points for manufacturing firms. Both the *Key High-tech Fields Supported by the State* and the *Twelfth Five-Year Plan for the National Development of Strategic Emerging Industries* establish plans for key developing-industry types beginning at the strategic level. These plans include the high-tech fields of energy saving and environmental protection; new-generation information technology; biotechnology; high-end equipment manufacturing; new energy; new materials; new energy vehicles; aeronautical and aerospace technology; and high-tech transformation of traditional industry. The plans stress the need to develop industries, in which technological breakthroughs and development requirements are of basic importance; these industries have large impacts on leading the way for overall and long-term socioeconomic development, are knowledge and technology intensive, have low material resource consumption, have significant development potential, and deliver good comprehensive benefits. In addition, *Made in China 2025* planned crucial components in the next 10 years for development of the manufacturing sector, namely, innovation incentives, quality orientation, green development, structural optimization, and talent-based characteristics to promote the integration of the internet and traditional industries, and to begin a new stage of development for China's manufacturing sector. These plans not only directly indicate key development orientation and high-tech development orientation, but also take into account issues of development and transformation faced by lagging industries, from the perspective of industrial development characteristics identified in the plans.

In summary, both industrial development trends and industrial development plans provide clear paths for considering selection of the most respectable listed manufacturing sector firms. That is, those high-tech industries or firms that meet development criteria are deserving of recognition and development priority. Furthermore, other industries or firms that consistently achieve transformation and upgrading in the direction of advanced technology merit a certain level of respect.

3. Distinguishing types of high-tech industries

For listed manufacturing sector firms, comparatively relevant classification standards include *Guidelines for the Industry Classification of Listed Companies* (2012 revision), *Guidelines for the Administration of the Recognition of Hi-tech Enterprises* (2008), and *High-tech Industry (Manufacturing Sector) Classification* (2013). All these include systems with formal classification standards, as well as a certain level of authority and practicality; thus, these standards can be used as the primary basis for distinguishing between listed manufacturing sector firms.

The *Guidelines for the Industry Classification of Listed Companies* (2012 revision) by China's Securities Regulatory Commission and the *High-Tech Industry* (*Manufacturing Sector*) *Classification* (2013) China's National Bureau of Statistics both establish the industrial classification of manufacturing sector firms based on the national economic industry classification standards (GB/T4754-2011). Herein, the *High-tech Industry* (*Manufacturing Sector*) *Classification* (2013) clearly distinguishes subdivisions of the high-tech manufacturing sector, including six main groups, 29 categories, and 42 subcategories. The six main groups are as follows: the manufacture of medicines (code #27), the manufacture of aircraft, spacecraft, and related equipment (code #374 + 4343), the manufacture of electronic equipment and communications equipment (code #39 − 391 + 3562 + 3832 + 3841), the manufacture of computers and office equipment (code #391 + 3474 + 3475), the manufacture of medical equipment and measuring instruments (code #358 + 401 + 402 + 409 + 4041), and the manufacture of photographic equipment (code #2664). The industry classification standards utilize intensity of R&D expenditure (i.e., R&D expenditure as a proportion of core operating revenue), and designate manufacturing sector industries with higher index values as high-tech industries. The industrial scope is basically consistent with, albeit more clearly defined than, the industries identified in *Key High-tech Fields Supported by the State* and strategic emerging industries; in addition, the delineation of scope has specific corresponding relationships with the national economic industry classification standards. The *Guidelines for the Industry Classification of Listed Companies* (2012 revision) utilized core operating revenue or core profit margin to determine the main group of listed firms (code #13–43); where subcategories are uncertain or inconsistent with high-tech industrial divisions, it is difficult to determine the subcategory of listed manufacturing sector firms, and therefore, it is difficult to judge which high-tech industry group in which listed manufacturing sector firms should be placed.

With regard to the problem of categorizing high-tech firms, the *Guidelines for the Administration of the Recognition of Hi-tech Enterprises* (2008) prepared by China's Ministry of Science and Technology, Ministry of Finance, and State Administration of Taxation remedied this problem with a clear certification standard system and certification structure. In these guidelines, high-tech firms are identified within "*Key High-tech Fields Supported by the State*," and further R&D is conducted and technological achievements are commercialized, forming core intellectual property for firms and allowing for business activities to be conducted on this basis. Resident firms of more than 1 year (excluding Hong Kong, Macau, and Taiwan) can be registered. The specific identification criteria indexes consist of patent information, technical personnel ratio, intensity of R&D expenditure, volume of independent intellectual property rights (patent-related index), level of organization and management of corporate R&D (including project approval, investment accounting system, R&D activities, structural institutions, and facilities), capacity for commercializing research findings, as well as the growth potential of sales and total assets. This emphasizes that firms that simply produce and process high-tech products but do not engage in R&D cannot be considered high-tech firms.

Generally speaking, according to the collation and analysis of the abovementioned documents, development of listed manufacturing sector firms is closely related to China's national socioeconomic development. The most respectable listed manufacturing sector firms should comply with national strategic industrial development requirements and their development should be oriented toward the new industrial model. In addition, they should possess knowledge, technology intensity, and significant growth potential, and deliver good comprehensive benefits; such firms deserve the respect of society, the public, and those within and outside their respective industries.

Based on the developing connotation of most respectable listed manufacturing sector firms, industrial development conditions, industrial development plans, and information on group differentiation standards, and in accordance with the requirements and principles of selection feasibility, our selection of most respectable listed manufacturing firms continues to employ the selection steps used in the preceding year. In other words, this involves carrying out a two-stage system: a preliminary screening and the ranking of comprehensive assessments. During this process, close attention is paid to such factors as growth potential for economies of scale and innovativeness, which reflect the value orientation of the most respectable listed manufacturing sector firms. For this reason, listed manufacturing firms that fail to perform their duties of social responsibility are eliminated during the preliminary screening process (e.g., specially treated firms, illegal businesses, firms with audit results indicating non-standard reservations of opinion, and firms with negative values for growth potential in economies of scale). In addition, we are inclined to consider a greater number of high-tech firms as candidates so that firms with good growth potential in economies of scale and innovativeness receive greater respect.

6.2.2 Design of Standards for Selecting Listed Manufacturing Sector Firms Most Deserving of Respect

Our selection of the most respectable listed manufacturing sector firms draws from the concept of new industry model development used in previous years' reports, combines the unique characteristics of manufacturing sector firms, and utilizes comprehensive subjective and quantitative methods. Borrowing from the standards used in previous reports, our ultimate goal is to select listed manufacturing sector firms that can serve as benchmarks for most respectable firms, thereby establishing suitable value benchmarks for firms' development in the new industrial model.

Thus, in designing standards, we utilize the analysis guidelines of the Wuli–Shili–Renli system (Gu 2011), which is a conceptualization of human life as a dynamic web of human relationships to things, conditions, and rules; hidden interaction between people and the world; and ethical and social values. We combine the selection experiences and related indicators of previous reports. Strict product quality standards are proposed as entry barriers, and conditions, such as environmental protection and tax payment, are observed by all listed manufacturing sector firms, especially high-tech ones. This is also true of punishment and exit mechanisms designed within institutions in order to deal with firms that violate laws and fail to meet standards. Because of these factors, all listed manufacturing firms complying with standards and institutions deserve a basic level of respect. In assuming other social responsibilities and duties, and because listed manufacturing firms are the most active economic entities in the market economy, the benefits of their behavior are reflected in financial indicators. This can reveal relative advantages for assessing the most respectable listed manufacturing firms. Thus, the final three-tier selection index for the most respectable listed manufacturing sector firms is outlined below. The physical index layer reflects basic real indicators, such as total assets, core operating revenue, net profits, R&D expenditure, and numbers of valid invention patents. The affairs index layer is primarily reflected by efficiency indexes; these include core operating profit margins, average 3-year core operating revenue growth rates, total R&D expenditure as a proportion of core operating revenue, numbers of valid invention patents as a proportion of core operating revenue (measured in 100 million yuan), and utilization of new industrial model patents as a proportion of core operating revenue (measured in 100 million yuan). The personal index layer combines expert assessments, or combinations of product quality, environmental protection, and other subjective quantitative methods for measuring social responsibility in order to carry out reasonable amendments to results during the preliminary screening stage and the overall assessment stage.

6.2.3 *Method and Process for Selecting Listed Manufacturing Sector Firms Most Deserving of Respect*

Based on the above analysis, our selection of the most respectable listed manufacturing sector firms comprises outstanding firms from a group of good candidates; it is appropriate to use a two-stage system here.

Stage 1: Selecting candidate firms for listed manufacturing sector firms "most deserving of respect"

In this stage, 1656 manufacturing sector firms listed before December 31, 2014 are included. This consists of 1646 firms with A-type stocks, and 10 firms with B-type stocks.

First, listed manufacturing firms that have failed to perform their social responsibility duties are filtered out (e.g., those that have been specially treated, illegal businesses, and firms with audit results indicating non-standard reservations of opinion). After this process, 1435 firms remain.

Next, we consider candidate firms that maintain good economies of scale and growth potential, filtering out those firms with negative values for net profits in the past 3 years, core operating revenue growth rates, or net profit growth rates. After this process, 635 firms remain.

Next, we consider high-tech firms of larger scale and with stronger innovativeness than other firms in the same industry; these firms are more recognizable as being "most respectable." High-tech firms with core operating revenues ranking in the top 15 of their respective industries are selected for the following industries: manufacture of medicines, manufacture of computers, communications, and other electronic equipment, manufacture of measuring instruments, manufacture of railway, ship, aerospace, and other transportation equipment, and manufacture of special and general purpose machinery. For a balanced consideration of each industry, listed manufacturing firms ranking in the top 10 of their respective industries in terms of core operating revenue are selected for all remaining industries.

The final sample of 227 listed manufacturing sector firms undergoes firm ranking analysis in the second stage.

Stage 2 Analyzing candidate firms to generate a priority ranking of listed manufacturing industry firms "most deserving of respect"
In this stage, the 227 listed manufacturing sector firms selected as described above undergo a comprehensive assessment analysis, primarily according to the following economic innovation capacity indicators and technological innovation capacity indicators: core operating profit margin, average 3-year core operating revenue growth rate, R&D expenditure as a proportion of core operating revenue, invention patents as a proportion of core operating revenue (in units of 100 million yuan), number of new industry model patents utilized as a proportion of core operating revenue (in units of 100 million yuan), R&D expenditure, number of valid invention patents, net profits, core operating revenue, and total assets. Finally, a shortlist of the 50 "most respectable" listed manufacturing sector firms was selected, and is shown in Table 6.16.

6.3 Assessment Analysis for Listed Manufacturing Sector Firms Most Deserving of Respect

The total core operating revenue of the 1656 listed Chinese manufacturing sector firms was about 9.36 trillion yuan, while the combined core operating revenue of the top 50 firms was about 2.1 trillion yuan, or 22.48% of the total core operating revenue of all 1656 firms. Considering industry divisions, the top 50 firms were concentrated in the following industries: manufacture of computers, communications, and other electronic equipment (9 firms), manufacture of measuring instruments (6 firms), and manufacture of medicines (6 firms). In terms of regional divisions, the top 50 firms were concentrated in Guangdong (13 firms), Beijing (6 firms), Zhejiang (6 firms), Henan (3 firms), Shanghai (3 firms), Jiangsu (2 firms), and Shandong (2 firms). This information is displayed in Figs. 6.7 and 6.8.

Next, we conduct an assessment of the 50 top ranked most respectable firms.

1. SAIC Motor (600104.SH)

SAIC Motor (SAIC Motor Corporation Limited) was established on November 20, 1997. The company primarily engages in the production and sale of motor vehicles, mechanical equipment, and parts and components for cars, motorcycles, and tractors. In 2010, the company ranked 8th in the international automobile industry in terms of sales volumes, and was the first group to exceed sales of 3 million vehicles annually within China. In 2014, the company sold 24.082 million vehicles in China, an increase of 8.3% over the previous year. At the same time, it established an online-to-offline electronic business platform, and together with Alibaba Group engaged in R&D for internet cars and promoted online automotive financial services. Furthermore, SAIC Motor signed a technology cooperation agreement with Volkswagen AG in Berlin to upgrade and transform Shanghai Volkswagen's Anting production base by advancing cooperation for pure electric

Table 6.16 Ranking of the 50 most respectable listed manufacturing industry firms in 2014

Rank	Stock symbol	Stock abbreviation	Industry	Provincial-level division
1	600104.SH	SAIC Motor	Manufacture of automobiles	Shanghai
2	000651.SZ	Gree Electric	Manufacture of electrical machinery and equipment	Guangdong
3	000333.SZ	Midea Group	Manufacture of electrical machinery and equipment	Guangdong
4	601299.SH	China CNR Corporation	Manufacture of railway, ship, aerospace, and other transportation equipment	Beijing
5	601766.SH	China CSR Corporation	Manufacture of railway, ship, aerospace, and other transportation equipment	Beijing
6	000625.SZ	Changan Automobile	Manufacture of automobiles	Chongqing
7	000100.SZ	TCL Corporation	Manufacture of computers, communications, and other electronic equipment	Guangdong
8	600519.SH	Kweichow Moutai Company	Manufacture of wine, drinks, and refined tea	Guizhou
9	600535.SH	Tasly Pharmaceutical	Manufacture of medicines	Guangdong
10	000039.SZ	China International Marine Containers	Manufacture of metal products	Guangdong
11	002415.SZ	Hikvision	Manufacture of computers, communications, and other electronic equipment	Zhejiang
12	600887.SH	Yili Group	Manufacture of foods	Inner Mongolia
13	600585.SH	Conch Cement	Manufacture of non-metallic mineral products	Anhui
14	600398.SH	Heilan Home	Manufacture of textile wearing and apparel	Jiangsu
15	300007.SZ	Hanwei Electronics	Manufacture of measuring instruments	Henan
16	002008.SZ	Han's Laser	Manufacture of special purpose machinery	Guangdong
17	600276.SH	Hengrui Medicine	Manufacture of medicines	Jiangsu
18	000977.SZ	Inspur	Manufacture of computers, communications, and other electronic equipment	Shandong

(continued)

Table 6.16 (continued)

Rank	Stock symbol	Stock abbreviation	Industry	Provincial-level division
19	300338.SZ	Kaiyuan Instruments	Manufacture of measuring instruments	Hunan
20	300203.SZ	Focused Photonics	Manufacture of measuring instruments	Zhejiang
21	002152.SZ	GRG Banking	Manufacture of general purpose machinery	Guangdong
22	600089.SH	TBEA	Manufacture of electrical machinery and equipment	Xinjiang
23	601238.SH	GAC Group	Manufacture of automobiles	Guangdong
24	002241.SZ	GoerTek	Manufacture of computers, communications, and other electronic equipment	Shandong
25	002456.SZ	O-Film	Manufacture of computers, communications, and other electronic equipment	Guangdong
26	300286.SZ	Acrel	Manufacture of measuring instruments	Shanghai
27	600100.SH	Tsinghua Tongfang	Manufacture of computers, communications, and other electronic equipment	Beijing
28	601179.SH	XD Group	Manufacture of electrical machinery and equipment	Shaanxi
29	300259.SZ	Suntront	Manufacture of measuring instruments	Henan
30	600332.SH	Baiyunshan Pharmaceutical	Manufacture of medicines	Guangdong
31	002236.SZ	Dahua Technology	Manufacture of computers, communications, and other electronic equipment	Zhejiang
32	600660.SH	Fuyao Group	Manufacture of non-metallic mineral products	Fujian
33	600005.SH	Wuhan Iron and Steel	Smelting and pressing of ferrous metals	Hubei
34	000761.SZ	Bengang Steel Plates	Smelting and pressing of ferrous metals	Liaoning
35	002202.SZ	Goldwind	Manufacture of general purpose machinery	Xinjiang
36	600267.SH	Hisun Pharmaceutical	Manufacture of medicines	Zhejiang
37	002050.SZ	Sanhua	Manufacture of general purpose machinery	Zhejiang
38	000550.SZ	Jiangling Motors	Manufacture of automobiles	Jiangxi

(continued)

Table 6.16 (continued)

Rank	Stock symbol	Stock abbreviation	Industry	Provincial-level division
39	002385.SZ	Da Bei Nong Group	Processing of food from agricultural products	Beijing
40	000513.SZ	Livzon Group	Manufacture of medicines	Guangdong
41	600893.SH	AVIC	Manufacture of railway, ship, aerospace, and other transportation equipment	Shaanxi
42	000016.SZ	Konka Group A	Manufacture of computers, communications, and other electronic equipment	Guangdong
43	600066.SH	Yutong Bus	Manufacture of automobiles	Henan
44	002121.SZ	Clou Electronics	Manufacture of measuring instruments	Guangdong
45	600597.SH	Bright Dairy	Manufacture of foods	Shanghai
46	000050.SZ	Tianma Micro-electronics	Manufacture of computers, communications, and other electronic equipment	Guangdong
47	601877.SH	CHINT Electric	Manufacture of electrical machinery and equipment	Zhejiang
48	600329.SH	Zhongxin Pharmaceuticals	Manufacture of medicines	Tianjin
49	300003.SZ	Lepu Medical	Manufacture of special purpose machinery	Beijing
50	600582.SH	Tiandi Science and Technology	Manufacture of special purpose machinery	Beijing

Note Manipulations and calculations are based on information from 2014 annual reports from the Shanghai Stock Exchange (www.sse.com.cn) and Shenzhen Stock Exchange (www.szse.com.cn) as well as information collected from China's State Intellectual Property Office (www.pss-system.gov.cn)

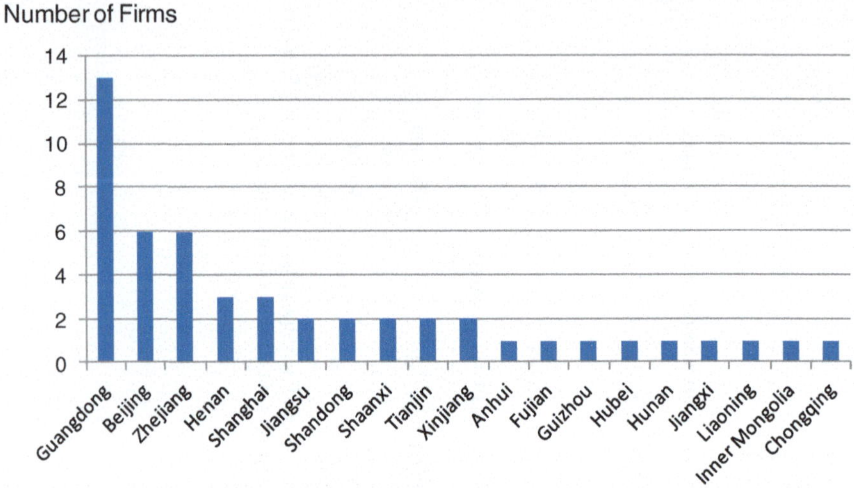

Fig. 6.7 Regional distribution of 50 listed manufacturing industry firms

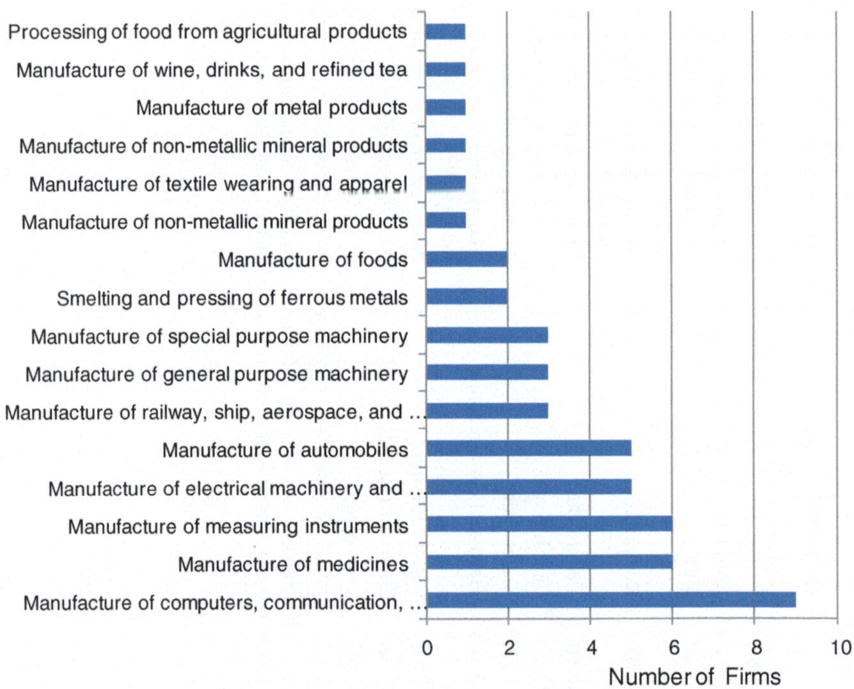

Fig. 6.8 Industry distribution of 50 listed manufacturing sector firms

vehicle development. In this study, SAIC Motor ranked first out of 50 firms in terms of total assets, core operating revenue, net profits, and R&D expenditure.

2. Gree Electric (000651.SZ)

Gree Electric (Gree Electric Appliances Inc.) was established on December 13, 1989. Its core operations are the production and sale of air conditioners, export of proprietary air conditioners, and import and export of related components. The company independently developed the R290 natural refrigerant air-conditioning unit, for which it earned a VDE certificate, a German national electrical safety certification. In addition, 29 of Gree's centrifugal and screw chillers passed certification by the American Air-Conditioning, Heating and Refrigeration Institute, making it the only Chinese firm currently producing air conditioners to have received this certification. The company has the world's largest air conditioner R&D Center, with two national technological research centers, one provincial-level key laboratory, six research institutes, and 52 research centers. In 2014, Gree Electric won China's State Scientific and Technological Progress Award, second class. Following the trend of smart home appliances, the company released a smart green-home system, utilizing photovoltaic and multi-connectivity technology, internet communication technology, preventative security technology, automatic control technology, and audio and video technology to integrate home appliances, thereby creating a high-efficiency energy-saving environmentally friendly energy management system, and a fast and convenient home task-management system. In this study of 50 firms, Gree Electric ranked near the top in terms of total assets, core operating revenue, and net profits; it ranked first in number of valid invention patents.

3. Midea Group (000333.SZ)

Midea Group (Midea Group Co., Ltd.) was established on April 7, 2000. It is a large-scale comprehensive enterprise group focused on the manufacture of home electronics, although it is also involved in electric motors and logistics. The company is divided into four business segments: large home appliances, small home appliances, electric motors, and logistics. In 2014, Midea Group won China's 16th National Patent Gold Prize, as well as the China Patent Excellence Award; it was the only air-conditioning firm to receive a gold medal, or to receive a second-class State Scientific and Technological Progress Award. It became China's first company to win a State Scientific and Technological Progress Award for the application of air-conditioning energy-saving technology, and won the US industrial design award, the Industrial Design Excellence Award. It also won the German iF Product Design Award, the Red Dot Design Concept Award, and the Chinese home appliance Aipu Lan award. The company has released the M-Smart home strategy, established a smart home research center, and rapidly promoted the establishment of a connective network for all classes of their home appliances. In 2014, Midea signed a strategic cooperation framework agreement with Xiaomi to develop a smart-home industrial chain and mobile internet supply. In 2014, the company achieved core operating revenue of 142.311 billion yuan, an increase of 17.36% over the previous year. In our study, the company ranked highly in the four indexes of total assets, net profits, R&D expenditure, and number of valid invention patents.

4. China CNR (601299.SH)

China CNR (China CNR Corporation Ltd.) was established on June 26, 2008. It is primarily engaged in the R&D and production of railroad locomotives, urban railroad, and electric equipment, as well as the provision of technological repair services and equipment leasing. Its primary products include railroad transportation, general electrical machinery, and modern services. It is the world leader in the railroad transportation equipment production industry, and ranks highly internationally in terms of scale of operation, core technological R&D, and production processes. According to a 2012 report released by the international railroad industry authority and strategic consulting firm, Germany's SCI Verkehr, China CNR ranked first in the global rolling stock industry 3 years in a row, and in 2011, retained its position as the world's top producer of railroad transportation equipment. The company currently has 4 national-level R&D organizations, 9 national-level firm technology centers, 27 provincial-level key R&D organizations, 3 international R&D centers, 9 specialized technology R&D centers, 5 postgraduate workstations, and 5 academic expert workstations. It has become a core firm for China's independent innovation in railroad transportation equipment, having independently developed the NECT traction electric drive system and network control system for loading operations. This shows that China CNR has completely mastered the design and production technology for high-power AC traction-drive systems and network-control systems. Its locomotive brake system has completed car-loading assessments, and has broken the foreign monopoly on locomotive braking technology. In 2014, China CNR's overseas operations grew rapidly, with export contracts totaling 2.994 billion US dollars, an increase of 73% from the previous year. The company's 2014 annual core operating revenue was 104.29 billion yuan, an increase of 7.3% from the previous year; profits reached 6.61 billion yuan, an increase of 29.7% from the previous year. In our study, the company ranked highly in terms of total assets, core operating profits, and net profits, and was second in terms of R&D expenditure.

5. China CSR (601776.SH)

China CSR (China South Locomotive and Rolling Stock Corporation Limited) was established on December 28, 2007. It is a transportation machinery and equipment manufacturing firm, primarily engaging in the R&D, production, sale, repair, and leasing of railroad locomotives, passenger cars, freight cars, electric multiple units (EMU), urban subway cars, and key components. The brands "South Locomotive" and "CSR" have a high degree of visibility and recognition in the international and domestic railroad transportation equipment manufacturing industry; the company actively explores overseas markets, and its newly signed overseas orders amount to 3.76 billion US dollars in 2014, a record high and increase of 68.6%. Its bid for a South African electric locomotive project worth more than 20 billion US dollars was accepted; this is China's largest ever export order for high-end railroad transportation equipment. The company spent EUR 290 million acquiring the 100-year-old BOGE company from Germany's ZF Group, helping China CSR to gradually become a world leader, as well as China's largest firm for the supply of automobile vibration and noise reduction products. Standardized Chinese EMUs, new model cold and sandstorm resistant EMUs, China's first Maglev train, China's first smart high-speed train, and the experimental train CRH380AM are among the outstanding products resulting from the company's R&D. In December 2014, a board meeting passed a motion leading to a merger of China CSR and China CNR. In 2014, the firm's annual core operating revenue was 119.724 billion yuan, an increase of 20.48% over the previous year. The parent company achieved net profit of 5.315 billion yuan in 2014, an increase of 27.61% from the previous year. In our study, China CSR ranked highly in terms of total assets, core operating revenue, and net profits, and ranked third in terms of R&D expenditure.

6. Changan Automobile (000625.SZ)

Changan Automobile (Changan Automobile Group Co. Ltd.) was established on October 31, 1996. It is a comprehensive firm engaged in the production and sale of automobiles, automobile engine lines, and auxiliary components. Its own car brand, the Benben Mini, was China's first selection for the energy-saving car promotion catalogue, the "energy-saving product serving people's work," and went on to win other awards, such as the 2010 Best New Chinese Car Build Quality Award, Own-Brand Sales Innovativeness Award, and Best Micro-Car. The Changan Taurustar won such awards as the Best Microvan of the Year, and Microvan of the Year. In 2010, Changan won the "National Innovation Award," following a comprehensive assessment by the Ministry of Science and Technology, the State-owned Assets Supervision and Administration Commission, and the All-China Federation of Trade Unions. Changan was awarded the second-class State Scientific and Technological Progress Award for a global collaboration project for independent development. Changan was awarded the Chinese automobile industry's top science and technology prize, the only winner of the award in the automobile industry, for an R&D project and industrialized application of passenger cars, and for a project on an electric car full-vehicle control method. In 2014, Changan had a market share of about 10.8% in China's automobile market, an increase of 1.2% from the previous year; Changan's sales continue to rank in the top four of China's automobile industry. Changan had annual core operating revenue of 52.913 billion yuan in 2014, an increase of 35.18% from the previous year, and net profits of 7.561 billion yuan, an increase of 124.46%. In our study, for the three indexes of net profit growth rate, net profit, and R&D expenditure, Changan ranked highly, and ranked third in terms of number of valid invention patents.

7. TCL Corporation (000100.SZ)

TCL Corporation (TCL Corporation Ltd.) was established on April 19, 2002. It is a corporate group with a core involvement in multimedia electronics, mobile communications, and digital electronics, including home electronics, core components (modules, chips, displays, and energy components), lighting, and cultural industries. The firm's business-wide smart television platform achieves a unified and open platform, and was awarded the 2010 Chinese Audio and Video Industry Product Innovation Award. At the 44th international Consumer Electronics Show in 2011, the TCL brand was ranked 25th in the Top 50 Global Consumer Electronics. It was also ranked the 6th top global television brand, and was selected for 3 consecutive years as the 5th "TOP 10 Chinese Consumer Electronics Brand," and in the Top 20 Global Television Brands. The company's awards also include the Annual Global Flat Screen Television Quality Award, and the Annual Best All-Round 3D Television Award. The company's cellphone, Alcatel OT-808, was awarded the Red Star Award at the Chinese Industrial Design "Oscars," and jointly with the Alcatel OT-806, won the Red Cotton Award in the Original Product Competition. In 2014, the corporate group had operating revenue of 23.31 billion yuan, an increase of 10.19% from the previous year. In the same period, net profit grew by 5.81% to 982 million yuan, and net profits attributable to listed company shareholders grew by 2.61%, rising to 754 million yuan. In our study, TCL Corporation ranked highly in terms of total assets, core operating revenues, net profits, R&D expenditure, and number of valid invention patents.

8. Kweichow Moutai Company (600519.SH)

Kweichow Moutai Company (Kweichow Moutai Company Limited) was established on November 20, 1999, and is an iconic symbol of Chinese alcohol companies. It is primarily engaged in the production and sale of one of the world's three largest alcohol brands, Moutai Baijiu. It also produces and sells beverages, food products, and packaging materials, as well as engaging in R&D on anti-counterfeiting technologies, and information industry-related products. The company's annual production volume has already exceeded 10,000 tons, and its unique age-tiered product-development model has formed a three-tier product series, low–mid–high, for the selection of 70+ products. Among the entire market, Moutai is the leader in the Baijiu industry, and is the only national brand that produces green food products, organic food products, and national geographical indication (GI) products that are part of China's intangible cultural heritage and were selected among the product list for Europe and China's mutual recognition of GI. In addition, the company continues to strengthen its environmental management and ecological development, and has committed to spending a total of 500 million yuan over 10 years on environmental protection and management of the Chishui River, beginning with 50 million yuan in 2014. In 2014, the company achieved operating revenue of 31.574 billion yuan, a growth of 2.11% from the previous year. It had net profit of 15.35 billion yuan, an increase of 1.41%. Of the 50 firms selected in our study, Kweichow Moutai ranked first and second in terms of core operating profit margins and net profits, respectively. It ranked highly in terms of total assets.

9. Tasly Pharmaceutical (600535.SH)

Tasly Pharmaceutical (Tasly Pharmaceutical Group Co., Ltd.) was established on April 30, 1998, and is a high-tech firm principally engaged in the pharmaceutical manufacturing industry. It is one of Tianjin's key supporting corporate groups, and is primarily engaged in the research, cultivation, extraction, and sale of modern Chinese medicine and chemical medicines. The firm is a national-level enterprise technology center, and has supported the Ninth Five-Year Plan, Tenth Five-Year Plan, Eleventh Five-Year Plan, the 863 Program, the 973 Program, and a key R&D scientific research project for national key new drug development. The company independently developed a large-scale automated drop-pill production line in line with advanced international standards, and is committed to the creation of a unified and modern Chinese medicine industrial chain conforming to serial standards. It was the first Chinese company to establish Good Engineering Practices for modern Chinese medicine and plant extracts, and established a modern Chinese medicine production base that passed national pharmaceutical Good Management Practice (GMP) certification. With its "high speed magnetic levitation pill machine" at the core of its advanced and modern Chinese medicine production system, it attained EU GMP certification, which formed the basis for the introduction of Chinese medicine to the EU market. The firm was honored as one of the CCTV Finance's 50 Index Top 10 Corporate Governance Constituent Stocks, and among the Top 50 Listed Asia Pacific Companies. In 2014, Tasly's operating revenue was 12.567 billion yuan, a growth rate of 13.24% from the previous year. In 2014, Tasly achieved net profit attributable to listed company shareholders, excluding non-recurring gains and losses, of 1.346 billion yuan, an increase of 31.91% from the previous year.

10. China International Marine Containers (000039.SZ)

China International Marine Containers (China International Marine Containers Group Co., Ltd.) was established on January 14, 1980. Its core products include shipping containers, road transport vehicles, energy and chemical products, liquid food equipment, marine engineering products, airport equipment, and logistics services and equipment. The company is one of China's earliest specialized shipping container production firms, as well as one of China's earliest foreign joint ventures. The firm is now the world's largest-scale and most diverse shipping container production group. In 2014, in order to ensure excellent integration of Zeigler, the group signed an asset transfer agreement with China Fire Safety Enterprise Group Ltd. When this agreement is finalized, a good foundation will be laid for fire and safety operations for the group. The group's technology center is a national-level firm technology center; the group has established 26 group-level technology centers, 5 of which research centers and 21 branches. In 2014, the firm achieved operating revenue of 70.071 billion yuan, an increase of 21.07% from the previous year. The firm achieved net profit attributable to parent company shareholders of 2.478 billion yuan, an increase of 13.64%, and earnings per share of 0.93 yuan, an increase of 13.41%. In this study, the firm ranked highly for the indexes of total assets and core operating revenue, and was fifth in terms of the number of valid invention patents.

11. Hikvision (002415.SZ)

Hikvision (Hangzhou Hikvision Digital Technology Co., Ltd.) was established on November 30, 2001. It is a high-tech firm specialized in the R&D, production, and sale of video surveillance security products, including data storage and processing equipment, such as hard-disk recorders and audio and video codec cards, as well as audio and video information acquisition and processing equipment, such as surveillance cameras, surveillance ball cameras, and video servers. It is China's leading firm in the audio and video surveillance industry; it has achieved the highest sales in China for many consecutive years. The company is involved in R&D for "Key Technologies for Active Object Mass Storage Systems," and won a second-class State Scientific and Technological Progress Award in 2014. In addition, through cooperation with internet companies and large enterprises, such as BAT, Letv, and JD.com, Hikvision developed an internet platform for the establishment of win–win cooperative ecosystems to accelerate the layout of internet business. In 2014, the firm achieved operating revenue of 17.233 billion yuan, an increase of 60.37% from the previous year, and net profit of 5.206 billion yuan, an increase of 53.76%. In this study, it ranked highly for the three indexes of net profits, core operating revenue growth rate, and R&D expenditure as a proportion of core operating revenue.

12. Yili Group (600887.SH)

Yili Group (Inner Mongolia Yili Industrial Group Company Limited) was established on June 4, 1993. It is primarily engaged in the production of liquid milk, milk products, and mixed feed. It is one of China's 520 key industrial enterprises, and is among eight Chinese ministries' first selection of 151 leading firms for agricultural industrialization. Its more than 1000 products have passed the national green-food product development center's green food product certification. According to 2014 sales statistics jointly released by the Chinese Chamber of Commerce and the Chinese Commercial Information Center, Yili has the largest market share in national markets for comprehensive milk products, milk powder, cold beverages, liquid milk, and children's milk. Yili has become a trusted Chinese brand for many customers. In Rabobank's 2014 Global Dairy Top-20, Yili was ranked number one in Asia, and 10th globally. The company achieved operating revenue of 54.436 billion yuan in 2014, an increase of 13.93% from the previous year, net profits attributable to listed company shareholders of 4.144 billion yuan, an increase of 30.03%, and earnings per share of 1.35 yuan, an increase of 22.73%. Of the 50 firms in our study, Yili ranked highly in terms of core operating revenue, and ranked fifth in terms of number of valid invention patents.

13. Conch Cement

Conch Cement (Anhui Conch Cement Company Limited) was established on September 1, 1997. It is primarily engaged in the production and sale of cement and clinker. It is currently the largest supplier of cement and clinker in Asia, with production and sales ranking first in China for 10 consecutive years. Its production line represents the most advanced in the global cement industry, and at the time of writing this report, 95% of its clinker production line had been completed, while its denitrification technical transformation was proceeding well. In addition, the firm is focused on increasing its environmental protection expenditure; it has implemented complete dust control reforms for the packaging systems at all of its subsidiaries, as well as actively developing electrical dust precipitation technological transformations to ensure conformance to dust emissions standards. The group had annual core operating revenues of 58.965 billion yuan in 2014, an increase of 8.79% from the previous year, while net profits attributable to listed company shareholders increased by 17.19% to 10.993 billion yuan and earnings per share were 2.07 yuan, an increase of 0.3 yuan per share. In this study, Conch ranked highly in terms of total assets and core operating revenue, and ranked fifth of the 50 firms listed here in terms of net profits.

14. Heilan Home (600398.SH)

Heilan Home (Heilan Home Co., Ltd.) was established on January 8, 1997. It is a wool clothing company in the textile garments industry. Its primary products and services include the production and sale of high-grade worsted wool fabric, high-grade suits, shirts, and professional clothing, as well as dyeing and finishing. Heilan is China's original corporate leader for worsted wool fabric, having China's best production capacity and quality for worsted wool fabric. Heilan's business focus has gradually shifted to higher value-added products, such as the production and sale of clothing; revenue from clothing sales now comprises more than 70% of Heilan's core operating revenue. At the time of writing this report, Heilan had 3348 retail outlets, an increase of 15.97% from 2013. Throughout China's 31 provincial-level divisions (including autonomous regions and municipalities), Heilan covered more than 80% of counties. In 2014, it was named a Wuxi Intellectual Property Rights Advantageous Enterprise, a Jiangsu Province Firm Model for Intellectual Property Rights Management, and a Chinese Business-wear Leading Enterprise. In 2014, the firm achieved operating revenue of 12.338 billion yuan, an increase of 72.56% over the previous year. Its net profits attributable to shareholders were 2.375 billion yuan, an increase of 75.83%. In this study, the company ranked first for the two indexes of net profit growth rate and average core operating revenue growth rate.

15. Hanwei Electronics (300007.SZ)

Hanwei Electronics (Henan Hanwei Electronics Co., Ltd.) was established on January 28, 2008. It researches and produces gas sensors, and its core business consists of the R&D, production, sale, and export of gas sensors, gas-detection instruments and meters, and gas-detection control systems. The company actively develops brand-building activities, and cooperates with other firms in the emerging home-health and smart-home market. In 2014, Hanwei achieved operating revenue of 127.5645 million yuan, an increase of 92.75% from the preceding year, and net profits attributable to the parent company owners were 114.006 million yuan, an increase of 40.72%. In this study, the company ranked highly in terms of core operating profit margin and number of invention patents as a proportion of core operating revenue (in units of 100 million yuan). For the two metrics of number of invention patents as a proportion of core operating revenue (in units of 100 million yuan) and number of new industry model patents utilized as a proportion of core operating revenue (in units of 100 million yuan), the company ranked first and third, respectively.

16. Han's Laser (002008.SZ)

Han's Laser (Han's Laser Technology Industry Group Co., Ltd.) was established on September 28, 2001. It is primarily engaged in the R&D, production, and sales of specialized equipment, such as laser processing equipment, printed circuit boards, photovoltaics, and light-emitting diode packaging. Han's Laser is China's leader in laser equipment, and is the largest laser processing-equipment production company in Asia, as well as being internationally recognized. In 2014, the high-powered laser equipment sales industry continued to innovate, and in terms of sales volume and revenue from high-powered laser cutting and welding equipment, Han's ranked first in China. The sales momentum for laser equipment-driven automation component systems has already started to materialize, with the company's fiber laser-cutting machine sales topping 1500 units, comfortably ranking it first globally. The combination of laser technology, automation technology, and a capital platform are the firm's inherent advantages in its expansion toward robotics, in which it already has made significant investments through the "Han's Yichuang" and "Han's Guoxin" robotics industry funds. In doing so, the company is striving to set aside investment income as reserves for its robotics program, and to increase its competitiveness through industrial chain integration. Han's Laser's project for key technology and application of post-semiconductor packaging of core equipment was awarded a second-class State Scientific and Technological Progress Award in 2014. The company achieved operating incomes of 5.566 billion yuan in 2014, an increase of 28.41% from the previous year, operating profits of 664 million yuan, an increase of 44.43%, and net profits attributable to listed company shareholders of 708 million yuan, an increase of 29.55%. In this study, Han's ranked highly in the following three indexes: number of valid invention patents, number of valid invention patents as a proportion of core operating revenue (in units of 100 million yuan), and utilization of new industry model patents as a proportion of core operating revenue (in units of 100 million yuan).

17. Hengrui Medicine (600276.SH)

Hengrui Medicine (Jiangsu Hengrui Medicine Co., Ltd.) was established on April 28, 1997. Its core products are pharmaceutical products, and the company is China's largest research and production base for anti-neoplastic and surgical use drugs, as well as being one of China's most innovative large-scale pharmaceutical companies. Hengrui Medicine was one of the first domestic pharmaceutical companies to pass China's new GMP certification, and was the first Chinese pharmaceutical company to receive approval to sell injection agents in European and US markets. In competing, Hengrui Medicine adheres to the use of scientific and technological innovation as the driving force behind its competitiveness, and is committed to building its core competitiveness. As of December 31, 2014, Hengrui had total share capital of 1.50399 billion shares, and total assets of 9.087 billion yuan; in 2014, it had net profits of 1.573 billion yuan, and the highest net profits in the chemical agent market.

18. Inspur (000977.SZ)

Inspur (Inspur Electronic Information Industry Co., Ltd.) is China's leader in computing platforms, and the supply of IT application solutions. In addition, it is China's largest producer of servers and supplier of server solutions; its core products are electronics, and it has been ranked first for 10 consecutive years in terms of sales volumes and sales revenue for domestically produced servers. Furthermore, it is the largest server production base in the Asia-Pacific region. As of December 31, 2014, Inspur had total share capital of 480 million shares, with total assets of 5.949 billion yuan; in 2014, its net profits were 339 billion yuan, ranking it sixth in its industry. In terms of net profit margin in our study, it was sixth of the 50 firms listed here.

19. Kaiyuan Instruments (300338.SZ)

Kaiyuan Instruments (Changsha Kaiyuan Instruments Co., Ltd.) is engaged in the development, production, and sale of coal quality testing instruments, and is the leading firm for Chinese coal quality-testing instruments. It is the only Chinese firm that simultaneously possesses scale production capacity for coal quality sampling, sample preparation, and chemical testing instruments. It has provided nearly 20,000 units of coal quality testing instruments to customers, encompassing primary energy source metering and conceptual energy-management systems solutions. Kaiyuan is continuously expanding its integrated system through product fields, such as online coal quality testing instruments and biological quality testing instruments. As of December 31, 2014, the company's total share capital was 126 million shares, with total assets of 879 million yuan, and net profits in 2014 of 49 million yuan. Operating revenue in this period was 307 million yuan, an increase of 8.86% from the previous year, while investment in technological development was 24 million yuan, an increase of 7.42% from the previous year.

20. Focused Photonics (300203.SZ)

Focused Photonics (Focused Photonics (Hangzhou) Inc.) is engaged in the R&D, production, and sale of instruments and meters for application in the fields of environmental testing, industrial process analysis, and safety testing. For 3 consecutive years, the firm has been the only analytics instrument company to be selected among the Top 100 Most Dynamic Enterprises in China, and for 3 consecutive years, was selected as one of Forbes' 100 Most Promising Enterprises. Furthermore, Focused Photonics' R&D organization was named a National Technology Center. As of the end of 2014, the company had 215 patents over its products, and 137 copyrights for its computing software. By the end of 2014, the company had total share capital of 453 million shares, with total assets of 2.906 billion yuan. In 2014, it had net profits of 197 million yuan, operating revenues of 1.231 billion yuan, an increase of 30.76% from the previous year, and profits of 228 million yuan, an increase of 27.14% from the previous year.

21. GRG Banking (002152.SZ)

The core products of GRG Banking (GRG Banking Equipment Co., Ltd.) are equipment and components for smart businesses, as well as equipment maintenance services and ATM operations. GRG Banking is China's only producer to grasp the core technologies for cash recycling machines, and is China's leader for domestically produced ATMs, the company's leading product, in terms of production capacity and scale of sales. In the traditional business market and the domestic ATM market, GRG Banking has consolidated and deepened its position; it has maintained the top market share for 7 consecutive years, and has continuously increased this share. It has already developed into the Chinese ATM industry's largest, most technologically capable firm, and has successfully competed alongside leading global ATM suppliers. In 2014, GRG Banking was named a National Standards Implementation Pilot Firm for Fusion of Informatization and Industrialization Management Systems, as well as a 2014 National Industrial Benchmarking Enterprise in Quality. The firm's operations have been growing steadily and in 2014, it achieved total operating revenue of 3.152 billion yuan, a growth of 25.28% from the previous year, and net profits of 817 million yuan, ranking it third in its industry.

22. TBEA (600089.SH)

TBEA (TBEA Co., Ltd.) was China's first listed company in the transformer industry. Its primary products are reactors, electrical wire and cable, international project contracting, solar wafers, and solar-powered system photovoltaic modules for transformers. The firm established the first international scientific R&D production base in India, an ultra-high voltage R&D and production base. In 2014, the firm was involved in one topic of the 863 Program, three National Science and Technology Major Projects, and two National Key Technology Research and Development programs. In 2014, 39 new TBEA products received national identification, 16 of which achieved international advanced level. In 2014, the firm received 177 new patents and had total assets of 59.292 billion yuan. It achieved operating revenues of 36.075 billion yuan, operating profits of 1.693 billion yuan, total profits of 2.051 billion yuan, and net profits of 1.811 billion yuan, increases of 23.65, 31.07, 32.01, and 31.42% from 2013, respectively.

23. GAC Group (601238.SH)

GAC Group (Guangzhou Automobile Group Co., Ltd.) is a leading automobile producer in China's domestic automobile industry. Based in southern China, its main business consists of R&D, production, and sales of automobiles and accessory products, as well as related services. It has formed a closed-loop industrial chain with the production of entire automobiles at its core, covering upstream automobile R&D, components, and downstream automobile services and financial investments. With its foothold in southern China extending to northern China, central China, eastern China, and the Bohai Rim, GAC has become a Chinese automobile group with one of the industry's most integrated industrial chains, and the most optimized industrial layout in the nation. The company's profitability is among the top in its industry, and as of December 31, 2014, its total share capital was 6.435 billion shares, with total assets of 62.318 billion yuan. Its operating revenue in 2014 was 22.376 billion yuan, an increase of 18.87% from the previous year, while net profits were 2.927 billion yuan.

24. GoerTek (002241.SZ)

GoerTek (GoerTek Electronics Co., Ltd.) is primarily engaged in the R&D, production, and sale of micro-electric acoustic components and consumer audio products. Currently, in the market for miniature electret microphones, GoerTek is the leading domestic firm in terms of market share, and third globally. For cell-phone micro-speaker/receivers, GoerTek ranks second among domestic firms. GoerTek has established many product R&D platforms for the integration of many technologies, and has already achieved the integration of acoustic electric technology, photoelectric technology, MEMS technology, wireless communication technology, automation technology, semiconductor technology, and software technology. Thus, it has formed overall technological advantages, and has developed innovative cross-platform products. In 2014, GoerTek applied for a total of 1093 patents, 467 of which were invention patents; this is a growth rate of 26.95% and 32.29% respectively, from the previous year. As of December 31, 2014, GoerTek had total assets of 17.758 billion yuan, and in 2014, net profits were 1.684 billion yuan, while operating revenue was 12.699 billion yuan, an increase of 26.37% from the previous year.

25. O-Film (002456.SZ)

O-Film (Shenzhen O-Film Tech Co., Ltd.) is a precision thin-film optical components manufacturing firm. Its key products include infrared cut filters and lens mount assemblies for digital imaging systems, as well as touch screens, optical fiber coatings, low-pass filters, and other precision thin-film optical components.

The company is a domestic leader in terms of scale, and its process technology capabilities for its precision optics optoelectronic components thin-film production line have reached advanced international standards and passed ISO 9001–2000 quality management system certification. The company has also received ISO 14001:2004 environmental management system certification, and in 2014, was named one of China's Top 100 Electronics and IT Companies. As of December 31, 2014, O-Film's main business was growing rapidly, achieving operating revenues of 19.482 billion yuan, an increase of 114% from the previous year, ranking it second in the industry. In 2014, the company had total assets of 14.156 billion yuan, net profits of 682 million yuan, and technological investments of 854 million yuan, an increase of 70% compared to the previous year.

26. Acrel (300286.SZ)

Acrel (Acrel Electric Co., Ltd.) is primarily an R&D, production, and sales company for client-end intelligent power meters. It is one of China's comparatively large-scale, technologically advanced, and independently innovative firms. It has integrated low-voltage signal acquisition, measurement, monitoring, and protection in a single system. It is one of a handful of Chinese firms providing customers with intelligent power monitoring, power management, and electrical safety systematic solutions. As of December 31, 2014, Acrel had five trademarks and 136 patents, 7 of which were invention patents, 61 new industry model patents, and 68 design patents. Its operating revenue was 283 million yuan in 2014, an increase of 34.13% from the previous year.

27. Tsinghua Tongfang

Tsinghua Tongfang (Tsinghua Tongfang Company Limited) was established in the two core industries of information technology, and energy and environmental protection. After restructuring, the company now represents the 11 following markets: computing systems, digital cities, security systems, networking applications, microelectronics and core components, multimedia, knowledge networks, military projects, digital television systems, construction energy efficiency, and semiconductor lighting. In 2014, it was named a Key Enterprise of National Secure and Reliable Computer Information System Integration by China's Ministry of Industry and Information Technology. In terms of intellectual property rights, as of the end of December, 2014, Tongfang had applied for 2441 Chinese patents, and had received 1492 Chinese patents. It had applied for a total of 935 foreign patents, and had received 343 foreign invention patents. The company's total share capital was 2.198 billion shares, and total assets were 49.725 billion yuan. In our study, by total assets, it was ranked first in the manufacture of computers, communications, and other electronic equipment industry, and with net profits of 1.213 billion yuan, it ranked highly within its industry in terms of net profits and operating revenues.

28. XD Group (601179.SH)

The core business of XD Group (China XD Electric Co., Ltd.) is the production, R&D, and testing of electrical transmission and distribution equipment. Its products have been used in national key projects, including the first northwest to north China back-to-back DC transmission network project, as well as the Three Gorges Project and the West–East Electricity Transfer Project. XD Group has 1386 valid patents, and its number of accepted invention patents increased by 26.32% in 2014 from the previous year. In addition, the company received its first foreign patents, and received three national patent excellence awards. In the international market, its electrical products and technologies have already been exported to more than 40 countries and regions, and its XD brand is highly recognizable within and outside of China. As of December 31, 2014, its total share capital was 5.126 billion shares, while their total assets were 30.905 billion yuan. In 2014, it achieved operating revenue of 12.87 billion yuan, total profits of 823 million yuan, and net profits attributable to the parent company of 684 million yuan, increases of 6.27, 109.01, and 104.95% from the previous year, respectively. In terms of net profits, XD Group ranked third in the manufacture of electric machinery and equipment industry, and ranked second in terms of operating revenue and total assets.

29. Suntront (300259.SZ)

Suntront (Suntront Tech Co., Ltd.) is primarily an R&D, production, sales, and service company for civil smart measurement instrument systems. Its products include supporting system equipment and software for its four main lines of smart water meters, heat meters, smart gas meters, and smart power meters. Suntront was an early entrant to the civil smart instrument and meter industry, and was one of the industry's most technologically advanced companies. The firm is recognized as a High Tech Enterprise, an Outstanding High-tech Enterprise, one of Henan's Top 50 Innovative Industrial Enterprises, one of Henan's Top 10 Technologically Innovative Companies, and a Henan Outstanding Industrial Innovation Enterprise. In 2014, Suntront had total operating revenues of 380 million yuan, an increase of 16.11% from the previous year. Its operating profits were 102 million yuan, an increase of 17.51% from the previous year, and net profits were 107 million yuan. Its total share capital was 272.448 million shares and assets totaled 965 million yuan.

30. Baiyunshan Pharmaceutical (600332.SH)

Baiyunshan Pharmaceutical (Guangzhou Baiyunshan Pharmaceutical Holding Co., Ltd.) is primarily involved in the production of pharmaceuticals; its core products are proprietary Chinese medicines, western medicines, and prepackaged foods. With regard to intellectual property rights, Baiyunshan has applied for 62 Chinese invention patents, and has received 33 Chinese invention patents. It has 4 well-known Chinese trademarks, 20 famous Guangdong province trademarks, and 27 famous Guangzhou trademarks. With respect to pharmaceutical formulations, it has 21 types of chemical pharmaceutical formulations, 16 types of proprietary

Chinese medicine formulations, and produces nearly 2000 varieties and specifications; these include 3 national first-class new drugs, 9 second-class new drugs, 71 protected Chinese medicines, and 31 exclusively produced varieties. As of December 31, 2014, Baiyunshan's total share capital was 1.29134 billion shares, and its total capital was 14.211 billion yuan. In 2014, it had net profits of 1.211 billion yuan and operating revenue of 18.799 billion yuan, an increase of 6.77% from the previous year. For the three indexes of net profits, operating revenue, and total assets, the company ranked highly within the manufacture of medicines industry.

31. Dahua Technology (002236.SZ)

Dahua Technology (Dahua Technology Co., Ltd.) is China's leading firm in the security video surveillance industry; its main product is the embedded DVR. The company has already formed a "total security" frontier product architecture consisting of the following five core video surveillance technology platforms: audio video codec algorithm technology, information storage and retrieval technology, integrated circuits application technology, network control and transmission technology, and embedded development technology. Dahua's products are used in the world's largest hydroelectric project, the Three Gorges Dam Hydroelectric Power Plant remote monitoring project, as well as in China's largest 500 kV DC conversion station, the Yichang Longquan converter station project. The company has two Chinese Torch Program projects, two national key new products, 15 patented technologies, and copyrights for 19 pieces of software. In 2014, Dahua maintained steady increasing operating revenues, achieving an operating revenue of 7.332 billion yuan, an increase of 35.52% from the previous year. Its net profits were 1.105 billion yuan, an increase of 1.04% from the previous year, and its annual R&D expenditure totaled 780 million yuan, an increase of 56.06% from the previous year.

32. Fuyao Group

Fuyao Group (Fuyao Glass Industry Group Co., Ltd.) is the world's only specialized supplier focusing on the automotive glass business. Its core products are automotive glass and float glass; it is China's largest-scale, most high-tech, largest exporting producer and supplier of automotive glass. Fuyao not only supports domestic automotive brands, but also has become a qualified supplier and high-volume supplier for Audi, Volkswagen, Hyundai, Holden, and Suzuki. The company's "FY" trademark is the only Chinese Brand and Famous Chinese Trademark in the Chinese automotive glass industry. As of December 31, 2014, Fuyao's total assets were 16.876 billion yuan. In 2014, its net profits were 2.217 billion yuan, earnings per share were 1.11 yuan, an increase of 15.63% from the previous year, and operating revenue was 12.928 billion yuan, an increase of 12.41% from the previous year. R&D expenditure as a proportion of operating revenue reached 4.01%, up 0.63% from the previous year.

33. Wuhan Iron and Steel (600005.SH)

Wuhan Iron and Steel (Wuhan Iron and Steel Corp.) is held by Wuhan Iron and Steel Group Corporation, and is China's second largest listed iron and steel company. Its primary businesses are the production and sale of iron and steel. The company's complete iron and steel production process for the smelting and rolling of iron and steel is in line with advanced international standards. It produces seven main classes of steel products, with more than 500 varieties; it currently has more than 100 products that have been awarded national, provincial, or departmental superior quality certificates. Wuhan Iron and Steel is listed in the top 20 of China's top 50 overall economically efficient companies. It is one of Shanghai and Shenzhen's top 10 listed companies, and has been named a national advanced equipment management enterprise eight times consecutively. In addition, it has received an AAA-rating certification for metering equipment systems, in addition to a plant engineering Golden Wrench Award for outstanding contributions to assessment activities. As of December 31, 2014, the company had total assets of 9.6064 billion yuan. In 2014, its net profits were 1.293 billion yuan, and it was ranked third in the smelting and pressing of ferrous metals industry for net ROE in our study.

34. Bengang Steel Plates (000761.SZ)

Bengang Steel Plates (Bengang Steel Plates Co., Ltd.) is primarily a steel smelting and pressing company, whose primary products are steel plates and steel billets. In recent years, Bengang has focused on R&D and production of high value-added, high technological content products, such as high-grade automobile plates and home appliance plates. It has aggressively engaged in the elimination and transformation of outdated production capabilities and high-energy consuming equipment; its main processing equipment is now up to advanced international standards. In 2014, Bengang produced a total of 10.093 million tons of pig iron, and 10.1038 million tons of crude steel, exhibiting steady growth from the previous year. In 2014, it engaged in R&D for continuous hot-dip steel sheet galvanization, and produced 52 new brand designations for hot-rolled dual-phase steel. It was engaged in 14 key scientific research products, 2 of which were awarded third-class awards for scientific and technological progress in Liaoning province. The company's Bengang cold-rolled dual-phase steel plates for automobile products was awarded a first-class new product award in Liaoning province. As of December 31, 2014, its total share capital was 3.136 billion shares, and it had total assets of 49.171 billion yuan. Net profits were 315 million yuan in 2014.

35. Goldwind (002202.SZ)

Goldwind (Xinjiang Goldwind Science and Technology Co., Ltd.) is currently China's largest original equipment manufacturer for complete wind turbine modules. Its core products are wind turbines and component development. For the past 3 years, Goldwind has sustained at least an 80% share of the Chinese-produced wind turbine market. In 2014, it was awarded the Large-scale Wind Turbine

Research and Demonstration Project award for Chinese machinery industry science by the China Machinery Industry Federation. With regard to intellectual property rights, as of December 31, 2014, Goldwind had 321 valid patents, 51 of which were invention patents, as well as 34 domestic trademarks. It had total assets of 45.777 billion yuan and operating revenue of 17.704 billion yuan in 2014, an increase of 43.84% from the previous year. Net profits were 1.854 billion yuan, an increase of 327.44% from the previous year. In our study, in terms of net profits, it was ranked second in its industry, and third for total assets and operating revenues.

36. Hisun Pharmaceutical (600267.SH)

Hisun Pharmaceutical (Zhejiang Hisun Pharmaceutical Co., Ltd.) was established on February 11, 1998. Its core products are chemical raw materials for medicines and intermediate chemicals. It is China's largest production base for antibiotics and anti-cancer drugs. In 2014, Hisun achieved a breakthrough in the fermentation of statin class drugs, and completed 35 formulation production R&D projects, while also developing a new respiratory drug delivery technology platform. The company's Hisun Pfizer Fuyang production base passed China's Food and Drug Administration's (FDA) new GMP certification in May 2014, becoming the only formulation GMP training base of the FDA's drug audit identification center. In addition, its first production line began production in May 2014, with 70 of Hisun's product designations entering the Hisun-Pfizer brand, and another 33 product types completing certification and initial production. In 2014, Hisun achieved operating revenue growth of 17.35% from the previous year, and a 20.69% increase in net profits in the same period. Net profits attributable to parent company shareholders increased by 2%. Hisun ranked highly in terms of operating revenue and total assets within the chemical and manufacture of medicines industries in our study.

37. Sanhua (002050.SZ)

Sanhua (Zhejiang Sanhua Co., Ltd.) was established on December 19, 2001. Its core products are household electronics control components and parts. The company is a leader in its industry, and has good brand image, sales channels, as well as R&D and production bases. It has a high market share among primary overseas air-conditioner manufacturers, and it has a strong competitive advantage. The company's products possess independent intellectual property rights and core technologies, and have established complete and strict quality control systems. Sanhua has won national quality awards, Chinese famous brand awards, Chinese well-known trademark awards, and has been named a national key high-tech enterprise. Furthermore, it was named one of Zhejiang province's top 100 high-tech enterprises, and one of China's 500 most valuable brands. For the company's leading products, four-way valves and a full range of shut-off valves, Sanhua has the largest global market share, and has a leading share in the world market for solenoid valves. Within its industry, and compared with world-class air-conditioner producers, it has advantages of scale and cost for collaborative development, and mutually beneficial operations. In 2014, Sanhua grew operating revenue 10.42% from the previous year and operating profit 64.88% from the previous year. Its net profits attributable to listed company shareholders rose by 50.83% from the previous year. It ranked first in terms of net profits compared to other white goods industry firms.

38. Jiangling Motors (000550.SZ)

Jiangling Motors (Jiangling Motors Co., Ltd.) was established on November 28, 1993. Its core products are commercial-use vehicles, SUVs, and related components. Jiangling forms the backbone of the domestic light commercial-use vehicle industry, and ranks among the top 100 Chinese listed companies in terms of comprehensive strength. In addition, it is a nationally recognized enterprise technology center, high-tech enterprise, and national vehicle export base, and has invested in the utilization of a new national R&D center. In addition, it possesses the advanced world-class technology of the DURATORQ JX4D24 engine, the renowned Chinese 4JB1 diesel engine, and the all new developed and produced 3 L diesel engine, in addition to the 1.5 L and 1.8 L GTDI gasoline engines. In 2014, Jiangling had a 1.17% share of China's overall car market, an increase of 0.13% from the previous year; it occupies a competitive position within the automobile manufacturing industry. In 2014, Jiangling's total sales revenue increased by 22% from the previous year, while net profits increased by 24%. It was ranked among the top of the automobile manufacturing industry in terms of net assets per share.

39. Da Bei Nong Group (002385.SZ)

Da Bei Nong Group (Beijing Da Bei Nong Technology Group Co., Ltd.) was established on October 18, 1994. Its core products are animal feed, animal protection, and seeds. The group has 3 national agricultural industrialization key enterprises and 12 national high-tech enterprises, and has been recognized as an enterprise technology center and a national innovative new industrial enterprise. It has established a marketing service network for national key breeding and growing areas. In 2014, it achieved three national certifications for new seed varieties, as well as seven provincial certifications for new seed varieties. In addition, it has several pending varietal certifications, including seven maize variety certifications (three national certifications and four provincial certifications), and nine rice variety certifications (three national certifications and six provincial certifications). In 2014, Da Bei Nong had 100 new patents, 43 of which were invention patents. In 2014, its operating revenue grew 10.71% from the previous year; net profits attributable to listed company shareholders grew 3.54% in the same period. Its consolidated gross profit margin was 21.79%, an increase of 1.13% from the previous year. The company ranked highly within the agricultural products processing industry in terms of net profits and net ROE in our study.

40. Livzon Group (000513.SZ)

Livzon Group (Livzon Pharmaceutical Group Inc.) was established on January 26, 1985. Its core products are drug formulations, pharmaceutical raw materials, intermediates, diagnostic reagents, and related equipment. With regard to intellectual property and international authorization, in 2014, Livzon applied for 37 domestic invention patents, 1 foreign patent, and received 18 domestic and international invention patents. In addition, it received two foreign and domestic patent subsidy funding totaling 179,200 yuan, as well as seven design patents. Furthermore, it received 2 trademark registration certificates and 30 domestic and international trademark renewals, and it filed for 12 trademark licenses. The company was awarded the major Ilaprazole Project, a part of the Twelfth Five-Year Plan, by the Ministry of Science and Technology, with funding of 30 million yuan, in addition to receiving many science and technology awards. In 2014, Livzon achieved operating revenue growth of 20.04% from the previous year and net profit growth of 5.66% over the same period. Net profit attributable to shareholders grew 5.84%, with benefits per share ranking among the top in the manufacture of medicines industry.

41. AVIC (600893.SH)

AVIC (AVIC Aviation Engine Corporation Plc.) was established on April 8, 1996. Its core products are aviation engines, non-aviation products, and civil services. It is China's top large-scale aviation engine production company. It has nearly 60 years of experience producing aviation engines, and has accumulated a large amount of data in this time. It has R&D capabilities for military and civilian aviation engines, and has aviation engine research and testing facilities and production processes that meet advanced international standards. The company has complete independent intellectual property rights over the Taihang and Kunlun jet engines, and provides the Qinling power plant for the Xian JH-7 Chinese fighter jet. AVIC fills a number of gaps in China's aviation technology, producing more than 30,000 sets of auxiliary equipment. AVIC is China's leader in the aviation engine industry, having China's largest propeller blade production line, as well as high-precision equipment, including a full set of spiral bevel gear processing modules, automatic welding machines, and laser-cutting equipment. In addition, it has more than 100 patents for related technologies, and has engaged in multiple 863 Project topics. In 2014, AVIC applied for 648 patents, 405 of which were invention patents; 445 patents were authorized. In 2014, it received more than 40 technological achievement awards at or above the provincial level. Its operating revenue grew by 3.74% in 2014 from the previous year, and net profits attributable to the listed company grew 14.57% over the same period. In terms of net profits, it ranked first in the aviation equipment industry.

42. Konka Group A (000016.SZ)

Konka Group A (Konka Group Co., Ltd.) was established on October 1, 1980. Its core products are home appliances and communications equipment. The company was the first foreign joint venture electronics corporation after China's reform and opening. In 2014, the firm's priority was process reform and establishing informatization, in order to optimize Konka's business management process and to reform its brand concept, namely, "refined products, beautiful life." In addition, it released new-brand advertisement videos, and implemented a new communications plan. In 2014, Konka's operating revenues decreased by 2.92% compared to the previous year, although net profits attributable to listed company shareholders increased by 14.85%. The company ranked highly for key indexes in the computer, communication, and other electronic equipment industry.

43. Yutong Bus (600066.SH)

Yutong Group (Zhengzhou Yutong Bus Co., Ltd.) was established on May 8, 1997. Its core products are busses and related products. In 2014, Yutong made groundbreaking advances in purely electric busses, cementing the place of its National Electric Bus Power Control and Safety Project Technology Research Center. Yutong's renewable energy bus technology is an industry leader, and while vigorously pushing renewable energy public busses, Yutong has also engaged in R&D on purely electric busses for the group tourism market, for which it has achieved market recognition. Under the backdrop of a downturn in the large and medium-sized bus industry in 2014, Yutong's operating revenues increased by 15.9% compared to the previous year. In 2014, its domestic sales of large and medium-sized busses increased by 4.6% from the previous year, while sales of renewable energy busses increased by 90.01% from the previous year. Yutong's total sales growth for large and medium busses was higher than that of the automobile manufacturing industry, and its key competitors. It continues to rank first in terms of sales and scale in the commercial passenger vehicle industry, and is the leading brand in China's bus industry.

44. Clou Electronics (002121.SZ)

Clou Electronics (Shenzhen Clou Electronics Co., Ltd.) was established on November 30, 2000. Its core products are electrical instruments and meters, and power automation products. Clou has already successfully entered the markets of 60 countries and regions, and is a leader in China's smart grid industry. Clou has completed many scientific and technological research and demonstration projects for China's national 863 Project. Clou has a unified industrial distribution chain integrating smart power generation, smart power storage, smart power distribution, and energy services. In addition, it has very strong core competitiveness, and has grasped the most advanced core technologies of the manufacture of measuring instruments industry. It is recognized within the industry as an A-type listed firm with the most complete industrial chain in the smart power grid field. As of the end of 2014, Clou had applied for 685 patents, and received 399 patents. In 2014, Clou's operating revenue grew 38.74% from the previous year, while total profits increased by 41.83% over the same period. Net profits attributable to the listed company increased 46.28% from the previous year.

45. Bright Dairy (600597.SH)

Bright Dairy (Bright Dairy & Food Co., Ltd.) was established on October 7, 1996. Its core products are liquid milk and other dairy products. The company has the national trademark, "Bright," and a series of well-known brands. It has a national key laboratory for dairy industry biological technology, and advanced dairy product processing techniques, technologies, and equipment. In addition, it has stable supplies and high-quality raw dairy material bases in China and New Zealand. In 2014, Bright Dairy achieved operating revenue growth of 25.13% from the previous year, while net profits attributable to the parent company owners increased by 39.87% from the same period. Earnings per share in 2014 increased by 39.39%, and the company ranked highly in the manufacture of foods industry in terms of earnings per share and net profits.

46. Tianma Micro-electronics A (000050.SZ)

Tianma Micro-electronics (Tianma Micro-electronics Co., Ltd.) was established on November 8, 1983. Its core products are electronic components and its main services are for the mobile terminal consumer display market and specialized display market. The company has a TFT-LCD key materials and technologies national project laboratory, a national enterprise technology center, and a postdoctoral mobile workstation. In addition, it has taken on multiple major national projects from the National Development and Reform Commission, the Ministry of Science and Technology, and the Ministry of Industry and Information Technology. With regard to the forward-looking technological layout of Tianma, the company has developed many internationally advanced, nationally leading technologies, and has independently developed the strength to lead the market in 3D displays, LTPS and AMOLED displays, flexible displays, transparent displays, and the integration of touch technology, in which it has achieved many positive results. The company has been awarded many prizes for innovative products and applications. In addition, Tianma's was the first to mass produce its naked-eye 3D technology. It was awarded many prizes, including best global partner, awards by core customers and a core supplier, the diamond award, best service award, and outstanding supplier award. In 2014, Tianma's operating revenue increased by 12% from the previous year. In terms of earnings per share and net profits, Tianma ranked highly in the computer, communication, and other electronic equipment industry.

47. CHINT Electric (601877.SH)

CHINT Electric (Zhejiang CHINT Electric Co., Ltd.) was established on August 5, 1997. Its core products are low-voltage electrical products. In 2014, it was ranked 92nd among China's Top 500 Private Enterprises, was placed 8th in the 15th China Electric Industry Top 100 list, and is the only firm specializing in low-voltage electrical products to be selected more than once as a top 10 firm among the China Electric Industry Top 100. In 2014, CHINT received the Chinese Industrial Award, was named a National Standards Implementation Pilot Firm for Fusion of Informatization and Industrialization Management Systems, as well as a National

Traditional Enterprise of Outstanding Integrity. In 2014, CHINT completed the development of 76 new products, completed 22 technological transformation projects, and received 173 patents. It took part in eight industry standard establishment and revision activities, and completed 353 domestic and international certifications (including 133 renewals). In 2014, the company's annual sales revenue grew by 6.78% from the previous year, and net profits attributable to the parent company increased by 18.88% in the same period. In 2014, its ROE reached 32.17%, and it was ranked first in the electronic equipment industry in terms of earnings per share.

48. Zhongxin Pharmaceuticals (600329.SH)

Zhongxin Pharmaceuticals (Tianjin Zhongxin Pharmaceutical Group Co., Ltd.) was established on December 20, 1992. Its core products are Chinese and western medicines. Zhongxin has one national enterprise technology center, five municipal enterprise technology centers, a municipal modernized technological progress center, and a postgraduate research workstation approved by the Chinese Ministry of Personnel. Through these structures, Zhongxin has formed an R&D platform encompassing the three fields of Chinese medicines, chemical medicines, and biotechnology medicines. All the group's medicine-producing subsidiaries have passed the new GMP certification, and some of its production workshops have also passed Australian and Japanese Current Good Manufacturing Practice (cGMP) certification. At the end of 2014, Zhongxin held 226 domestic trademarks, 21 international trademarks, and 4 famous trademarks. The company has strong brand advantages, especially in its distinctive traditional brand names dating to the Ming and Qing dynasties that are still in use today; these include the world-renowned "Darentang" and "Lerentang," which rank among the top 10 Chinese time-honored brands, as named by the China Time-Honored Brand Network. Zhongxin has a wide range of resources for its formulations, and as of the end of 2014, it had 17 formulation types and 601 drug approvals. Among these, it had four National Treasure-grade Chinese medicines, one National Confidential Variety, and three State Secret Varieties. In addition, it had eight varieties of protected Chinese medicines, 101 exclusively produced varieties, 85 varieties included in the National Essential Drugs List, and 267 varieties included in the National Medical Insurance Catalogue. In 2014, Zhongxin's annual sales revenue grew 8.87% from the previous year, operating revenue grew 17.92%, net profits attributable to the parent company grew 1.71%, and operating profits grew 10.06%. Zhongxin ranked among the top of the manufacture of medicines industry in terms of operating revenue in this study.

49. Lepu Medical (300003.SZ)

Lepu Medical (Lepu Medical Technology (Beijing) Co., Ltd.) was established on June 11, 1999. Its core products are medical devices and drug products, and it is one of the few Chinese companies capable of competing strongly with foreign firms in the field of high-end medical devices. Since its establishment, Lepu has completed the R&D and industrialization processes of many core interventional medical products, such as stents and catheters. In the manufacture of special purpose machinery industry, it was the first to receive the Coronary Stent Delivery System product registration card (Class III) from China's State Food and Drug Administration, and the first to successfully develop and test an anti-infection central-venous catheter drug delivery system. Lepu's dual-chamber implantable pacemaker is expected to break the current market structure in which foreign dual-chamber pacemakers hold a monopoly. As of the end of 2014, Lepu had received 86 medical device product registration certificates, and 63 drug approvals. In 2014, Lepu received 54 patents, 6 of which were invention patents; and approved and pending patents totaled 284, of which 188 have already been granted (including 34 invention patents). In 2014, Lepu's operating revenue grew 28.03% from the previous year, and operating profit grew 22.78% over the same period. Net profits attributable to listed company shareholders grew by 17.02% from the previous year, while end-of-period total assets grew by 22.43% from the start of the period; the company's owner equity attributable to the parent company grew by 11.14% in the same period. In the manufacture of special purpose machinery industry, Lepu ranked first in terms of net profits.

50. Tiandi Science and Technology (600582.SH)

Tiandi Science and Technology (Tiandi Science and Technology Co., Ltd.) was established on March 24, 2000. Its core products are related to coal production machinery. Tiandi has contributed to many projects within China's national 863 Project and 973 Project; it was a key work unit in completing the Complete Technology and Equipment Set R&D for Ultra-Thick, High Seam Caving Mining project, which solved the difficult global problem of 14–20 m ultra-thick seam mining. This led to the development of international coal-mining technology, with Tiandi awarded a first-class 2014 State Scientific and Technological Progress Award. Beijing China Coal, a subsidiary of Tiandi, released China's first ultra large-scale raise borer, and won the China Coal Industry Association's 2014 first-class Science and Technology Award. The company is the first in China to achieve remote-controlled surface coal mining, supplementing China's progress toward fully automated pilotless smart coal mining. As of December 31, 2014, Tiandi had total share capital of 1.21392 billion, total assets of 28.156 billion yuan, and in 2014, net profits were 1.862 billion yuan. The company ranked first in terms of net profits and operating revenue within the manufacture of special purpose machinery industry.

6.4 Chapter Summary

This chapter collects authoritative data and statistics published by relevant state departments. It primarily utilizes quantitative statistical and analytical methods to assess and analyze the numerical distribution, scale, efficiency growth potential, and innovativeness of listed Chinese manufacturing sector firms, as well as assessing the selection of the most respectable firms. This section arrives at the following conclusions.

Considering regional distribution of listed Chinese manufacturing sector firms, listed firms are primarily concentrated in the eastern coastal region, where Guangdong, Zhejiang, Jiangsu, and Shandong provinces remain the top four provinces in this regard. Considering industry distribution, the following three industries contain a total of 568 listed firms, accounting for 34.3% of the industry total: manufacture of computers, communications, and other electronic equipment; manufacture of raw chemical materials and chemical products; and manufacture of electrical machinery and equipment. The three manufacturing sector industries with the fewest listed firms are manufacture of leather, fur, feather, and related products and footware (six firms, or 0.36% of the total); manufacture of furniture (five firms, or 0.3% of the total); and utilization of waste resources (one firm, or 0.06% of the total).

With regard to firm scale, according to an analysis of the scale of core operating revenues, there was a clear expansion of the scale of listed Chinese manufacturing sector firms in 2014 compared to the previous year, with an average scalar increase of 4.15%. The total average scale of core operating revenues of listed manufacturing sector firms was highest in the following six provincial-level divisions: Shanghai, Guangdong, Beijing, Shandong, Jiangsu, and Zhejiang. Considering China's eastern, central, and western regions, the total scale of listed firms was largest in China's east, followed by central China, and was smallest in the west. With regard to industry differences, the total scale of listed manufacturing sector firms was largest in the communications equipment, computers, and other electronic equipment manufacturing industry, while the only listed firm in the utilization of waste resources industry had the smallest scale. Using total assets to conduct a statistical analysis of the scale of listed manufacturing sector firms, we can learn that the general trend of listed Chinese manufacturing sector firms is a continuous increase in scale. Very large-scale firms with total assets greater than 5 billion yuan account for more than two-thirds of the overall scale of the manufacturing sector, at 79.92%. This proportion of the overall sector comprised of very large-scale firms is clearly increasing, having increased by 1.64% in the period of this study. Where industry concentration of listed manufacturing sector firms is more pronounced, market competitiveness is stronger.

Considering efficiency, this chapter utilizes the three indexes of ROE, ROA, and core operating profit margins to assess the profitability of listed manufacturing sector firms. The statistics obtained here include 1640 manufacturing sector firms, and from the perspective of industries, average core operating profit margin was 22.32%, average ROA was 3.1%, and average ROE was 3.04%. Here, the average core operating profit margins of the 147 listed pharmaceutical manufacturing firms was highest, at 47.77%, while that of the 75 listed non-metallic mineral production firms was lowest, at −25.38%. Fifteen industries had core operating profit margins higher than the industry average, including the following: manufacture of computers, communications, and other electronic equipment; manufacture of electrical machinery and equipment; manufacture of medicines; manufacture of special purpose machinery; and manufacture of general purpose machinery. The seven listed firms in the printing, reproduction of recording media industry had the highest average ROA, at 7.97%, while the 19 listed firms in processing of petroleum, coking, processing of nuclear fuel had the lowest average ROA, at −1.25%. Fourteen industries had average ROA levels higher than the industry average, including manufacture of computers, communications, and other electronic equipment; manufacture of electrical machinery and equipment; manufacture of medicines; and manufacture of automobiles. The 17 listed firms in other manufacturing industries had the highest average ROE, at 12.12%, while the 50 listed firms in the manufacture of rubber and plastic industry had the lowest average ROE, at −38.75%. Meanwhile, 22 industries had ROE levels higher than the sector average, including the manufacture of computers, communications, and other electronic equipment; manufacturing of raw chemical materials and chemical products; manufacture of electrical machinery and equipment; manufacture of medicines; manufacture of special purpose machinery; manufacture of general purpose machinery; and manufacture of automobiles. From a regional point of view, in 2014, the average core operating profit margin for China's 31 provincial-level divisions was 24.61%. The five listed firms in Tibet had the highest average core operating profit margin, of 38.07%, while the 18 listed firms in Inner Mongolia had the lowest average core operating profit margin, at −1.84%. In 2014, the average ROA of China's 31 provincial-level divisions was 3.41%. The five listed firms in Tibet had the highest average ROA of 8.17%, while the nine listed firms in Ningxia had the lowest ROA at only 0.1%. In 2014, the average ROE of China's 31 provincial-level divisions was 2.63%. The eight listed firms in Qinghai had the highest average ROE of 18.63%, while the 37 listed firms of Hebei had the lowest average ROE, at only −53.63%.

With regard to growth, this chapter utilizes the 3-year average core operating revenue growth rate to measure the growth of listed manufacturing sector firms. Our analysis leads us to consider that the overall growth of each industry in the manufacturing sector is comparatively good. Two industries, the manufacture of non-metallic mineral products and utilization of waste resources, have relatively high revenue growth rates, both of which are higher than 0.7%. However, smelting and pressing of ferrous metals and processing of petroleum, coking, processing of nuclear fuel have negative values for the growth capacity index; their future development will be difficult, and they must accelerate their transformation, upgrading, and transition to the new industry model. In addition, we believe that most overall regions have relatively good growth, including Gansu, Heilongjiang, and Shanghai. These three provincial-level divisions have relatively high revenue growth rates, all above 0.5%. The two provincial-level divisions of Ningxia and Shanxi have negative revenue growth rates, showing the difficulties that listed manufacturing sector firms in these regions face while pursuing growth. However, when considering the individual firm, the overall growth of industries and regions as described above is highly susceptible to the influence of outlying firms, such as Shanghai Zhongji Investment Holding Co., Ltd., Gansu Shangfeng Cement Co.,

Ltd., and Harbin Electric Corporation Jiamusi Electric machine Co., Ltd., which increased the measured overall growth of their respective industries and regions.

Considering innovativeness, this chapter utilizes intensity of R&D expenditure and number of valid invention patents as a proportion of core operating revenue (in units of 100 million yuan) to reflect the technological input–output levels of listed Chinese manufacturing sector firms. In 2014, the average R&D intensity of China's 29 manufacturing sector industries was 2.96%. The 214 listed firms in the manufacture of computers, communications, and other electronic equipment industry had the highest average R&D intensity, at 7.69%, while the 17 listed firms in processing of petroleum, coking, processing of nuclear fuel had the lowest average R&D intensity, at 0.46%. In 2014, the average R&D intensity for the manufacturing industries of China's 31 provincial-level divisions was 3.54%. The 23 listed firms in Shaanxi had the highest average R&D intensity, at 11.15%, while the 8 listed firms in Qinghai had the lowest average R&D intensity, at only 0.86%. At the same time, individual listed manufacturing sector firms with comparatively high levels of investment in innovation, such as Caihong Group Corporation, increased the overall levels of innovation investment of their respective region and industry. In 2014, the average level of technological innovation output for the 29 manufacturing sector industries was 1.2836 patents/100 million yuan. The 72 listed firms of the non-metallic mineral products industry had the highest level of average innovation output, at 4.9864 patents/100 million yuan. The 17 listed firms in processing of petroleum, coking, processing of nuclear fuel had the lowest level of average innovation output, at 0.142 patents/100 million yuan. In 2014, the average level of technological innovation output for the manufacturing sector in China's 31 provincial-level divisions was 2.1857 patents/100 million yuan. The 17 listed firms of Yunnan had the highest average level of technological innovation output, at 19.6332 patents/100 million yuan, while the 8 listed firms of Qinghai had the lowest average technological innovation output, at only 0.0442 patents/100 million yuan. In addition, listed manufacturing sector firms with relatively high levels of technological innovation output, such as Caihong Group Corporation and Bowin Technology Industry, increase the levels of technological innovation output of their respective regions and industries.

The selection of the most respectable listed manufacturing sector firms involved a selection of outstanding firms from a group of good candidates, and utilized a two-stage assessment process. This process resulted in the selection of 50 outstanding "most respectable" listed manufacturing sector firms; the top 5 firms were SAIC Motor, Gree Electric, Midea Group, China CNR Corporation, and China CSR Corporation. Considering distribution across industries, the top 50 firms were concentrated in the following industries: manufacture of computers, communications, and other electronic equipment (9 firms); manufacture of measuring instruments (6 firms); and manufacture of medicines (6 firms). With regard to regional distribution, the top 50 firms were concentrated in Guangdong (13 firms), Beijing (6 firms), Zhejiang (6 firms), Henan (3 firms), Shanghai (3 firms), Jiangsu (2 firms), and Shandong (2 firms).

References

Aimin, Qian, Zhang Xinmin, and Zhou Ziyuan. 2009. Quality of Earnings Structure, Core Profitability and Earnings Persistence: Evidence from China's A Share Listed Companies in Manufacturing Industry. *China Soft Science* 08: 108–118.

Chengyong, Yu., Shi Jianjun, and Fang Hong. 2013. Control Right, Scale and M&A Performance: An Empirical Study Based on Listed Manufacturing Corporations in Shanghai and Shenzhen. *Journal of International Trade* 05: 128–142.

China Securities Regulatory Commission. *Guidelines for the Industry Classification of ListedCompanies* (2012 Revision). http://csrc.gov.cn/. 26 Oct, 2012.

China State Intellectual Property Office. Patent Search and Analysis. http://www.pss-system.gov.cn/. 1 May, 2015.

Fuming, Zhang, and Meng Xianzhong. 2011. An Empirical Study of the Relationship BetweenProfitability and Growth Based on the GMM System. *Modern Management Science* 04: 31–33.

Haiyun, Ren, Shi Ping, and Zhang Lin. 2010. Empirical Research on the Relationship BetweenFirm Size and R&D Input. *Science and Technology Progress and Policy* 04: 68–71.

Hongya, Li, Xuegui, Shi and Yinjie, Zhang 2014. Study On Firm Scale Distribution andFinancial Constraints—An Analysis Based on Listed Chinese Manufacturing Firm Data. *Modern Economic Science* 02 (95–109): 127–128.

Jifa, Gu. 2011. Practice on Wuli-Shili-Renli System Approach. *Chinese Journal of Management* 8 (3): 317–322.

Pingfang, Zhu, Hong, Liu, and Jiang Guolin. 2002. A Reflection on the Patent Output Problem ofShanghai's High Tech Industry.. *Quantitative and Technical Economics.*

Qiang, He, and Chen Song. 2013. Innovative Development, Director's Innovative Preference andR&D Expenditure: Empirical Evidence from Listed Manufacturing Enterprises in China. *Industrial Economics Research* 06: 99–110.

Shanghai Stock Exchange. Annual Report of Listed Company 2011–2014. http://www.sse.com.cn 30 Apr 2015.

Shenzhen Stock Exchange. Annual Report of Listed Company 2011–2014. http://www.szse.com.cn. 30 Apr, 2015.

Weiqiu, Yuan. 2013. Financial Constraints, Working Capital Management Efficiency and Company Profitability: Evidence from Chinese Listed Manufacturing Companies. *Shanghai Journal of Economics* 10 (22–34): 133.

Wenyi, Wang, and An Tongliang. 2014a. Firm Size and R&D Performance of Chinese Manufacturing Listed Companies. *Forum on Science and Technology in China* 05 (62-67): 73.

Wenyi, Wang, and An Tongliang. 2014b. Industrial Agglomeration, Innovation and Knowledge Spillover: An Empirical Research Based on the Chinese Listed Manufacturing Companies. *Industrial Economics Research* 04: 22–29.

Xin, Liu, and Chen Song. 2014. Effect of Firm Scale on Process Innovation—An Empirical Study Based on Chinese Listed Manufacturing Sector Firms. *Science and Technology Progress and Policy* 21: 83–86.

Yan, Cao. 2011. Study of the Profitability of Listed Home Appliance Manufacturing Firms—Based Data from 2009 Annual Reports. *Journal of Shanxi Finance and Economics University* S1: 134–135.

Yuanqiong, He, and Chen Yun 2009. An Empirical Study of the Relationship between TMT Size and Firm Performance in the Uncertain Environment:Evidence from the Chinese Listed Companies in the Manufacturing Sector. Science of Science and Management of S. & T., 2009, 02: 123–128.

Yuejun, Tang, and Song Yuanyang. 2008. The Impacts of Size and Age of Chinese Companies on Firm Growth—Panel Data of Manufacturing Listed Companies. *Industrial Economics Research* 06: 28–35.

The manufacturer's authorised representative in the EU is Springer Nature Customer Service Centre GmbH, Europaplatz 3, 69115 Heidelberg, Germany. If you have any concerns regarding our products, please contact ProductSafety@springernature.com

Printed and bound by CPI Group (UK) Ltd, Croydon, CR0 4YY
25/09/2026
02222884-0002